AF356549

Use of Essential Oils and Volatile Compounds as Biological Control Agents

Use of Essential Oils and Volatile Compounds as Biological Control Agents

Editors

Marie-Laure Fauconnier
Haïssam Jijakli
Caroline De Clerck

MDPI • Basel • Beijing • Wuhan • Barcelona • Belgrade • Manchester • Tokyo • Cluj • Tianjin

Editors
Marie-Laure Fauconnier
University of Liège
Belgium

Haïssam Jijakli
University of Liège
Belgium

Caroline De Clerck
University of Liège
Belgium

Editorial Office
MDPI
St. Alban-Anlage 66
4052 Basel, Switzerland

This is a reprint of articles from the Special Issue published online in the open access journal *Foods* (ISSN 2304-8158) (available at: https://www.mdpi.com/journal/foods/special_issues/Use_Essential_Oils_Volatiles_Compounds_Biocontrol_Agents).

For citation purposes, cite each article independently as indicated on the article page online and as indicated below:

LastName, A.A.; LastName, B.B.; LastName, C.C. Article Title. *Journal Name* **Year**, *Volume Number*, Page Range.

ISBN 978-3-0365-4127-3 (Hbk)
ISBN 978-3-0365-4128-0 (PDF)

Contents

About the Editors

Marie-Laure Fauconnier

Marie-Laure Fauconnier, Full Professor, Head of the Laboratory of Chemistry of Natural Molecules. She is specialized in extraction, chemical characterization and use of plant secondary metabolites including essential oils for applications in agronomy.

Haïssam Jijakli

Haïssam Jijakli, Full Professor, Professor in Urban agriculture and Plant Pathology, Director of Research Center in Urban Agriculture, Director of Integrated and Urban Plant Pathology Laboratory, co-funder of APEO, he is developing and implementing biocontrol methods against plant diseases.

Caroline De Clerck

Caroline De Clerck, Assistant Professor, in Charge of the AgricultureIsLife Platform, she is specialized in plant bio-assays in controlled and field conditions, in the implementation of new agricultural practices and new biocontrol methods against insects and plant diseases.

Editorial

Use of Essential Oils and Volatile Compounds as Biological Control Agents

Caroline De Clerck [1,†], Manon Genva [2,†], M. Haissam Jijakli [3] and Marie-Laure Fauconnier [2,*]

1 AgricultureIsLife, Gembloux Agro-Bio Tech, Liege University, Passage des Déportés 2, 5030 Gembloux, Belgium; Caroline.declerck@uliege.be
2 Laboratory of Chemistry of Natural Molecules, Gembloux Agro-Bio Tech, Liege University, Passage des Déportés 2, 5030 Gembloux, Belgium; m.genva@uliege.be
3 Integrated and Urban Plant Pathology Laboratory, Gembloux Agro-Bio Tech, Liege University, Passage des Déportés 2, 5030 Gembloux, Belgium; MH.Jijakli@uliege.be
* Correspondence: marie-laure.fauconnier@uliege.be
† These authors contributed equally to this work.

Citation: De Clerck, C.; Genva, M.; Jijakli, M.H.; Fauconnier, M.-L. Use of Essential Oils and Volatile Compounds as Biological Control Agents. *Foods* **2021**, *10*, 1062. https://doi.org/10.3390/foods10051062

Received: 8 May 2021
Accepted: 10 May 2021
Published: 12 May 2021

Publisher's Note: MDPI stays neutral with regard to jurisdictional claims in published maps and institutional affiliations.

Plants containing essential oils have been used for centuries as spices, remedies or for their pleasant odor. In the Middle Ages, the development of distillation techniques made it possible to obtain essential oils, which have continued to be used in their historical applications in food, medicine or cosmetics [1]. However, over the last few decades, the essential oil sector has entered a new dimension, as its fields of application are constantly increasing, largely due to the biocidal properties of its constituents.

The emergence of the resistance of targeted populations, ecological concern and impact on human health paved the way to the development of more sustainable alternatives to synthetic conventional biocides. Essential oils that combine highly biocidal properties with a specific or broad spectrum of action as well as a high volatility, thus limiting residues in foodstuff or the environment, are perfect candidates for a new generation of biocides. Used in plant protection as bactericides, fungicides or insecticides in both pre- and post-harvest treatments; as food ingredients to increase shelf-life; or incorporated in innovative packaging, research in the field of essential oils has a bright future ahead of it.

Three major subjects were discussed in the present Special Issue entitled "Use of Essential Oils and Volatile Compounds as Biological Control Agents": stored product insecticides, plant protection and food additives-food packaging.

Six research articles were published on the first topic, focusing on the insecticidal properties of essential oils, with the challenging perspective of replacing chemical insecticides that are widely used during crop cultivation and post-harvest treatments and therefore reducing the quantities of residues in foods. Oftadeh et al. first highlighted the high level of interest in essential oil from flowers of *Artemisia annua* L. in the control of *Glyphodes pyloalis* Walker, which damages mulberry leaves and induces the transmission of plant pathogenic agents [2]. In the second paper, the authors described the interesting contact toxicity of *Satureja intermedia* C.A.Mey essential oil against *Aphis nerii* Boyer de Fonscolombe, which is an insect pest in many ornamental plant cultures causing direct plant damage and transmitting pathogenic viruses. Interestingly, *Coccinella septempunctata* L., which is a predator of *A. nerii* and is used as biocontrol agent, was less susceptible to the essential oil. Moreover, the authors also described the elevated fumigant toxicity of *S. intermedia* essential oil against *Trogoderma granarium* Everts, *Rhyzopertha dominica* Fabricius, *Tribolium castaneum* Herbst, and *Oryzaephilus surinamensis* L., which are all common insect pests in stored products [3]. Loss during food storage due to insect infestation is a huge problem, both in developing and in developed countries. Contact chemical insecticides are therefore traditionally used to reduce food losses, with the problems of resistance appearance and the persistence of chemical residues in food. Essential oils, along with their complex composition, their low mammal toxicity and their high volatility, have emerged as promising alternatives to chemical insecticides in stored products. In the next article, Demeter et al. studied the insecticidal activity of 25 essential oils against *Sitophilus granarius* L., which is one of the main insect pests during grain storage. The authors showed a high potential in different essential oils, such as those from *Allium sativum* L., *Mentha arvensis* L. and *Eucalyptus dives* Schauer for the control of *S. granarius* in stored products [4]. Tanoh et al. also showed the toxicity of the newly described essential oils from *Zanthoxylum leprieurii* Guill. & Perr. against the same insect [5]. Thereafter, Owolabi et al. described the high toxicity of essential oils from a Nigerian plant, *Launaea taraxacifolia* (Willd.) Amin ex C. Jeffrey, against *Sitophilus oryzae* L., the rice weevil that causes high food losses during grain storage [6]. Finally, Liang et al. showed the high insecticidal properties of essential oil from *Elsholtzia ciliata* (Thunb.) Hyl. and of its major components, carvone and limonene, in the control of *Tribolium castaneum* Herbst, a common beetle affecting many stored products, such as cereals and flours [7].

In the second topic of the present Special Issue, two articles described the high interest of essential oils in crop plant protection. De Clerck et al. firstly screened 90 commercially available essential oils for their in vitro antifungal and antibacterial activities against 10 phytopathogens that particularly attack plant crops and decrease food production yields. The authors highlighted that several essential oils, such as that from *Allium sativum* L., are active on diverse pathogens and thus have a "generalist" effect, while other essential oils such as that from *Citrus sinensis* (L.) Osbeck have an action on one to three pathogens, and thus a more "specific" effect [8]. In the review from Werrie et al., the authors described the high interest of essential oils for the development of biopesticides, but they also underlined the different restrictions on their use, as some of them display phytotoxicity on untargeted crops. The authors mentioned the different parameters that need to be taken into account to limit that risk, such as the mode of application, the phenological state and the product formulation [9].

In the last topic of this Special Issue, different authors studied the potential of essential oils as food additives or for their incorporation into food packaging. Siroli et al. firstly showed that the incorporation of essential oils into the marinade increased the sensorial perception of the marinated pork loin [10]. In the next article, Licon et al. showed that the incorporation of essential oils from *Thymus vulgaris* L. in milk used for the production of pressed ewes' cheese had an interesting antimicrobial effect, with a decrease in the growth of exogenous detrimental microorganisms without affecting the cheese natural flora [11]. Ben-Fadhel et al. then highlighted the antimicrobial interest of essential oils for the treatment of ready-to-eat carrots. Indeed, their incorporation into emulsions that were applied to the carrot surface allowed the lengthening of the carrot shelf-life by two days [12]. Ruengvisesh et al. studied the antimicrobial activities of micelles formed from sodium dodecyl sulfate. The authors showed that eugenol-loaded micelles were particularly effective in inhibiting *Escherichia coli* and *Salmonella enterica* when applied on fresh spinach surfaces [13]. Essential oils also emerged as interesting bioactive additives for their incorporation into active packaging. In their article, Díaz-Galindo et al. showed that the incorporation of cinnamon essential oil emulsions into thermoplastic starch leads to a decrease in the growth rate of *Botrytis cinerea* without affecting the thermal stability of the packaging [14]. As essential oil volatility may limit their applications when the release is too fast, Maes et al. studied the potential of biosourced dendrimers to encapsulate essential oils. Their results show that stirring time and stirring rate are crucial parameters that need to be optimized for an efficient encapsulation, which paves the way to numerous essential oil applications when a slower release is needed [15]. Bleoancă et al. studied two different treatments for the formation of edible films containing thyme extracts. Both high-pressure-thermally treated films and thermally treated ones display different structures with different abilities to retain volatile compounds [16]. Finally, Kostoglou et al. showed the promising potential of three plant terpenoids—carvacrol, thymol and eugenol—as anti-biofilms agents, as they showed significant anti-biofilm activities against *Staphylococcus aureus* and *Staphylococcus epidermidis.* Those two microorganisms are notably the cause of foodborne diseases and nosocomial infections [17].

The success of this Special Issue demonstrates clearly the scientific interest around the use of volatile compounds, especially essential oils, as biological control agents in food products. In addition, with controversial products being removed from the market, alternative products such as essential oils are expected to rise.

While this topic seems to have a bright future, some questions and difficulties remain. One of the first challenges encountered in the development of biopesticides using volatile molecules is their short persistence (volatility, degradation, etc.) in comparison to synthetics. This can be positive in terms of environmental impacts and in terms of food residues, but the release kinetic of the compounds and their molecular dynamics have to be known and controlled to ensure the product's efficacy. The formulation thus plays an important role, and technology is evolving, as highlighted in several papers of this Special Issue, with the development of nano-emulsions and encapsulation, among others. These formulations are also important to avoid the apparition of any adverse tastes or odors on stored food products. The authors also pointed out the need for an upscaling of the tests, which will help to assess the practical applicability of the treatments. A number of compounds have proven their efficacy in vitro and seem promising. However, in vitro tests will always need to be confirmed in vivo.

Essential oils and volatile compound activities are often attributed to mixtures of compounds. While this could be an advantage to prevent the development of resistances if they present different modes of action, as has been shown in [18], with two constituents of essential oils with distinct chemical structure interacting differentially with plant plasma membrane, this complex composition presents challenges to regulatory standards, where regulations are generally designed for synthetic substances that contain a single, highly concentrated and persistent molecule. This is leading to difficulties regarding market approval by the different regulatory agencies throughout the world, as well as economic considerations. Even if procedures are sometimes available for plant-based products, few active substances have been registered so far, especially in the pre- and post-harvest fields. Uses as ingredients in food products are less problematic, as only a few essential oils have restricted regulation concerns (e.g., mint essential oils).

More investigations need to be performed to decipher the mechanism of action of these volatile compounds, including the role of minor components and the synergic or additive effect among them. This will be crucial to evaluate the risks on the environment (plants, beneficial organisms (insects, worms . . .), soil microbiota, etc.), and human health, as well as to secure their industrial use.

To conclude, the use of volatile compounds and essential oils in particular for sustainable agricultural practices or as food ingredients seems promising, and extensive research will probably clarify or deny their relevance in diverse applications. They can be an efficient alternative to synthetic plant protection products when properly formulated and integrated with other pest management strategies; they can also be valuable food ingredients or innovative packaging constituents

The works collected in this Special Issue will certainly contribute to the field by increasing the knowledge on volatile compounds used as biological control agents, their efficiency and formulation in a large panel of situations related to the food sector.

Author Contributions: Conceptualization, C.D.C., M.G., M.H.J. and M.-L.F.; writing—original draft preparation, C.D.C., M.G., M.H.J. and M.-L.F.; writing—review and editing, C.D.C., M.G., M.H.J. and M.-L.F. All authors have read and agreed to the published version of the manuscript.

Funding: This research was funded by the Education, Audiovisual and Culture Executive Agency (EACEA) trough EOHUB project 600873-EPP-1-2018-1ES-EPPKA2-KA.

Conflicts of Interest: The authors declare no conflict of interest.

References

1. *Handbook of Essential Oils*, 3rd ed.; Başer, K.H.C.; Buchbauer, G. (Eds.) CRC Press: Boca Raton, FL, USA, 2020; ISBN 9781351246460.
2. Oftadeh, M.; Sendi, J.J.; Ebadollahi, A.; Setzer, W.N.; Krutmuang, P. Mulberry Protection through Flowering-Stage Essential Oil of Artemisia annua against the Lesser Mulberry Pyralid, Glyphodes pyloalis Walker. *Foods* **2021**, *10*, 210. [CrossRef] [PubMed]
3. Ebadollahi, A.; Setzer, W.N. Evaluation of the Toxicity of Satureja intermedia C. A. Mey Essential Oil to Storage and Greenhouse Insect Pests and a Predator Ladybird. *Foods* **2020**, *9*, 712. [CrossRef] [PubMed]
4. Demeter, S.; Lebbe, O.; Hecq, F.; Nicolis, S.C.; Kenne Kemene, T.; Martin, H.; Fauconnier, M.-L.; Hance, T. Insecticidal Activity of 25 Essential Oils on the Stored Product Pest, Sitophilus granarius. *Foods* **2021**, *10*, 200. [CrossRef] [PubMed]
5. Tanoh, E.A.; Boué, G.B.; Nea, F.; Genva, M.; Wognin, E.L.; Ledoux, A.; Martin, H.; Tonzibo, Z.F.; Frederich, M.; Fauconnier, M.-L. Seasonal Effect on the Chemical Composition, Insecticidal Properties and Other Biological Activities of Zanthoxylum leprieurii Guill. & Perr. Essential Oils. *Foods* **2020**, *9*, 550. [CrossRef]
6. Owolabi, M.S.; Ogundajo, A.L.; Alafia, A.O.; Ajelara, K.O.; Setzer, W.N. Composition of the Essential Oil and Insecticidal Activity of Launaea taraxacifolia (Willd.) Amin ex C. Jeffrey Growing in Nigeria. *Foods* **2020**, *9*, 914. [CrossRef] [PubMed]
7. Liang, J.-Y.; Xu, J.; Yang, Y.-Y.; Shao, Y.-Z.; Zhou, F.; Wang, J.-L. Toxicity and Synergistic Effect of Elsholtzia ciliata Essential Oil and Its Main Components against the Adult and Larval Stages of Tribolium castaneum. *Foods* **2020**, *9*, 345. [CrossRef] [PubMed]
8. De Clerck, C.; Dal Maso, S.; Parisi, O.; Dresen, F.; Zhiri, A.; Jijakli, M.H. Screening of Antifungal and Antibacterial Activity of 90 Commercial Essential Oils against 10 Pathogens of Agronomical Importance. *Foods* **2020**, *9*, 1418. [CrossRef]
9. Werrie, P.-Y.; Durenne, B.; Delaplace, P.; Fauconnier, M.-L. Phytotoxicity of Essential Oils: Opportunities and Constraints for the Development of Biopesticides. A Review. *Foods* **2020**, *9*, 1291. [CrossRef] [PubMed]
10. Siroli, L.; Baldi, G.; Soglia, F.; Bukvicki, D.; Patrignani, F.; Petracci, M.; Lanciotti, R. Use of Essential Oils to Increase the Safety and the Quality of Marinated Pork Loin. *Foods* **2020**, *9*, 987. [CrossRef] [PubMed]
11. Licon, C.C.; Moro, A.; Librán, C.M.; Molina, A.M.; Zalacain, A.; Berruga, M.I.; Carmona, M. Volatile Transference and Antimicrobial Activity of Cheeses Made with Ewes' Milk Fortified with Essential Oils. *Foods* **2020**, *9*, 35. [CrossRef] [PubMed]
12. Ben-Fadhel, Y.; Maherani, B.; Aragones, M.; Lacroix, M. Antimicrobial Properties of Encapsulated Antimicrobial Natural Plant Products for Ready-to-Eat Carrots. *Foods* **2019**, *8*, 535. [CrossRef] [PubMed]
13. Ruengvisesh, S.; Kerth, C.R.; Taylor, T.M. Inhibition of Escherichia coli O157:H7 and Salmonella enterica Isolates on Spinach Leaf Surfaces Using Eugenol-Loaded Surfactant Micelles. *Foods* **2019**, *8*, 575. [CrossRef] [PubMed]
14. Díaz-Galindo, E.P.; Nesic, A.; Bautista-Baños, S.; Dublan García, O.; Cabrera-Barjas, G. Corn-Starch-Based Materials Incorporated with Cinnamon Oil Emulsion: Physico-Chemical Characterization and Biological Activity. *Foods* **2020**, *9*, 475. [CrossRef] [PubMed]
15. Maes, C.; Brostaux, Y.; Bouquillon, S.; Fauconnier, M.-L. Use of New Glycerol-Based Dendrimers for Essential Oils Encapsulation: Optimization of Stirring Time and Rate Using a Plackett—Burman Design and a Surface Response Methodology. *Foods* **2021**, *10*, 207. [CrossRef] [PubMed]
16. Bleoancă, I.; Enachi, E.; Borda, D. Thyme Antimicrobial Effect in Edible Films with High Pressure Thermally Treated Whey Protein Concentrate. *Foods* **2020**, *9*, 855. [CrossRef] [PubMed]
17. Kostoglou, D.; Protopappas, I.; Giaouris, E. Common Plant-Derived Terpenoids Present Increased Anti-Biofilm Potential against Staphylococcus Bacteria Compared to a Quaternary Ammonium Biocide. *Foods* **2020**, *9*, 697. [CrossRef] [PubMed]
18. Lins, L.; Dal Maso, S.; Foncoux, B.; Kamili, A.; Laurin, Y.; Genva, M.; Jijakli, M.H.; De Clerck, C.; Fauconnier, M.L.; Deleu, M. Insights into the Relationships Between Herbicide Activities, Molecular Structure and Membrane Interaction of Cinnamon and Citronella Essential Oils Components. *Int. J. Mol. Sci.* **2019**, *20*, 4007. [CrossRef] [PubMed]

 *foods*

MDPI

Article

Mulberry Protection through Flowering-Stage Essential Oil of *Artemisia annua* against the Lesser Mulberry Pyralid, *Glyphodes pyloalis* Walker

Marziyeh Oftadeh [1], Jalal Jalali Sendi [1,2,*], Asgar Ebadollahi [3,*], William N. Setzer [4,5] and Patcharin Krutmuang [6,7,*]

1 Department of Plant Protection, Faculty of Agricultural Sciences, University of Guilan, Rasht 416351314, Iran; marziye.oftade@gmail.com
2 Department of Silk Research, Faculty of Agricultural Sciences, University of Guilan, Rasht 416351314, Iran
3 Department of Plant Sciences, Moghan College of Agriculture and Natural Resources, University of Mohaghegh Ardabili, Ardabil 5697194781, Iran
4 Department of Chemistry, University of Alabama in Huntsville, Huntsville, AL 35899, USA; wsetzer@chemistry.uah.edu
5 Aromatic Plant Research Center, 230 N 1200 E, Suite 100, Lehi, UT 84043, USA
6 Department of Entomology and Plant Pathology, Faculty of Agriculture, Chiang Mai University, Chiang Mai 50200, Thailand
7 Innovative Agriculture Research Center, Faculty of Agriculture, Chiang Mai University, Chiang Mai 50200, Thailand
* Correspondence: jjalali@guilan.ac.ir (J.J.S.); ebadollahi@uma.ac.ir (A.E.); patcharink26@gmail.com (P.K.)

Citation: Oftadeh, M.; Sendi, J.J.; Ebadollahi, A.; Setzer, W.N.; Krutmuang, P. Mulberry Protection through Flowering-Stage Essential Oil of *Artemisia annua* against the Lesser Mulberry Pyralid, *Glyphodes pyloalis* Walker. *Foods* 2021, 10, 210. https://doi.org/10.3390/foods 10020210

Academic Editors: Marie-Laure Fauconnier, Haïssam Jijakli and Caroline De Clerck

Received: 26 November 2020
Accepted: 19 January 2021
Published: 20 January 2021

Publisher's Note: MDPI stays neutral with regard to jurisdictional claims in published maps and institutional affiliations.

Abstract: In the present study, the toxicity and physiological disorders of the essential oil isolated from *Artemisia annua* flowers were assessed against one of the main insect pests of mulberry, *Glyphodes pyloalis* Walker, announcing one of the safe and effective alternatives to synthetic pesticides. The LC_{50} (lethal concentration to kill 50% of tested insects) values of the oral and fumigant bioassays of *A. annua* essential oil were 1.204 % W/V and 3.343 µL/L air, respectively. The *A. annua* essential oil, rich in camphor, artemisia ketone, β-selinene, pinocarvone, 1,8-cineole, and α-pinene, caused a significant reduction in digestive and detoxifying enzyme activity of *G. pyloalis* larvae. The contents of protein, glucose, and triglyceride were also reduced in the treated larvae by oral and fumigant treatments. The immune system in treated larvae was weakened after both oral and fumigation applications compared to the control groups. Histological studies on the midgut and ovaries showed that *A. annua* essential oil caused an obvious change in the distribution of the principal cells of tissues and reduction in yolk spheres in oocytes. Therefore, it is suggested that the essential oil from *A. annua* flowers, with wide-range bio-effects on *G. pyloalis*, be used as an available, safe, effective insecticide in the protection of mulberry.

Keywords: essential oil; sweet wormwood; mulberry pyralid; mulberry; immunity; reproductive system; digestive system

1. Introduction

The mulberry (*Morus* sp. (Rosales: Moraceae)) leaves are used for rearing silkworm (*Bombyx mori* L. (Lepidoptera: Bombycidae)). The importance of lesser mulberry pyralid *Glyphodes pyloalis* Walker (Lepidoptera: Pyralidae)) is from the larvae damaging mulberry leaves and the transmission of plant pathogenic agents [1]. The extensive use of synthetic chemical pesticides has led to many concerns about the safety of humans, beneficial insects, and the environment [2,3]. Thus, management of insect pest through eco-friendly and biodegradable agents is critical in sericulture.

The essential oils obtained from several parts of plants, including leaves, flowers, fruits, twigs, bark, seeds, wood, rhizomes, and roots, are made as secondary metabolites in

5

the plant and possess diverse chemical compositions [4]. The effectiveness of essential oils as a more sustainable pest management tool has been noted previously [5–7]. It can easily be inferred from their biodegradable nature and safety compared to many of the synthetic insecticides. Since they have multiple target sites in insects, their application is less likely to result in resistance in comparison with synthetic insecticides [8]. It was indicated that plant-derived essential oils may have several effects, including ovicidal, ovipositional deterrents, feeding deterrents, growth retardants, and inhibition in detoxification enzymes [9–11].

The annual wormwood, *Artemisia annua* L. (Asterales: Asteraceae), native to temperate Asia, has been naturalized in many countries [12]. The *A. annua* has traditionally been used to treat certain diseases of humans, including asthma, fever, malaria, skin diseases, jaundice, circulatory disorders, and hemorrhoids [13]. Although our previous findings of the essential oil or extracts in the vegetative stage of *A. annua* showed the high potential of this medicinal plant species on insect pest control [14–18], the insecticidal effects of its floral essential oil were evaluated against *G. pyloalis* in the present study.

The evaluation of lethal (acute) and sublethal (chronic) effects of essential oil extracted from *A. annua* flowers on *G. pyloalis* was the main objective of the current study, recommending a biorational and available agent as a possible replacement for synthetic insecticides. Fumigant toxicity is considered to be a non-residual treatment in which no residue will commonly remain for future contaminants. In oral toxicity, the pest is eliminated by swallowing infested food, and it is a suitable method for controlling leaf-eating pests. Therefore, fumigant and oral toxicity and the effect on some key enzymes and biochemical compounds, immunology, digestive system in the larvae, and the ovary of emerged adults of insects, along with the chemical analysis of the essential oil, were evaluated.

2. Materials and Methods

2.1. Insects' Rearing

The larvae of *G. pyloalis* were handpicked from a mulberry orchard within the University of Guilan campus, Rasht (37.2682° N, 49.5891° E), Iran. The larvae were maintained on fresh leaves of 'Shin Ichinoise' mulberry variety in disposable transparent containers (high-density polyethylene plastic containers, $10 \times 20 \times 5$ cm) in a rearing room set at 25 ± 1 °C, 75 ± 5% RH (Relative Humidity), and 16:8 L:D (Light:Dark). The emerging adults were reserved in glass jars ($18 \times 7 \times 5$ cm), in which fresh leaves were positioned for egg laying, and 10% honey-soaked cotton wool was provided for feeding.

2.2. Essential Oil

2.2.1. Extraction of the Essential Oil

The mature and immature flowers of *A. annua* (autumn 2018) were collected on the University of Guilan campus. Samples were dried on a table out of direct sunlight for about a week until sufficiently dry to form a powder when ground. The dried flowers were made into a fine powder by a grinder (354, Moulinex, Normandy, France), and a solution was made with distilled water (50 g/750 mL). The solution was let to stand in the dark at laboratory room temperature for 24 h to maximum essential oil extraction. The mixture was distilled to extract the essential oil using a Clevenger apparatus (J3230, Sina glass, Tehran, Iran). The distillation process was run for two hours and the obtained essential oil was dried over anhydrous sodium sulfate. The obtained essential oil was stored in dark glass vials at 4 °C in a refrigerator until used.

2.2.2. Determination of Essential Oil Composition

The essential oil was analyzed through gas chromatography (Agilent Technologies 7890B) coupled with a mass spectrometer (Agilent Technologies 5977A), which was armed with an HP-5MS ((5%-phenyl)-methylpolysiloxane) capillary column with a 30-m length, 0.25-mm width, and an internal thickness of 0.25 μm. Helium gas at a 1 mL/min flow rate was used, while the column temperature started from 50 and reached to 280 °C at a rate of 5 °C/min. A 10% *A. annua* essential oil solution in methanol (*v/v*) was prepared, and

1 µL of solution was injected. Spectra were obtained in the electron impact mode with 70 eV of ionization energy. The scan range was between 30–600 m/z. The identification of components was performed by comparing mass spectral fragmentation patterns and retention indices with those described in the databases [19,20].

2.3. Insecticidal Activity

2.3.1. Oral Toxicity

Initial tests were conducted to assist in selecting the appropriate range of concentrations. Bioassays were carried out on 4th instar larvae, which were deprived of nutrition for 4 h before the onset of experiments. The essential oil concentrations of 0.5, 0.7, 1, 1.4 and 2% (W/V) in acetone as solvent (Merck, Darmstadt, Germany) were selected. For bioassays, mulberry leaf disks (8 cm in diameter) were immersed in desired concentrations for 10 s and then air-dried at room temperature for 30 min. Ten 4th instar *G. pyloalis* were placed on each disk. The mortality was documented after 24 h. Control groups were placed on disks treated with acetone. The control and treated groups were replicated four times.

2.3.2. Fumigant Activity

In order to carry out fumigation bioassays, two transparent polyethylene plastic containers (Pharman polymer company, Rasht, Iran) were used. A 250-mL container was used to place 10 4th instar larvae of mulberry pyralid. They were provided with fresh mulberry leaf disks, and the container top was covered with fine cotton fabric for aeration. The container was then placed inside a 1000-mL container. The desired amount of pure essential oil was poured onto filter papers (Whatman No. 1) cut to 2 cm in diameter using a micro applicator. It was then placed in the corner of the larger container, and its lid tightly sealed using Parafilm. The concentrations of 2, 3, 4, 5 and 6 µL/L air were used for this bioassay based on the initial tests. The controls were treated in the same way without any treatments of the filter papers. All tests were replicated four times.

2.4. Digestive Enzymes' Assays

In order to evaluate digestive enzymes activity, the larvae that were treated with LC_{50}, LC_{30}, and LC_{10} (Lethal Concentration to kill 50, 30, and 10% of insects, respectively) dosages of essential oil obtained from oral and fumigant bioassays and the controls were dissected in ringer's solution (9% v/v NaCl and isotonic) 24 h after treatment and their digestive systems (only midguts) were dissected out. Five midguts for each treatment and control were first homogenized in 500 µL of universal buffer (50 mM sodium phosphate-borate at pH 7.1) in a tissue homogenizer (DWK885300-0001-1EA, Merk, Darmstadt, Germany). The supernatant was then kept at $-20\,^{\circ}$C until analyzed.

2.4.1. The α-Amylase Activity

The reagent dinitrosalicylic acid (DNS, Sigma, St. Louis, MI, USA) in 1% soluble starch was used to estimate α-amylase activity according to the method of Bernfeld (1955) [21]. Briefly, 20 µL of the enzyme was poured into 40 µL of soluble starch and 100 µL of universal buffer (pH 7). The mixture was incubated for 30 min at 35 °C, and DNS (100 µL) was then added to stop the reaction. The absorbance was read at 540 nm in an ELISA reader (Awareness, Temecula, CA, USA).

2.4.2. Protease Assay

The protease activity was assessed by addition of 200 µL of casein solution casein (1%) to 100 µL of enzyme and 100 µL universal buffer (pH 7). Then, the obtained mixture was incubated at 37 °C for 60 min [22]. The mixture was centrifuged at $8000 \times g$ within 15 min and the absorbance was read at 440 nm.

2.4.3. Lipase Estimation

The method of Tsujita et al. (1989) [23] was adopted to estimate lipase. Concisely, 10 μL enzyme, 18 μL p-nitrophenyl butyrate (50 mM), and 172 μL universal buffer (pH 7) were mixed and incubated at 37 °C for 30 min. The absorbance was recorded at 405 nm in the ELISA reader.

2.4.4. The α- and β-Glucosidase Estimation

Here, we used Triton X-100 in order to hydrolyze glucosidases (α- and β-) for 20 h at 40 °C in a ratio of 10 mg of Triton X-100/mg protein. Then, we incubated 75 mL p-nitrophenyl-α-D-glucopyranoside (pNaG, 5 mM), p-nitrophenyl-β-D-glucopyranoside (pNbG, 5 mM), 125 mL universal buffer (made of 2%Mol MES (2-(N-morpholino)ethanesulfonic acid), glycine, and succinate, 100 mM, pH 5.0), and 50 mL enzyme solution. In order to stop the reaction, 2 mL of sodium carbonate (1 M) was used and the absorbance was read at 450 nm [24].

2.5. Detoxifying Enzymes' Assays

Quantitative analyses of biochemical constituents were carried out on insects remaining after treatments with LC_{10}, LC_{30}, and LC_{50} and controls. To quantify the whole body protein, the method of Bradford (1976) [25], using the kit (GDA01A, Biochem Co., Tehran, Iran), was incorporated, while glucose and triglyceride were measured by Siegert (1987) [26] method and the triglyceride diagnostic kit, respectively (Pars Azmoon Co., Tehran, Iran). Key enzymes including esterase (general esterases with α- and β-naphthyl acetate substrates), glutathione S-transferase (GST), and phenol oxidase (PO) were assessed by the method described by van Asperen (1962) [27], Habing et al. (1974) [28], and Parkinson and Weaver (1999) [29], respectively.

2.6. Hematological Study

The amount of various circulating blood cells in mm^{-3} of larval lesser mulberry pyralid treated with sublethal doses of *A. annua* oil and in controls were assessed. The hemolymph was drawn from one of the larval prolegs, cutting by a fine scissor, using a capillary glass tube (10 μL for each treatment). Then, the blood was diluted five times with a solution of anticoagulant (0.017 M EDTA, 0.186 M NaCl, 0.098 M NaOH, and 0.041 M citric acid at pH 4.5). An improved Neubauer hemocytometer (mlabs, HBG, Giessen, Germany) [30] was used to assess the total cells using the formula of Jones (1962) [31]. A drop of hemolymph was collected from cut proleg of treated and control larvae. A smear was formed and stained with diluted Giemsa (Merck, Darmstadt, Germany) in distilled water (1:9) for 25 min, then just dipped in a saturated solution of lithium carbonate, and, finally, washed with distilled water. Permanent slides were prepared in Canada balsam (Merck Darmstadt, Germany). The percentage profile of different cells was done after identification and counting of 200 cells per slide [32].

Immunity Responses

Initially the treated or control larvae were made immobile by keeping them on ice cubes for five minutes. Then, they were surface sterilized and injected with 1×10^4 spores/mL in 0.01% Tween-80 of *Beauveria bassiana* (IRAN403C isolate) or latex beads (1:10 dilution for each suspension and Tween-80, respectively) on the second abdominal sternum using a 10-μL Hamilton syringe. The treated larvae were then transferred to glass jars and were given fresh leaves of mulberry. The control larvae were injected with 1 μL of distilled water comprising 0.01% of Tween-80 only. The hemolymph was collected 24 h post-injection from each larva, and the number of nodules formed was scored in a hemocytometer [33]. The counting was repeated four times for each group.

2.7. Histological Studies of Larvae Midgut and Adults' Ovary

The larvae midguts were separated from the whole dissected gut in insect ringer and were immediately fixed in aqueous Buine solution for 24 h [10]. Also, the ovary of adults (2 days old), emerging from either treated or control larvae, were separated and fixed. The tissues were processed for embedding in paraffin after being dehydrated in grades of ethanol alcohol and also cleaned by xylene. The fixed tissues were then cut by 5-μM thickness through a rotary microtome (Model 2030; Leica, Wetzlar, Germany). The hematoxylin and eosin were used for staining and then permanent slides were thus prepared, observed, and photographed under a light microscope (M1000 light microscope; Leica, Wetzlar, Germany) armed with an EOS 600D digital camera (Canon, Tokyo, Japan).

2.8. Statistical Analysis

LC values were determined using the Polo-Plus software (2002) [34]. All the data were analyzed by ANOVA (SAS Institute, Cary, Cary, NC, USA, 1997) [35], and the comparison of means was performed using Tukey's multiple comparison test ($p < 0.05$).

3. Results

3.1. A. annua Essential Oil Analysis

The chemical composition of extracted *A. annua* essential oil is presented in Table 1. We identified 55 compounds in flowers of this plant, which represent 93.0% of the total composition. Camphor (13.1%), artemisia ketone (11.8%), β-selinene (10.7%), pinocarvone (7.4%), 1,8-cineole (6.8%), and α-pinene (5.9%) were considered as the major compounds detected, all of which are terpenes. However, other groups such as ester and phenylpropene were also recognized (Table 1).

Table 1. Chemical composition of the of *Artemisia annua* floral essential oil.

RI_{calc}	RI_{db}	Compound	%	RI_{calc}	RI_{db}	Compound	%
923	926	Tricyclene [MH]	0.2	1258	1259	Lepalone [OM]	0.1
938	939	α-Pinene [MH]	5.9	1281	1278	Lepalol [OM]	0.3
978	975	Sabinene [MH]	0.3	1299	1290	p-Cymen-7-ol [OM]	0.2
982	979	β-Pinene [MH]	0.1	1337	1327	p-Mentha-1,4-dien-7-ol [OM]	0.2
992	990	Myrcene [MH]	0.4	1361	1359	Eugenol [PP]	0.6
1013	999	Yomogi alcohol [OM]	1.2	1374	1376	α-Copaene [SH]	1.0
1021	1024	p-Cymene [MH]	0.8	1391	1392	Benzyl 2-methylbutanoate [E]	0.3
1026	1026	o-Cymene [MH]	0.8	1402	1392	(Z)-Jasmone [OC]	0.1
1030	1031	1,8-Cineole [OM]	6.8	1420	1419	(E)-β-Caryophyllene [SH]	3.1
1061	1062	Artemisia ketone [OM]	11.8	1426	1432	β-Copaene [SH]	0.2
1074	1070	cis-Sabinene hydrate [OM]	0.5	1448	1454	α-Humulene [SH]	0.3
1082	1083	Artemisia alcohol [OM]	1.4	1455	1456	(E)-β-Farnesene [SH]	1.0
1104	1114	3-Methyl-3-butenyl 3-methylbutanoate [E]	0.8	1471	1477	β-Chamigrene [SH]	0.2
1119	1126	α-Campholenal [OM]	0.7	1478	1485	Germacrene D [SH]	0.7
1131	1144	trans-Pinocarveol [OM]	0.4	1489	1490	β-Selinene [SH]	10.7
1144	1146	Camphor [OM]	13.1	1510	1516	Isobornyl isovalerate [OM]	0.1
1161	1164	Pinocarvone [OM]	7.4	1517	1523	δ-Cadinene [SH]	0.1
1169	1169	Borneol [OM]	1.5	1547	1555	iso-Caryophyllene oxide [OS]	0.3
1179	1177	Terpinene-4-ol [OM]	2.2	1585	1583	Caryophyllene oxide [OS]	5.4
1192	1188	α-Terpineol [OM]	0.9	1588	1590	β-Copaene-4α-ol [OS]	0.2
1199	1195	Myrtenol [OM]	2.6	1594	1594	Salvial-4(14)-en-1-one [OS]	0.2
1211	1205	Verbenone [OM]	0.3	1643	1640	Caryophylla-4(12),8(13)-dien-5β-ol [OS]	1.3
1219	1216	trans-Carveol [OM]	0.6	1700	1695	Germacra-4(15),5,10(14)-trien-1β-ol [OS]	1.7
1227	1230	cis-p-Mentha-1(7),8-dien-2-ol [OM]	0.2	1765	1767	β-Costol [OS]	1.3
1229	1235	(3Z)-Hexenyl 3-methylbutanoate [E]	0.2	1854	1847	Phytone [OC]	0.4
1234	1236	n-Hexyl 2-methylbutanoate [E]	0.1	1984	1960	Palmitic acid [OC]	1.2
1240	1241	Cuminaldehyde [OM]	0.2	2087	2106	Phytol [DT]	0.3
1244	1243	Carvone [OM]	0.1			Total identified	93.0

RIcalc = retention index determined with respect to a homologous series of n-alkanes on a HP-5 ms column; RIdb = retention index from the databases [19,20]; MH = monoterpene hydrocarbone; OM = oxygenated monoterpene; SH = sesquiterpene hydrocarbone; OS = oxygenated sesquiterpene; DT = diterpene; PP = phenylpropene; E = ester; OC = other components.

3.2. Insecticidal Activity

Based on oral and fumigant bioassays, *A. annua* essential oil was toxic to 4th instar larva of *G. pyloalis* 24 h post treatments. Probit analysis revealed that the LC_{50} values were 1.204 % W/V and 3.343 µL/L air for oral and fumigant toxicity, respectively. The mortality of tested larvae was augmented with increasing concentration (Table 2). Besides LC_{50}, the LC_{10} and LC_{30} values were used to evaluate sublethal bio-activities, including effects on energy reserves, digestive and detoxifying enzymes activity, and hematological and immunity responses and histological study of midgut and ovary of larvae (Table 2).

Table 2. Probit analysis of the oral and fumigant toxicity of *Artemisia annua* floral essential oil on 4th instar larva of *Glyphodes pyloalis*.

Bioassay	LC_{10} (95% CL)	LC_{30} (95% CL)	LC_{50} (95% CL)	LC_{90} (95% CL)	Slope ± SE	X^2 (df = 3)
Oral toxicity	0.593 (0.395–0.735)	0.901 (0.725–1.058)	1.204 (1.024–1.466)	2.445 (1.882–4.128)	4.165 ± 0.631	3.2567
Fumigant toxicity	1.945 (1.568–2.240)	2.678 (2.347–2.948)	3.343 (3.048–3.632)	5.745 (5.112–6.825)	5.449 ± 0.788	2.976

LC: lethal concentration (% W/V for oral toxicity and µL/L for fumigant toxicity), CL: confidence limits, X^2: Chi-square value, and df: degrees of freedom. According to Chi-square values, no heterogeneity factor was used in the calculation of confidence limits. Concentration rates were 0.5–2% (W/V) and 2–6 µL/L air for oral and fumigant toxicity, respectively.

3.3. Energy Reserves

The essential oil of *A. annua* flowers on the energy reserves of *G. pyloalis* larvae is shown in Table 3. As can be seen, for all macromolecules, increasing dose of essential oil decreased the concentrations of protein, glucose, and triglycerides. For example, doubling the essential oil concentration (LC_{10} to LC_{50}) reduced glucose by 29% in oral tests, while a 1.7-fold increase in fumigant concentration resulted in a 32% drop in glucose levels. The protein was also affected but the decrease in protein with increasing essential oil levels was insufficient to detect given background variability.

Table 3. Effect of *Artemisia annua* flowers' essential oil on macromolecules in 4th instar larvae of *Glyphodes pyloalis*.

Bio-assay	Concentrations	Protein (mg/dL)	Glucose (mg/dL)	Triglyceride (mg/dL)
Oral toxicity (% W/V)	Control	1.0200 ± 0.0360 [a]	1.7733 ± 0.0247 [a]	1.8800 ± 0.0145 [a]
	LC_{10}	0.9833 ± 0.0088 [a]	1.6666 ± 0.0033 [a]	1.8033 ± 0.0617 [a]
	LC_{30}	0.9700 ± 0.0057 [a]	1.6533 ± 0.0290 [a]	1.6557 ± 0.0531 [a]
	LC_{50}	0.9533 ± 0.0088 [a]	1.1733 ± 0.0783 [b]	1.1700 ± 0.0577 [b]
	F-Value	2.16	29.51	19.65
	Pr	0.0170	0.0001	0.0005
Fumigant toxicity (µL/L)	Control	1.0400 ± 0.0208 [a]	1.8100 ± 0.0655 [a]	1.9200 ± 0.0964 [a]
	LC_{10}	0.9900 ± 0.0057 [ab]	1.7266 ± 0.0384 [a]	1.7533 ± 0.0635 [ab]
	LC_{30}	0.9700 ± 0.0032 [b]	1.6900 ± 0.0208 [a]	1.433 ± 0.2185 [ab]
	LC_{50}	0.9366 ± 0.0088 [b]	1.1633 ± 0.0317 [b]	1.3000 ± 0.0765 [b]
	F-Value	12.94	47.80	5.04
	Pr	0.0019	0.0001	0.0300

In each separate column, means followed by different letters designate significant differences at $p < 0.05$ according to Tukey's test.

3.4. Digestive and Detoxifying Enzymes

The effects of *A. annua* floral essential oil on digestive enzymes' activity of *G. pyloalis* larvae was manifested by a decrease in protease, α-glucosidase, β-glucosidase, α-amylase, and lipase contents. The difference was significant between the LC_{50} versus the control in both oral and fumigant applications while other concentrations of the essential oil produced intermediate responses (Table 4).

The effect of essential oil of *A. annua* flowers on the activity of esterase and glutathione S-transferase (GST) of *G. pyloalis* larvae is shown in the Table 5. Glutathione S-transferase and esterase contents were reduced significantly when LC_{50} was applied in both oral and fumigation methods compared to the controls (Table 5).

3.5. Hematological Study and Immunity Responses

The essential oil affected the immune system, which included cellular quantity and quality, phenol oxidase activity, and the immune responses after *B. bassiana* and latex beads' injection (Figures 1–4). Total hemocyte counts (THC), plasmatocytes and granular cells, nodule formation, and phenol oxidase activity was recorded the lowest in LC_{50} both in oral and fumigation assays, respectively.

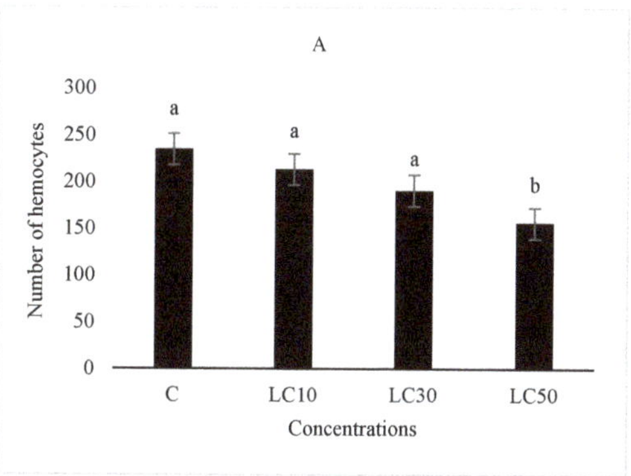
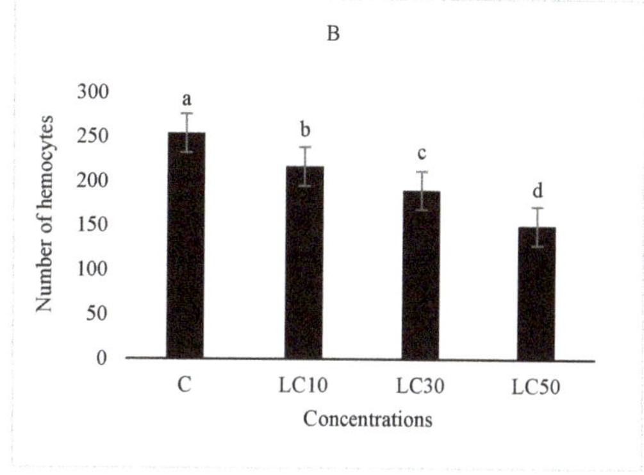

Figure 1. The effect of *Artemisia annua* floral essential oil on total hemocyte counts (THC) of *Glyphodes pyloalis* larvae treated with oral (**A**) and fumigant (**B**) assays. Bars with different letters above them indicate significant differences between means at $p < 0.05$, Tukey's test. Number of hemocytes $\times 10^4$.

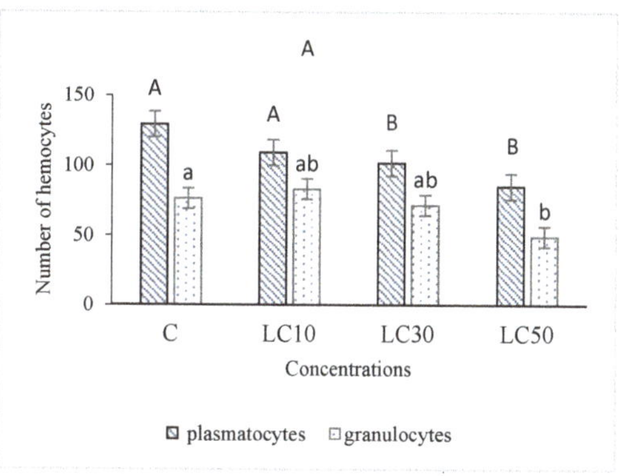
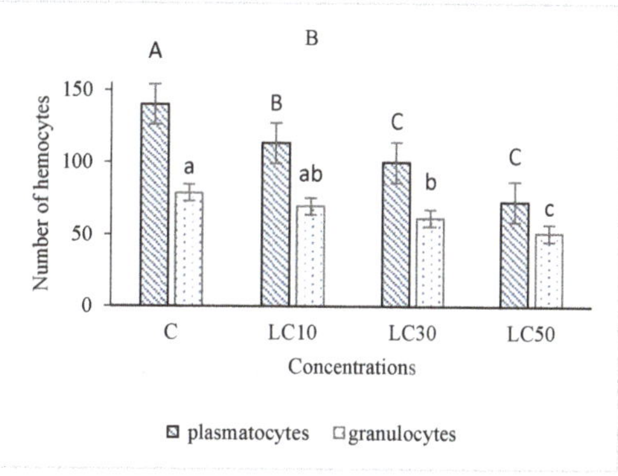

Figure 2. The effect of *Artemisia annua* floral essential oil on the plasmatocytes and granular cells of *Glyphodes pyloalis* larvae treated with oral (**A**) and fumigant (**B**) assays. Bars with different letters indicate significant differences among means of each hemocyte at $p < 0.05$, Tukey's test. The number of hemocytes $\times 10^4$.

Table 4. Effect of *Artemisia annua* floral essential oil on digestive enzyme activities in 4th instar larvae of *Glyphodes pyloalis*.

Bio-assay	Digestive Enzymes (U/mg Protein)	Control	LC_{10}	LC_{30}	LC_{50}	F-Value	Pr
Oral toxicity (% W/V)	Protease	1.9467 ± 0.3525 [a]	1.7833 ± 0.1201 [ab]	1.5433 ± 0.0876 [ab]	1.0667 ± 0.0437 [b]	3.96	0.0531
	α-glucosidase	1.374 ± 0.192 [a]	1.046 ± 0.0825 [ab]	0.7119 ± 0.0333 [b]	0.5640 ± 0.0360 [b]	9.31	0.0055
	β-glucosidase	1.4451 ± 0.1165 [a]	1.1635 ± 0.0955 [a]	0.8757 ± 0.05365 [b]	0.6873 ± 0.0515 [b]	15.61	0.0010
	α-amylase	0.3066 ± 1.732 [a]	0.2633 ± 0.0202 [ab]	0.2333 ± 0.01763 [b]	0.0833 ± 0.0120 [c]	41.32	0.0001
	Lipase	0.0571 ± 0.032 [a]	0.0387 ± 0.064 [ab]	0.03806 ± 0.089 [b]	0.03700 ± 0.059 [b]	22.75	0.0003
Fumigant toxicity (μL/L)	Protease	1.8333 ± 0.1244 [a]	0.8967 ± 0.1197 [b]	0.7167 ± 0.1591 [b]	0.4067 ± 0.1591 [b]	15.83	0.0010
	α-glucosidase	1.2034 ± 0.039 [a]	1.1083 ± 0.266 [a]	0.8870 ± 0.064 [b]	0.6921 ± 0.038 [b]	20.80	0.0004
	β-glucosidase	1.3451 ± 0.0330 [a]	1.3183 ± 0.1830 [a]	0.9537 ± 0.0282 [ab]	0.7591 ± 0.0717 [b]	8.07	0.0084
	α-amylase	0.2800 ± 0.0057 [a]	0.2700 ± 0.01731 [ab]	0.2300 ± 0.11541 [ab]	0.1333 ± 0.0145 [b]	37.49	0.0001
	Lipase	0.0559 ± 0.0010 [a]	0.0436 ± 0.0012 [b]	0.0378 ± 0.0027 [b]	0.02620 ± 0.0025 [c]	37.68	0.0001

In each separate row, means followed by different letters designate significant differences at $p < 0.05$ according to Tukey's test.

Table 5. Effect of the different concentrations of *Artemisia annua* flowers' essential oil on the activity of glutathione S-transferase (GST) and esterase in 4th instar larvae of *Glyphodes pyloalis*.

Bio-assay	Concentrations	GST (U/mg Protein)	Esterase (U/mg Protein)
Oral toxicity (% W/V)	Control	0.02300 ± 0.001 [a]	0.0953 ± 0.004 [a]
	LC$_{10}$	0.01733 ± 0.0032 [a]	0.08266 ± 0.007 [ab]
	LC$_{30}$	0.0065 ± 0.0025 [b]	0.07366 ± 0.002 [ab]
	LC$_{50}$	0.0001 ± 0.00001 [b]	0.06700 ± 0.001 [b]
	F-Value	23.46	14.13
	Pr	0.0003	0.0483
Fumigant toxicity (µL/L)	Control	0.02266 ± 0.0008 [a]	0.09566 ± 0.004 [a]
	LC$_{10}$	0.01533 ± 0.0006 [a]	0.07966 ± 0.0005 [ab]
	LC$_{30}$	0.0010 ± 0.0001 [b]	0.06066 ± 0.0063 [ab]
	LC$_{50}$	0.0001 ± 0.0000 [b]	0.04600 ± 0.0024 [b]
	F-Value	30.13	22.27
	Pr	0.0001	0.0003

In each separate column, means followed by different letters indicate significant differences at $p < 0.05$ according to Tukey's test.

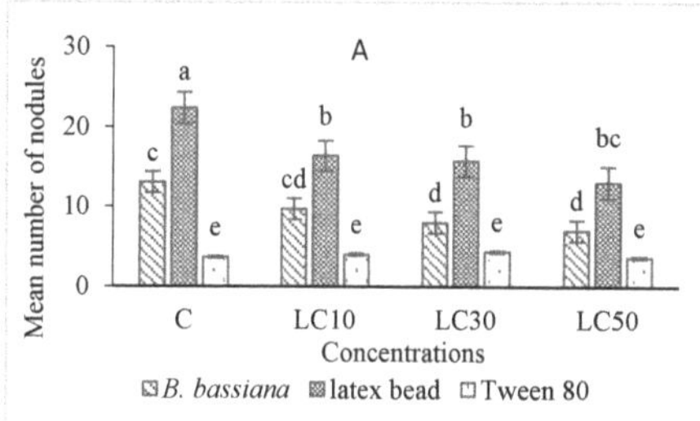
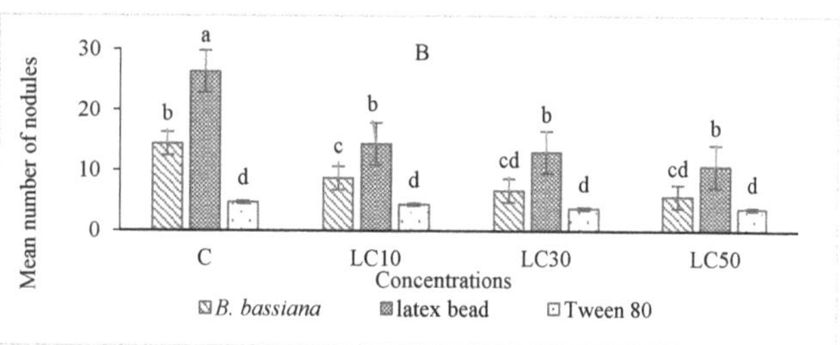

Figure 3. Effects of *Artemisia annua* floral essential oil on the nodule formation of *Glyphodes pyloalis* larvae treated with oral (**A**) and fumigant assays (**B**) and inoculated with *Beauveria bassiana* spores or latex beads. Bars with different letters indicate significant differences between means at $p < 0.05$. Tukey's test. The number of hemocytes $\times 10^4$.

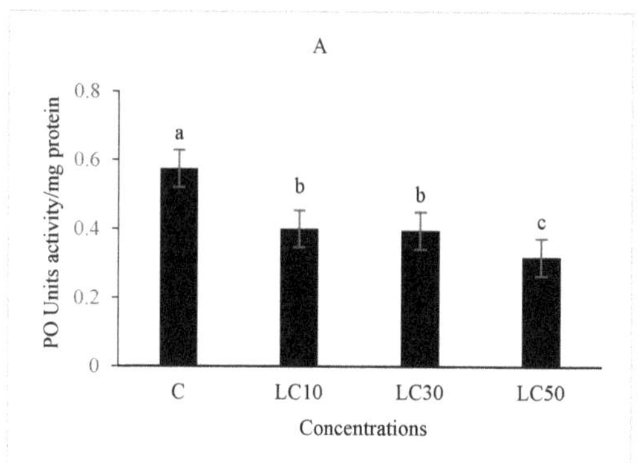
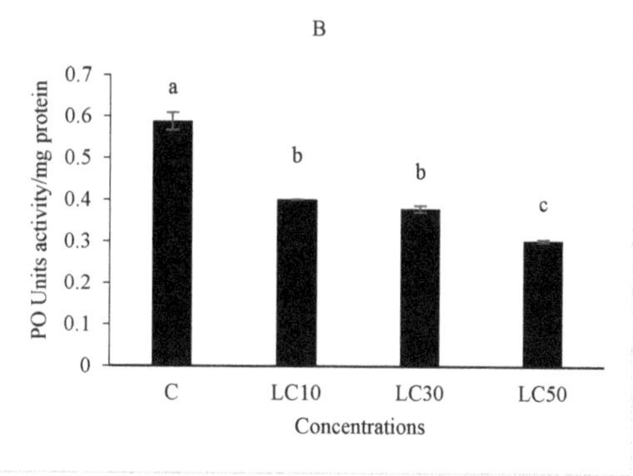

Figure 4. The effect of *Artemisia annua* floral essential oil on phenol oxidase (PO) activity of *Glyphodes pyloalis* larvae treated with oral (**A**) and fumigant (**B**) assays. Bars with different letters above them indicate significant differences between means at $p < 0.05$, Tukey's test. The number of hemocytes $\times 10^4$.

3.6. Histological Studies

The histological texture of larval midgut upon treatment with *A. annua* essential oil revealed significant differences with the controls, the most significant of which was the elongation and separation of epithelial cells losing the compactness (Figure 5). The most significant changes in ovarian structure was thinning of epithelial cells around each follicle compared with that of control. Also, the significant reduction in cytoplasm was seen after vacuolization in yolk spheres of the oocytes (Figure 6).

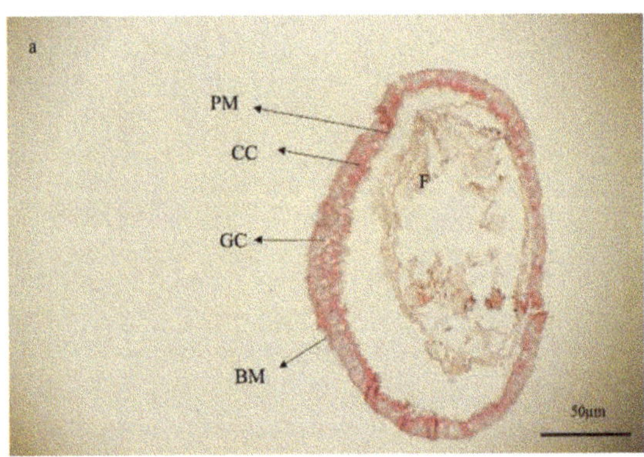

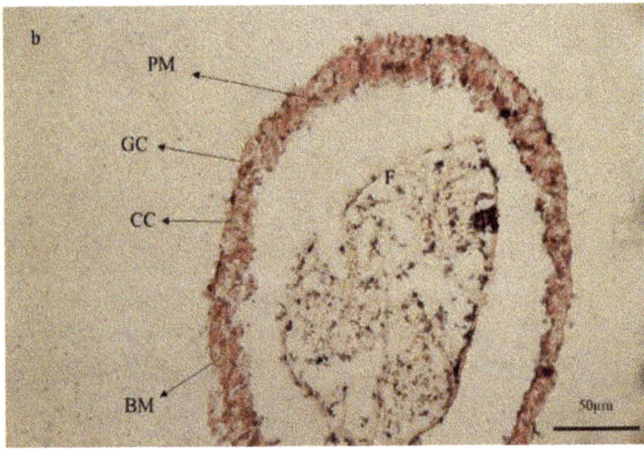

Figure 5. Light microscopy of the larval midgut of *Glyphodes pyloalis* in control (**a**) and after oral treatment with *Artemisia annua* floral essential oil (**b**). Normal texture of all cell types (**a**) was contrasted to changes in size and texture in treated larvae (**b**). In the midgut of insects treated with essential oil from *A. annua* the cohesion of the columnar epithelial layer was damaged. (**BM**) basement membrane, (**CC**) columnar cell, (**GC**) goblet cell, and (**PM**) peritrophic membrane.

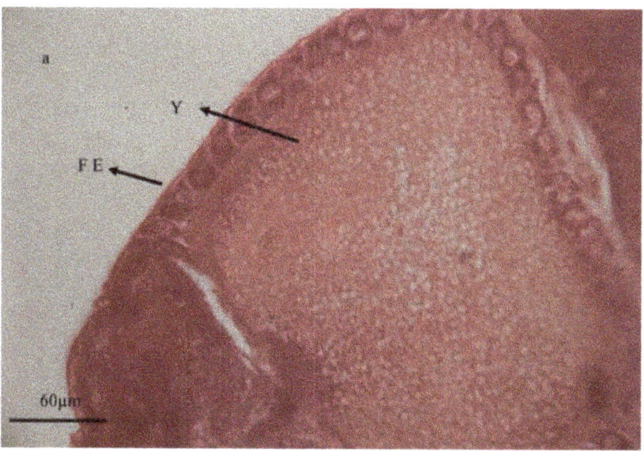

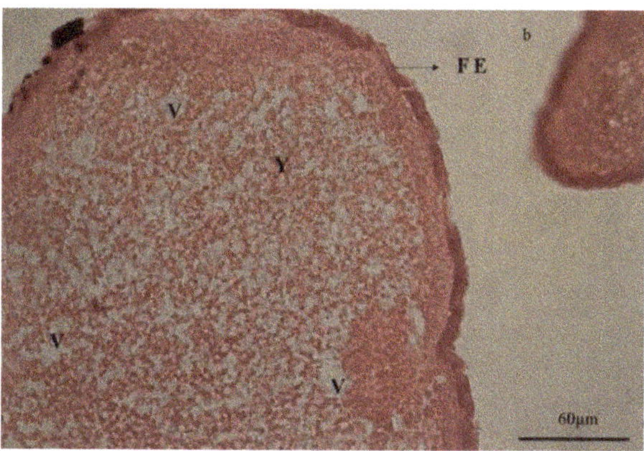

Figure 6. Histology of ovaries in adults of *Glyphodes pyloalis* emerging from untreated (**a**) and treated larvae by *Artemisia annua* floral essential oil (**b**). In treatments of the ovarian sheath significant changes and yolk granules were reduced under the influence of vacuolization in cytoplasm compared to the control. (**FE**) follicular epithelium, (**V**) vacuole, and (**Y**) yolk granules.

4. Discussion

The chemical composition of *A. annua* essential oil in the vegetative stage was investigated in the previous studies [15,36–39], in which terpenes such as 1,8-cineole, camphor, and artemisia ketone were introduced as major constituents. Although 1,8-cineole (6.8%), camphor (13.1%), and artemisia ketone (11.8%) were also identified as main compounds in the essential oil extracted from *A. annua* flowers, some other terpenes such as β-selinene (10.7%), pinocarvone (7.4%), and α-pinene (5.9%) had high amounts. However, a range of minor constituents, including compounds from ester and phenylpropene groups, were also recognized. Such differences can be caused by exogenous and endogenous factors, including geographic location, harvesting time, and the growth stage of plants [40]. The chemical composition of each essential oil has a significant impact on its insecticidal activity. For example, the promising insecticidal effects of terpenes like camphor and 1,8-cineole identified and extracted from essential oils were reported [41,42].

Our study clearly showed decreased enzymatic activity in *G. pyloalis* larvae related to ingestion of *A. annua* essential oil-treated mulberry leaves. Our findings support earlier findings where disruption in insects' physiology and their inability to digest food was reported [43,44]. Reduction in α-amylase, protease, and α- and β-glucosidase, and disruptions on immunology and digestive system in the larvae and the ovary of emerged adults of *G. pyloalis* were described in our results. Such activities are common for botanical insecticides against several insect pests [45–47]. Also, there were further supports for the interference or even deformation of midgut cells, which were responsible for the production of key enzymes in insects [15,48].

Protein plays a key role in digestion, metabolism, and also energy conversion. Klowden (2007) [49] believes that reduction in the insect's protein content after applying biopesticides may stem from the reduction of growth hormone level. We observed a reduction in protein content and also retardation in growth; however, growth hormone level was not worked out. Lipids are other important macromolecules that help the insect reserve energy from feeding. They play a key role in insects' intermediary metabolism and, therefore, they are essential in insect physiology [49]. Significant reduction in the triglyceride content of *G. pyloalis* larvae treated with *A. annua* essential oil was observed in the present study. There are several reasons for reducing insect lipid content after treatments by toxins, alteration in lipid synthesis patterns, and hormonal dysfunction to control its metabolism [49]. Glucose as a key carbohydrate (monosaccharide) was also decreased following treatment with *A. annua* essential oil. This reduction could be related to reduced feeding following treatment, since the essential oil acts as a deterrent [2]. Any disruption causing reducing resources at larval stages could affect insects' survival and reproduction in their later generations. A reduction in protein, lipid, and glucose contents may have adverse effects on the reproductive parameters such as egg production, fertility, and fecundity [50].

Detoxifying enzymes, including esterases and glutathione *S*-transferases, are involved in reducing the impacts of exogenous compounds [51]. In the current study, the activity of detoxifying enzymes, including esterases and glutathione *S*-transferases, was reduced by essential oil of *A. annua* flowers. Certainly, the reduced activity of these enzymes is related to their production halt somewhere in the process of production [15].

Insect cellular immunity is considered as the main system challenging natural enemies entering the insect body [52]. The immunocytes provide the insect ability to combat invading organisms by several means including phagocytosis, nodulation, and encapsulation [53]. So, the reduced immunocytes, as shown for *G. pyloalis* larvae treated with *A. annua* essential oil in the present study, could cause larvae to become susceptible to any invasion [54,55]. The reduced number of hemocytes is mostly due to cytotoxic effect of the botanicals used [56]. We do believe this toxic effect of botanicals to be more reliable as a reasoning for the reduction of immunocytes [57–59].

Phenol oxidase system is considered as the key component in the immune system of insect and a bridge in the gap between cellular and humeral insect immunity. Its action is critically required in the last stage of cellular defense in order to form melanization, a process that terminates the action and kills the pathogenic agent. Phenol oxidase inhibition, documented for *G. pyloalis* larvae treated with *A. annua* essential oil in the present study, probably helps to make the insects susceptible to pathogenic agents if they have not received the toxic concentration [45,58,60].

The insect midgut principal cells are the main cells taking the role of producing the enzymes needed for digestion and then absorbing the nutrients. Therefore, any damages to these cells will lower the activities in digestive enzymes already reported by other researchers [15,31,61]. The elongation and separation of midgut epithelial cells of *G. pyloalis* larvae treated by *A. annua* essential oil were observed in the present study.

Inhibiting insect reproduction has long been the subject of many studies. In lepidopterans, obtaining all nutrients at larval stages is necessary for reproductive development [62]. So, if larval nutrition is disrupted by any means, it will be reflected in adult reproductive function. Our previous findings and the current study display the changes in morphology and histology of emerging adults [15,31]. Our study showed the essential oil of *A. annua* brought about subtle changes in ovarian tissue, such as disruption of follicular cells. As the insect tries to compromise to reduce nutrients in detoxification processes, follicles' cells deplete its content into the oocytes, which then disrupts the cell texture [63].

5. Conclusions

Plant-derived allelochemicals are beneficial agents in controlling pests. As we know, the plant kingdom mainly depends on secondary metabolites to defend against herbivores. With this knowledge in mind, scientists exploit the use of secondary plant chemicals for pest control. One of the main reasons for this increased demand is that the plant-originated chemicals are comparatively safer for humans and the environment. Our study's results clearly document that the essential oil of *A. annua* flowers is toxic to larval mulberry pyralid and disrupt its various physiological systems in a way that the insect can hardly get resistance to it. Consequently, this wild-growing plant in Iran can be considered an efficient natural source capable of controlling insect pests. To apply the research results, it is recommended to evaluate the possible side effects of essential oil on mulberry and the biological control agents in future research. Regarding the insect pest's resistance, identifying specific modes of action of essential oil active components and their overlapping with other insecticides should also be assessed.

Author Contributions: Conceptualization, M.O., J.J.S., and A.E.; methodology, M.O. and J.J.S.; formal analysis, M.O., J.J.S., A.E., and W.N.S.; investigation, M.O.; writing—original draft preparation, M.O., J.J.S., A.E. and P.K.; writing—review and editing, M.O., J.J.S., A.E., P.K. and W.N.S.; supervision, J.J.S. and A.E.; funding acquisition, P.K. All authors have read and agreed to the published version of the manuscript.

Funding: This research received no external funding.

Institutional Review Board Statement: Not applicable.

Informed Consent Statement: Not applicable.

Data Availability Statement: The data that support the findings of this study are available upon request from the authors.

Acknowledgments: This research was financially supported by the University of Guilan, Rasht, Iran, and was partially supported by Chiang Mai University, Thailand, which is greatly appreciated. W.N.S. participated in this work as part of the activities of the Aromatic Plant Research Center (APRC, https://aromaticplant.org/).

Conflicts of Interest: The authors declare no conflict of interest.

References

1. Khosravi, R.; Sendi, J.J. Biology and demography of *Glyphodes pyloalis* Walker (Lepidoptera: Pyralidae) on mulberry. *J. Asia-Pacific Èntomol.* **2010**, *13*, 273–276. [CrossRef]
2. Isman, M.B. Plant essential oils for pest and disease management. *Crop. Prot.* **2000**, *19*, 603–608. [CrossRef]
3. Lamichhane, J.R.; Dachbrodt-Saaydeh, S.; Kudsk, P.; Messéan, A. Toward a reduced reliance on conventional pesticides in european agriculture. *Plant. Dis.* **2016**, *100*, 10–24. [CrossRef] [PubMed]
4. El Asbahani, A.; Miladi, K.; Badri, W.; Sala, M.; Addi, E.H.A.; Casabianca, H.; El Mousadik, A.; Hartmann, D.; Jilale, A.; Renaud, F.N.R.; et al. Essential oils: From extraction to encapsulation. *Int. J. Pharm.* **2015**, *483*, 220–243. [CrossRef] [PubMed]
5. Isman, M.B.; Grieneisen, M.L. Botanical insecticide research: Many publications, limited useful data. *Trends Plant. Sci.* **2014**, *19*, 140–145. [CrossRef]
6. Ebadollahi, A.; Ziaee, M.; Palla, F. Essential oils extracted from different species of the Lamiaceae plant family as prospective bioagents against several detrimental pests. *Molecules* **2020**, *25*, 1556. [CrossRef]
7. Basaid, K.; Chebli, B.; Mayad, E.H.; Furze, J.N.; Bouharroud, R.; Krier, F.; Barakate, M.; Paulitz, T. Biological activities of essential oils and lipopeptides applied to control plant pests and diseases: A review. *Int. J. Pest. Manag.* **2020**, 1–23. [CrossRef]
8. Campos, E.V.; Proença, P.L.; Oliveira, J.L.; Bakshi, M.; Abhilash, P.; Fraceto, L.F. Use of botanical insecticides for sustainable agriculture: Future perspectives. *Ecol. Indic.* **2019**, *105*, 483–495. [CrossRef]
9. Nathan, S.S.; Choi, M.-Y.; Paik, C.-H.; Seo, H.-Y. Food consumption, utilization, and detoxification enzyme activity of the rice leaffolder larvae after treatment with *Dysoxylum* triterpenes. *Pestic. Biochem. Physiol.* **2007**, *88*, 260–267. [CrossRef]
10. Lazarevic, J.; Jevremović, S.; Kostić, I.; Kostić, M.; Vuleta, A.; Jovanović, S.M.; Jovanović, D. Šešlija Toxic, oviposition deterrent and oxidative stress effects of *Thymus vulgaris* essential oil against *Acanthoscelides obtectus*. *Insects* **2020**, *11*, 563. [CrossRef]
11. Afraze, Z.; Sendi, J.J.; Karimi-Malati, A.; Zibaee, A. Methanolic extract of winter cherry causes morpho-histological and immunological ailments in mulberry pyralid *Glyphodes pyloalis*. *Front. Physiol.* **2020**, *11*, 908. [CrossRef] [PubMed]

12. Konovalov, D.; Khamilonov, A.A. Biologically active compounds of *Artemisia annua* essential oil. *Pharm. Pharmacol.* **2016**, *4*, 4–33. [CrossRef]

13. Xiao, L.; Tan, H.; Zhang, L. *Artemisia annua* glandular secretory trichomes: The biofactory of antimalarial agent artemisinin. *Sci. Bull.* **2016**, *61*, 26–36. [CrossRef]

14. Shekari, M.; Sendi, J.J.; Etebari, K.; Zibaee, A.; Shadparvar, A. Effects of *Artemisia annua* L. (Asteracea) on nutritional physiology and enzyme activities of elm leaf beetle, *Xanthogaleruca luteola* Mull. (Coleoptera: Chrysomellidae). *Pestic. Biochem. Physiol.* **2008**, *91*, 66–74. [CrossRef]

15. Hasheminia, S.M.; Sendi, J.J.; Jahromi, K.T.; Moharramipour, S. The effects of *Artemisia annua* L. and *Achillea millefolium* L. crude leaf extracts on the toxicity, development, feeding efficiency and chemical activities of small cabbage *Pieris rapae* L. (Lepidoptera: Pieridae). *Pestic. Biochem. Physiol.* **2011**, *99*, 244–249. [CrossRef]

16. Zibaee, A. Botanical Insecticides and Their Effects on Insect Biochemistry and Immunity. In *Pesticides in the Modern World—Pests Control and Pesticides Exposure and Toxicity Assessment*; IntechOpen: London, UK, 2011.

17. Mojarab-Mahboubkar, M.; Sendi, J.J.; Aliakbar, A. Effect of *Artemisia annua* L. essential oil on toxicity, enzyme activities, and energy reserves of cotton bollworm *Helicoverpa armigera* (Hübner) (Lepidoptera: Noctuidae). *J. Plant. Prot. Res.* **2015**, *55*, 371–377. [CrossRef]

18. Oftadeh, M.; Sendi, J.J.; Ebadollahi, A. Toxicity and deleterious effects of *Artemisia annua* essential oil extracts on mulberry pyralid (*Glyphodes pyloalis*). *Pestic. Biochem. Physiol.* **2020**, *170*, 104702. [CrossRef]

19. Adams, R.P. *Identification of Essential Oil Components by Gas. Chromatography/Mass Spectrometry*, 4th ed.; Allured Publishing Corporation: Carol Stream, IL, USA, 2007.

20. NIST. *NIST17*; National Institute of Standards and Technology: Gaithersburg, MD, USA, 2017.

21. Bernfeld, P. Amylases alpha and beta. *Methods Enzymol.* **1955**, *1*, 140–146.

22. García-Carreño, F.L.; Haard, N.F. Characterization of proteinase classes in Langostilla (*Pleuroncodes planipes*) and Crayfish (*Pacifastacus astacus*) extracts. *J. Food Biochem.* **1993**, *17*, 97–113. [CrossRef]

23. Tsujita, T.; Ninomiya, H.; Okuda, H. P-nitrophenyl 865. buyrate hydrolyzing activity of Hormone-senditive lipase 866 from bovine adipose tissue. *J. Lipid. Res.* **1989**, *30*, 867–997. [CrossRef]

24. Ferreira, C.; Terra, W.R. Physical and kinetic properties of a plasma-membrane-bound β-d-glucosidase (cellobiase) from midgut cells of an insect (*Rhynchosciara americana* larva). *Biochem. J.* **1983**, *213*, 43–51. [CrossRef] [PubMed]

25. Bradford, M.M. A rapid and sensitive method for the quantitation of microgram quantities of protein utilizing the principle of protein-Dye binding. *Anal. Biochem.* **1976**, *72*, 248–254. [CrossRef]

26. Siegert, K.J. Carbohydrate metabolism in *Manduca sexta* during late larval development. *J. Insect Physiol.* **1987**, *33*, 421–427. [CrossRef]

27. Van Asperen, K. A study of housefly esterases by means of a sensitive colorimetric method. *J. Insect Physiol.* **1962**, *8*, 401–416. [CrossRef]

28. Habing, W.H.; Pabst, M.J.; Jakoby, W.B. Glutathione S-transferases. The first step in mercapturic acid formation. *J. Biol. Chem.* **1974**, *249*, 7130–7139.

29. Parkinson, N.M.; Weaver, R.J. Noxious Components of Venom from the Pupa-Specific Parasitoid *Pimpla hypochondriaca*. *J. Invertebr. Pathol.* **1999**, *73*, 74–83. [CrossRef]

30. El-Aziz, N.M.A.; Awad, H.H. Changes in the haemocytes of *Agrotis ipsilon* larvae (Lepidoptera: Noctuidae) in relation to dimilin and Bacillus thuringiensis infections. *Micron* **2010**, *41*, 203–209. [CrossRef]

31. Jones, J.C. Current concepts concerning insect hemocytes. *Am. Zoöl.* **1962**, *2*, 209–246. [CrossRef]

32. Arnold, J.W.; Hinks, C.F. Haemopoiesis in Lepidoptera. I. The multiplication of circulating haemocytes. *Can. J. Zool.* **1976**, *54*, 1003–1012. [CrossRef]

33. Seyedtalebi, F.S.; Safavi, S.; Talaei-Hasanloui, A.R.; Bandani, A.R. Quantitative comparison for some immune responses among *Eurygaster integriceps*, *Ephestia kuehniella* and *Zophobas morio* against the entomopathogenic fungus *Beuveria bassiana*. *Invert. Surviv. J.* **2017**, *14*, 174–181. [CrossRef]

34. LeOra Software. *Polo Plus, a User's Guide to Probit or Logit Analysis*; LeOra Software: Berkeley, CA, USA, 2002.

35. SAS Institute. *SAS/STAT User's Guide for Personal Computers*; SAS Institute: Cary, NC, USA, 1997.

36. Mojarab-Mahboubkar, M.; Sendi, J.J. Chemical composition, insecticidal and physiological effect of methanol extract of sweet wormwood (*Artemisia annua* L.) on *Helicoverpa armigera* (Hübner) (Lepidoptera: Noctuidae). *Toxin Rev.* **2016**, *35*, 106–115. [CrossRef]

37. Bedini, S.; Flamini, G.; Cosci, F.; Ascrizzi, R.; Echeverria, M.C.; Guidi, L.; Landi, M.; Benelli, G.; Conti, B. *Artemisia* spp. essential oils against the disease-carrying blowfly *Calliphora vomitoria*. *Parasit. Vectors* **2017**, *10*, 1–10. [CrossRef] [PubMed]

38. Pandey, A.K.; Singh, P. The Genus *Artemisia*: A 2012–2017 Literature review on chemical composition, antimicrobial, insecticidal and antioxidant activities of essential oils. *Medicines* **2017**, *4*, 68. [CrossRef]

39. Nigam, M.; Atanassova, M.; Mishra, A.P.; Pezzani, R.; Devkota, H.P.; Plygun, S.; Salehi, B.; Setzer, W.N.; Sharifi-Rad, J. Bio-active compounds and health benefits of *Artemisia* species. *Nat Prod. Commun* **2019**. [CrossRef]

40. Isman, M.B. Commercial development of plant essential oils and their constituents as active ingredients in bioinsecticides. *Phytochem. Rev.* **2020**, *19*, 235–241. [CrossRef]

41. Kumar, P.; Mishra, S.; Malik, A.; Satya, S. Repellent, larvicidal and pupicidal properties of essential oils and their formulations against the housefly, *Musca domestica*. *Med. Veter Èntomol.* **2011**, *25*, 302–310. [CrossRef] [PubMed]

42. Filomeno, C.A.; Barbosa, L.C.; Teixeira, R.R.; Pinheiro, A.L.; Farias, E.D.S.; Ferreira, J.S.; Picanço, M.C. Chemical diversity of essential oils of Myrtaceae species and their insecticidal activity against *Rhyzopertha dominica*. *Crop. Prot.* **2020**, *137*, 105309. [CrossRef]

43. Shannag, H.K.; Capinera, J.L.; Freihat, N.M. Effects of neem-based insecticides on consumption and utilization of food in larvae of *Spodoptera eridania* (Lepidoptera: Noctuidae). *J. Insect Sci.* **2015**, *15*, 152. [CrossRef]

44. Zou, C.; Wang, Y.; Zou, H.; Ding, N.; Geng, N.; Cao, C.; Zhang, G. Sanguinarine in Chelidonium majus induced antifeeding and larval lethality by suppressing food intake and digestive enzymes in Lymantria dispar. *Pestic. Biochem. Physiol.* **2019**, *153*, 9–16. [CrossRef]

45. Zibaee, A.; Bandani, A.R. A study on the toxicity of the medicinal plant, *Artemisia annua* L. (Astracea) extracts the Sunn pest, *Eurygaster integriceps* Puton (Heteroptera: Scutelleridae). *J. Plant. Prot. Res.* **2010**, *50*, 48–54. [CrossRef]

46. Bezzar-Bendjazia, R.; Kilani-Morakchi, S.; Maroua, F.; Aribi, N. Azadirachtin induced larval avoidance and antifeeding by disruption of food intake and digestive enzymes in *Drosophila melanogaster* (Diptera: Drosophilidae). *Pestic. Biochem. Physiol.* **2017**, *143*, 135–140. [CrossRef] [PubMed]

47. Magierowicz, K.; Górska-Drabik, E.; Sempruch, C. The effect of Tanacetum vulgare essential oil and its main components on some ecological and physiological parameters of *Acrobasis advenella* (Zinck.) (Lepidoptera: Pyralidae). *Pestic. Biochem. Physiol.* **2020**, *162*, 105–112. [CrossRef] [PubMed]

48. Mishra, M.; Sharma, A.; Dagar, V.S.; Kumar, S. Effects of β-sitosterol on growth, development and midgut enzymes of *Helicoverpa armigera* Hübner. *Arch. Biol. Sci.* **2020**, *72*, 271–278. [CrossRef]

49. Klowden, M.J. *Physiological Systems in Insects*, 2nd ed.; Academic Press: Cambridge, MA, USA, 2007.

50. Wu, M.-Y.; Ying, Y.-Y.; Zhang, S.-S.; Li, X.-G.; Yan, W.-H.; Yao, Y.-C.; Shah, S.; Wu, G.; Yang, F.-L. Effects of diallyl trisulfide, an active substance from garlic essential oil, on energy metabolism in male moth *Sitotroga cerealella* (Olivier). *Insects* **2020**, *11*, 270. [CrossRef]

51. Nattudurai, G.; Baskar, K.; Paulraj, M.G.; Islam, V.I.H.; Ignacimuthu, S.; Duraipandiyan, V. Toxic effect of *Atalantia monophylla* essential oil on *Callosobruchus maculatus* and *Sitophilus oryzae*. *Environ. Sci. Pollut. Res.* **2016**, *24*, 1619–1629. [CrossRef]

52. Perazzolo, L.M.; Gargioni, R.; Ogliari, P.; Barracco, M.A. Evaluation of some hemato-immunological parameters in the shrimp *Farfantepenaeus paulensis* submitted to environmental and physiological stress. *Aquac.* **2002**, *214*, 19–33. [CrossRef]

53. Kraaijeveld, A.R.; Limentani, E.C.; Godfray, H.C.J. Basis of the trade-off between parasitoid resistance and larval competitive ability in Drosophila melanogaster. In *Proceedings of the Royal Society B: Biological Sciences*; The Royal Society: London, UK, 2001; Volume 268, pp. 259–261.

54. Ghasemi, V.; Yazdi, A.K.; Tavallaie, F.Z.; Sendi, J.J. Effect of essential oils from *Callistemon viminalis* and *Ferula gummosa* on toxicity and on the hemocyte profile of *Ephestia kuehniella* (Lep.: Pyralidae). *Arch. Phytopathol. Plant. Prot.* **2014**, *47*, 268–278. [CrossRef]

55. Sadeghi, R.; Eshrati, M.R.; Mortazavian, S.M.M.; Jamshidnia, A. The Effects of the essential oils isolated from four ecotypes of cumin (*Cuminum cyminum* L.) on the blood cells of the pink stem borer, *Sesamia cretica* Ledere (Lepidoptera: Noctuidae). *J. Kans. Èntomol. Soc.* **2019**, *92*, 390–399. [CrossRef]

56. Clark, K.D.; Strand, M.R. Hemolymph melanization in the silkmoth *bombyx mori* involves formation of a high molecular mass complex that metabolizes tyrosine. *J. Biol. Chem.* **2013**, *288*, 14476–14487. [CrossRef]

57. Khanikor, B.; Bora, D. Effect of plant based essential oil on immune response of silkworm, *Antheraea assama* Westwood (Lepidoptera: Saturniidae). *Int. J. Ind. Èntomol.* **2012**, *25*, 139–146. [CrossRef]

58. Ghoneim, K. Disturbed hematological and immunological parameters of insects by botanicals as an effective approach of pest control: A review of recent progress. South Asian. *J. Exp. Biol.* **2018**, *1*, 112–144.

59. Rahimi, V.; Hajizadeh, J.; Zibaee, A.; Sendi, J.J. Changes in immune responses of *Helicoverpa armigera* Hübner followed by feeding on Knotgrass, *Polygonum persicaria agglutinin*. *Arch. Insect Biochem. Physiol.* **2019**, *101*, e21543. [CrossRef] [PubMed]

60. Cerenius, L.; Lee, B.L.; Söderhäll, K. The proPO-system: Pros and cons for its role in invertebrate immunity. *Trends Immunol.* **2008**, *29*, 263–271. [CrossRef] [PubMed]

61. Murfadunnisa, S.; Vasantha-Srinivasan, P.; Ganesan, R.; Senthil-Nathan, S.; Kim, T.-J.; Ponsankar, A.; Kumar, S.D.; Chandramohan, D.; Krutmuang, P. Larvicidal and enzyme inhibition of essential oil from *Spheranthus amaranthroids* (Burm.) against lepidopteran pest *Spodoptera litura* (Fab.) and their impact on non-target earthworms. *Biocatal. Agric. Biotechnol.* **2019**, *21*. [CrossRef]

62. Riddiford, L.M. How does juvenile hormone control insect metamorphosis and reproduction? *Gen. Comp. Endocrinol.* **2012**, *179*, 477–484. [CrossRef]

63. Reis, T.C.; Soares, M.A.; Dos Santos, J.B.; Dos Santos, C.A.; Serrão, J.E.; Zanuncio, J.C.; Ferreira, E.A. Atrazine and nicosulfuron affect the reproductive fitness of the predator *Podisus nigrispinus* (Hemiptera: Pentatomidae). *Anais da Academia Brasileira de Ciências* **2018**, *90*, 3625–3633. [CrossRef]

 *foods*

Article

Use of New Glycerol-Based Dendrimers for Essential Oils Encapsulation: Optimization of Stirring Time and Rate Using a Plackett—Burman Design and a Surface Response Methodology

Chloë Maes [1,2,*], Yves Brostaux [3], Sandrine Bouquillon [1,†] and Marie-Laure Fauconnier [2,†]

[1] Institut de Chimie Moléculaire de Reims, UMR CNRS 7312, Université Reims-Champagne-Ardenne, UFR Sciences, BP 1039 boîte 44, CEDEX 2, 51687 Reims, France; sandrine.bouquillon@univ-reims.fr

[2] Laboratoire de Chimie des Molécules Naturelles, Gembloux Agro-Bio Tech, Université de Liège, 2 Passage des Déportés, 5030 Gembloux, Belgium; marie-laure.fauconnier@uliege.be

[3] Unité de Statistique, Informatique et Modélisation appliquées, Gembloux Agro-Bio Tech, Université de Liège, 2 Passage des Déportés, 5030 Gembloux, Belgium; y.brostaux@uliege.be

* Correspondence: chloe.maes@uliege.be; Tel.: +32-495-879-737

† Contributed equally to the work.

Citation: Maes, C.; Brostaux, Y.; Bouquillon, S.; Fauconnier, M.-L. Use of New Glycerol-Based Dendrimers for Essential Oils Encapsulation: Optimization of Stirring Time and Rate Using a Plackett—Burman Design and a Surface Response Methodology. *Foods* **2021**, *10*, 207. https://doi.org/10.3390/foods10020207

Academic Editor: Qin Wang

Received: 18 December 2020

Accepted: 18 January 2021

Published: 20 January 2021

Abstract: Essential oils are used in an increasing number of applications including biopesticides. Their volatility minimizes the risk of residue but can also be a constraint if the release is rapid and uncontrolled. Solutions allowing the encapsulation of essential oils are therefore strongly researched. In this study, essential oils encapsulation was carried out within dendrimers to control their volatility. Indeed, a spontaneous complexation occurs in a solution of dendrimers with essential oils which maintains it longer. Six parameters (temperature, stirring rate, relative concentration, solvent volume, stirring time, and pH) of this reaction has been optimized by two steps: first a screening of the parameters that influence the encapsulation with a Plackett–Burmann design the most followed by an optimization of those ones by a surface response methodology. In this study, two essential oils with herbicide properties were used: the essential oils of *Cinnamomum zeylanicum* Blume and *Cymbopogon winterianus* Jowitt; and four biosourced dendrimers: glycerodendrimers derived from polypropylenimine and polyamidoamine, a glyceroclikdendrimer, and a glyceroladendrimer. Meta-analysis of all Plackett–Burman assays determined that rate and stirring time were effective on the retention rate thereby these parameters were used for the surface response methodology part. Each combination gives a different optimum depending on the structure of these molecules.

Keywords: essential oil; encapsulation; controlled release; biosourced; surface response methodology

1. Introduction

For the last 70 years, industrial countries intensively used chemical pesticides in order to increase agricultural yields to feed a constantly growing population. Unfortunately, with time passing, controversies and the knowledge about their harmful effects on human health and environment have blown up quickly [1]. In this context, biopesticides are priceless candidates to create new weeds- and crops-managing strategies. Among natural compounds from plant origin, essential oils (EOs) are increasingly used for their various biological properties [2,3].

Essential oils are natural mixtures of volatile compounds frequently used in cosmetics, perfume, and sanitary products for both their fragrance and biological activities [4–7]. Another principal characteristic of EOs is their volatility, which limits residues after treatment. Unfortunately this can be a constraint for their utilization as biopesticide because their spread is not controlled [8]. To counter this, scientists developed several different encapsulation techniques. Depending on their properties, emulsion, coacervation, spray drying,

Foods **2021**, *10*, 207. https://doi.org/10.3390/foods10020207 — https://www.mdpi.com/journal/foods

complexation, ionic gelation, and nanoprecipitation help maintain a controlled release of EOs, either quick or slow [9]. EOs encapsulation may appear useless to enhance herbicidal activities on plants, because shoot death occurs after 1 h to 1 day of application [10]. However, an actual interest exists for the improvement of the seed's germination inhibition effects because this one occurs for longer periods (up to 30 days) thus EO encapsulation with controlled release allows to use a lower concentration. Lethal dose depends on the target plant/seed [11].

Cinnamon and Java citronella essential oils are of particular interest for herbicidal applications in a context where the replacement of conventional herbicides is increasingly wanted [12–14]. In a previous study [12], we determined that the major constituents in cinnamon essential oil are trans-cinnamaldehyde (70%) followed by eugenol, caryophyllene, cinnamyl acetate, and linalool in decreasing concentration order. Java citronella EO is constituted of 57 different molecules; among them citronellal (40%), geraniol (20%), citronellol (15%), limonene (5%), and eugenol (2%) are the main representatives [12,15,16]. The modes of action of the main constituents of these EOs as herbicides are not fully characterized but their interaction with respectively the lipid and protein fraction of the plant plasma membrane might be involved [12].

In the present research, glycerol-based dendrimers (GDs) are proposed as new and original matrix to encapsulate EOs. GDs are macromolecules synthesized from glycerol carbonate (a side product from biofuel production) which already showed good encapsulation ability of contrast agent for medical sectors, metals (nanoparticles), and organic pollutants of used water. Indeed, their tree structure allows intern cavities (Figure 1), from various sizes depending of the dendrimer generation, to retain molecules [17–20]. Glyceroclikdendrimer (GAD) and glyceroladendrimer (GCD) have been recently developed and described in two patents with specific encapsulation abilities toward organic pollutants and metallic salts [21,22]. Beyond the agronomic field, EOs encapsulation within dendrimers can be used in a wide range of applications, including food industry (active packaging) and pharmaceutical (drug delivery system) through their bactericidal, viricidal, and fungicidal activities [23,24].

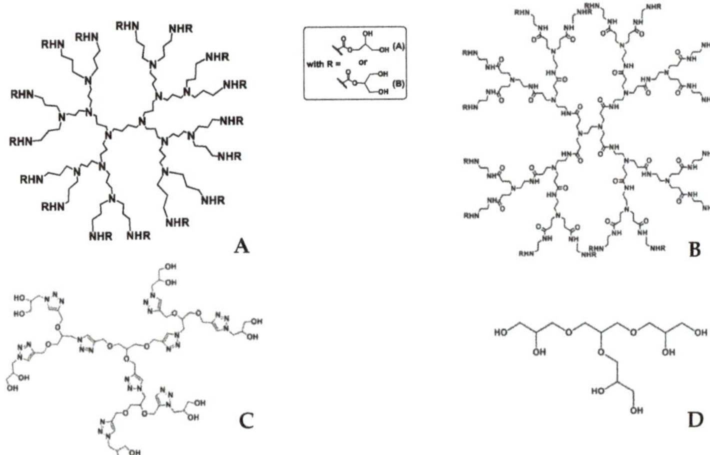

Figure 1. Structures of dendrimers: (**A**) Glycerodendrimers polypropylenimine 3rd generation (GD-PPI-3). (**B**) Glycerodendrimers polyamidoamine 2nd generation (GD-PAMAM). (**C**) Glyceroclikdendrimers 2nd generation (GCD-2). (**D**) Glyceroladendrimers 1st generation (GAD).

The goal of this study is to optimize the encapsulation reaction of two essential oils by four selected dendrimers by maximizing the retention of two GDs, a GCD and a GAD using a Plackett–Burman design (PBD) and response surface methodology (RSM) in order to eventually create an effective biosourced herbicide or for other applications where a slow release of EOs is required. PBDs are a screening design that takes into account a large number of factors with a minimal number of trials, while RSMs are an experimental design intended to optimize factors and their combinations [25]. Obviously, since this study highlights the statistical optimization of the encapsulation, these results can be applied in other fields cited before such as food preservatives creations [26].

2. Materials and Methods

2.1. Chemicals and Reagents

The essential oils of *Cinnamomum zeylanicum* Blume bark (Cinnamon, CAN) and *Cymbopogon winterianus* Jowitt leaves (citronella, CIT) were purchased from Pranarom (Belgium).

Glycerodendrimers-polypropilenImine (GD-PPI) and glycerodendrimers-polyamidoa mine (GD-PAMAM) were synthesized according the previously described work related to the decoration of dendrimers [17,18].

GlycerolADendrimers (GAD) and GlyceroClickDendrimers (GCD) were synthesized following the procedures described in two patents [21,22].

2.2. Essential Oils Encapsulation

Essential oils encapsulation take place by a spontaneous complexation; the dendrimers were dissolved in H_2O (8 mL) and EOs were dissolved in ethanol (various concentrations). EOs solutions or pure ethanol was added to dendrimers solution (3/1 v/v) in a 22 mL glass vial which was directly hermetically sealed with a Teflon cap and covered with an aluminum foil to avoid light interference. Solutions were then stirred for at least 10 min at 100 rpm. According on the stirring settings, an emulsion of EOs occurs in the dendrimer solution, which provides a liquid phase EOs retention. This retention leads to a change in dynamic balance between solution and headspace compared to free EO solution (control), which is quantified by the following analysis.

2.3. Dynamic-Headspace Gas Chromatography–Mass Spectrometry (DHS-GC–MS) Analysis

The percentage of retention (r) of EOs by GDs was determined by dynamic head- space sampling (DHS, Gerstel, Germany) coupled to a thermal desorption unit (TDU, Gerstel, Germany), a gas chromatograph (Agilent Technologies 7890A), and a mass spectrometer (MS, Agilent Technologies 5975C). During treatment in the DHS unit, the vials were conditioned at 25 °C for 30 min with stirring (500 rpm). The head-space sampling was performed on Gerstel TDU desorption tubes (OD 6.00 mm, filled with 60 mg of Tenax TA, Gerstel, Germany), on 200 mL at 20 mL/min, followed by 200 mL at 60 mL/min of drying phase. Desorption then occurred for 10 min at 300 °C and coupled to a cooled injection system (CIS, Gerstel, Germany) set at −80 °C. EOs were then transferred to the GC column (VF-WAXms, Agilent technologies USA; 30 m length, 0.250 mm I.D, 0.25 l m film thickness) for separation with temperature program as follow: Java citronella—from 70 °C (5 min) to 100 °C at a rate of 8 °C/min, then 2 °C/min to 160 °C, and then 20 °C/min to 260 °C (10 min); Cinnamon—from 40 °C (4 min) to 80 °C at a rate 3.5 °C/min, then 5 °C/min to 160 °C, and then 20 °C/min to 220 °C (10 min) with helium as carrier gas at a flow rate of 1.5 mL/min. The MS were recorded in electron ionization mode at 70 eV (scanned mass range: 35 to 300 m/z); source and quadrupole temperature at 230 °C and 150 °C respectively. The component identification was performed by comparison of the recorded spectra with two data libraries (Pal 600K® and Wiley275) and injection of pure commercial standards in the same chromatographic conditions.

The percentage of retention (r) of EOs by GDs was calculated by the equation [27]:

$$r(\%) = \left(1 - \frac{\sum A_D}{\sum A_0}\right) \times 100 \tag{1}$$

$\sum A_D$: sum of peak areas of EO component in the presence of dendrimers, $\sum A_0$: sum of peak areas of EO component in free EO solution (control).

2.4. Screening of Six Encapsulation Parameters with Plackett–Burman Design

Plackett–Burman design was used to select the significant parameters for essential oils encapsulation. This design was applied to four combinations of dendrimers and EOs previously selected owing to their noticeable essential oil retention capacity (preliminary assays, data not shown but published soon). The combinations are: GD-PPI-3/CAN EO, GD-PAMAM-2/CIT EO, GAD-1/CAN EO, GCD-1/CIT EO. The independent parameters were set on the basis of those preliminary analyses, which considered the properties of the dendrimers for relative concentration and pH, the technical feasibility for rate of stirring, the solvent volume, and stirring time and the temperature which can be found in realistic agronomical conditions.

For each combination, a 12-run PBD was applied to evaluate six factors. Each variable was examined at two levels: −1 for the low level and +1 for the high level. Table 1 illustrates these parameters and the corresponding levels used. The values of two levels were set according to our previous preliminary experimental results. In Table 2, representing PBD and experimental results, data listed indicate the variations in retention rate between each combination of dendrimers-Eos, depending on the treatment. Negative values indicated that the opposite effect is observed: presence of dendrimers increase the volatility of EOs.

Table 1. Factors and their levels selected for the Plackett–Burman design.

Factors	Symbol	Levels	
		−1	+1
Temperature (°C)	T	4	20
Rate of stirring (rpm)	R	150	800
Relative concentration (mg/mmol)	C	500	1500
Solvent volume (mL)	V	3	10
Stirring time (min)	D	10	240
pH	P	4	7

Table 2. Experimental setting (12 runs) generated by Minitab® 19 and retention rate for the fourth combinations of dendrimers and essential oils (Eos) (%, experimental).

Run	T	C	V	R	D	P	r (GD-PPI-3/CAN)	r (GD-PAMAM-2/CIT)	r (GAD-1/CAN)	r (GCD-2/CIT)
1	1	−1	1	−1	−1	−1	−30.79	2.29	32.04	54.11
2	1	1	−1	1	−1	−1	22.82	−95.35	18.69	−69.69
3	−1	1	1	−1	1	−1	−5.14	−23.76	−37.17	18.85
4	−1	1	1	1	−1	1	0.74	−10.22	−0.46	−9.39
5	−1	1	−1	−1	−1	1	9.22	3.44	−11.03	37.81
6	1	1	1	−1	1	1	−28.86	−29.84	21.52	−4.44
7	1	1	−1	1	1	−1	17.70	35.88	−40.48	−36.38
8	1	−1	−1	−1	1	1	−36.31	−26.39	−5.58	−19.96
9	1	−1	1	1	−1	1	27.33	−42.67	−3.88	22.40
10	−1	−1	1	1	1	−1	5.40	−104.65	−67.95	22.02
11	−1	−1	−1	−1	−1	−1	21.00	−29.09	30.83	59.72
12	−1	−1	−1	1	1	1	−24.11	−52.64	9.97	9.86

2.5. Optimization of Two Encapsulation Parameters by Response Surface Methodology

Based on the results of the PDB design, only the most influential parameters on the encapsulation reaction have been selected for further optimization through response surface methodology. Experiments were performed according to a design with two parameters and three levels for each parameter [25]. Two blocks have been used to cover the potential heterogeneity during the course of the experiment. The selected independent variables were stirring rate (R) and stirring duration (D). The experimental design in the actual levels is shown in Table 3. As for PBD, variations in retention rate between each couple dendrimers-EOs were recorded. In RSM experimental results (Table 4), negative percentage of retention notifies an increase in EOs volatility in presence of dendrimers.

Maximums were represented with contour plots.

Table 3. Factors and their levels selected for the Box–Behnken design (response surface methodology).

Factors	Symbol	Levels		
		−1	0	+1
Stirring time (min)	D	10	60	240
Rate of stirring (rpm)	R	150	1000	2000

Table 4. Experimental setting (28 runs) generated by Minitab® 19 and retention rate for the fourth combinations of dendrimers and EOs (%, experimental).

Run	D	R	r (GD-PPI-3/CAN)	r (GD-PAMAM-2/CIT)	r (GAD-1/CAN)	r (GCD-2/CIT)	r (GD-PPI-3/CAN (2))
1	1	1	−15.17	4.64	−22.80	−10.01	6.69
2	0	0	12.85	14.96	2.91	23.56	25.64
3	0	0	13.55	9.56	−1.67	15.67	19.47
4	−1	1	3.93	−7.55	7.03	8.09	5.57
5	1	1	39.55	0.78	−12.61	−0.64	18.32
6	0	0	16.69	11.21	6.94	7.49	26.96
7	−1	−1	−30.53	20.49	−4.51	−0.65	−4.92
8	0	−1.4	−23.54	6.90	4.43	−11.46	6.19
9	−1.4	0	−6.82	8.72	−0.08	4.07	3.41
10	1.4	0	12.00	4.71	−19.16	−19.77	22.07
11	0	1.4	21.45	3.10	3.42	0.24	15.74
12	0	0	3.56	18.83	5.36	8.54	29.07
13	0	0	8.25	12.36	−1.60	9.18	32.35
14	0	0	13.59	17.92	7.21	1.67	28.94
15	1	1	34.07	−15.68	−13.38	−0.17	16.49
16	0	0	19.15	11.85	−2.39	8.04	17.92
17	−1	−1	1.79	24.67	−3.37	−53.77	4.54
18	0	0	21.78	13.71	4.26	7.78	23.28
19	1	−1	6.78	5.63	−14.68	−21.57	4.76
20	0	0	19.84	8.08	−9.13	4.06	15.35
21	−1	1	15.20	0.36	10.89	−15.47	−0.43
22	0	0	21.18	13.71	−1.12	1.40	15.45
23	0	1.4	36.87	3.64	−9.70	8.28	26.11
24	1.4	0	34.60	5.06	−19.58	4.00	21.34
25	−1.4	0	15.16	10.99	−0.38	−39.21	−1.98
26	0	−1.4	1.84	17.76	−6.11	−8.88	9.50
27	0	0	29.50	13.46	−6.81	8.14	14.77
28	0	0	28.67	10.72	7.25	7.90	16.70

2.6. Data Analysis

PBD and RSM were designed and processed using Minitab® 19 software [25].

3. Results and Discussion

3.1. Volatiles Profiles and Major Components of EOs

Chromatograms obtained by DHS-GS-MS for encapsulation optimizations show the volatile profiles of both EOs in Figures 2 and 3. Major compounds have been identified as it was previously mentioned [12]. On these figures, chromatograms of control and encapsulation solutions are overlaid which show that the only difference found is in the height (and peak area) of all compounds. Therefore, profiles were similar in the presence and absence of dendrimers. A thorough examination of the retention rate of each compound in Table 5 allows to observe that chemical structures and volumes of the major components of cinnamon EOs (volumes from 210 to 377 Å^3) are more variable than in citronella EOs (volumes from 270 to 303 Å^3), which seems to affect somewhat the profile (12% retention rate variations between eugenol and β-caryophyllene)

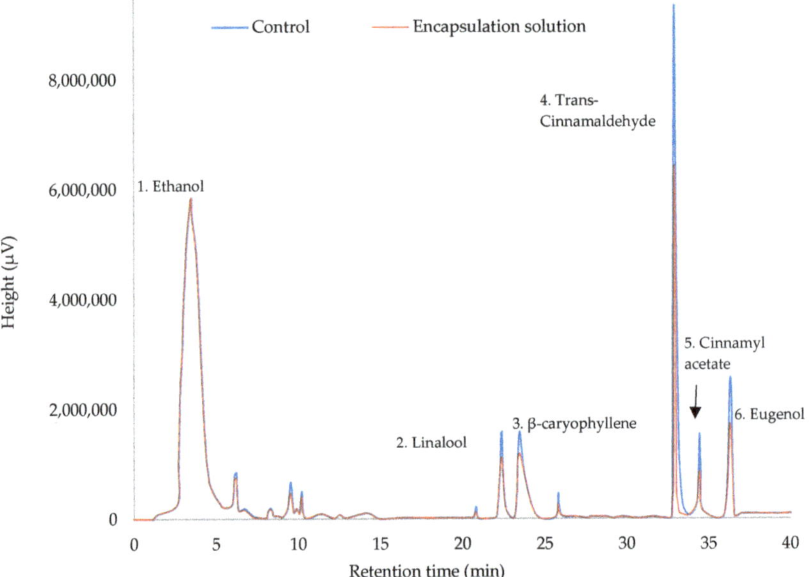

Figure 2. Overlaying of chromatographic analysis of free cinnamon EO (control) and cinnamon EO encapsulated within GD-PPI-3 under optimized conditions—(1) ethanol (sample solvent), (2) linalool, (3) β-caryophyllene, (4) trans-cinnamaldehyde, (5) cinnamyl acetate, (6) eugenol.

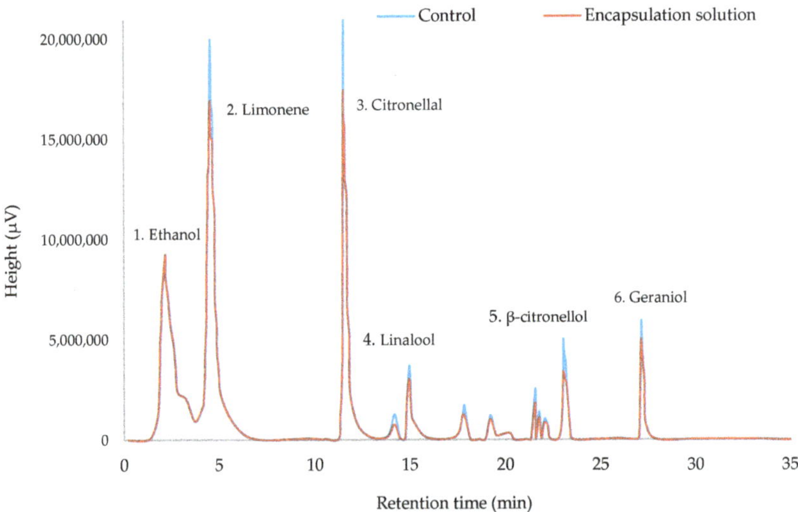

Figure 3. Overlaying of chromatographic analysis of free Java citronella EO (control) and Java citronella EO encapsulated within GD-PAMAM-2 under optimized conditions—(1) ethanol (sample solvent), (2) limonene, (3) citronellal, (4) linalool, (5) β-citronellol, (6) geraniol.

Table 5. Chemical structures and calculated molecular volumes of the major compounds of cinnamon and Java citronella EOs; and their individual retention rate in the optimized encapsulation within dendrimers.

Cinnamon EO	Linalool	β-Caryophyllene	Trans-Cinnamaldehyde	Cinnamyl Acetate	Eugenol	Mean All Components
Volume (Å^3) *	294	377	210	279	257	
r (GAD-1)	9.18%	8.76%	10.93%	11.73%	12.21%	10.89%
r (GD-PPI-3)	28.21%	26.59%	35.99%	32.82%	38.97%	32.35%

Citronella EO	Limonene	Citronellal	Linalool	β-citronellol	Geraniol	Mean All Components
Volume (Å^3) *	270	297	294	303	291	
r (GCD-2)	11.76%	13.22%	12.98%	14.45%	13.65%	13.56%
r (GD-PAMAM-2)	22.99%	23.67%	21.92%	24.01%	23.51%	24.67%

* V = M/dN$_A$ with M: molecular weight; d: density; N$_A$: Avogadro's number [27].

3.2. Influence of Parameters with PBD

In the present study, the dendrimer/EOs complexes were successfully prepared by a simple solubilization and stirring in controlled conditions. To minimize the experimental runs and times for the screening of the encapsulation parameters, the PBD was applied on the basis of two coded levels of the six independent variables, resulting in twelve experiments (Table 2).

Analysis of PBD has been done for each couple dendrimer/EO (Table 6) which showed that almost no one had a variable influencing significantly the encapsulation rate ($p < 0.05$). However, the meta-analysis of all results and a particular attention at the ranking of variables show that time and rate of stirring appeared important in the encapsulation process. Considering that, it seems the lack of significance of these results reveals that the influence had been attenuated by the variability among repetitions in the manipulations. Both parameters (duration and rate of stirring) were selected for further optimization both with RSM.

3.3. Rate and Duration Stirring Optimization with Response Surface Methodology
3.3.1. GD-PPI-3/CAN

For the first studied combination of dendrimer/EO, initially settled parameters were not optimal to find a maximum (Figure 4A) so new ones were defined in Table 7. Figure 5A shows that the model with those parameters was significant, with F-value equal to 10.34 and *p*-value < 0.001. Despite a slight rejection of the lack-of-fit test ($p = 0.022$) the applied model presented a good fitting to the encapsulation efficiency response (Figure 5B).

Table 6. Analyses of variance (ANOVA) of Plackett–Burman screening design experiments.

	Effect Size	**Coefficient**	**Std Error**	**F-Value**	*p*-**Value**
GD-PPI-3/CAN					
Constant		−2.86	7.28	0.74	0.643
T	3.64	−1.82	7.28	0.74	0.643
C	11.22	5.61	7.28	0.06	0.813
V	−4.71	−2.36	7.28	0.59	0.476
R	17.9	8.95	7.28	0.1	0.759
D	−13.36	−6.68	7.85	1.51	0.274
P	−16.06	−8.03	7.28	0.72	0.434
GD-PAMAM−2/CIT					
Constant		−31.6	14.9	0.3	0.913
T	11.1	5.6	14.9	0.14	0.723
C	23.2	11.6	14.9	0.61	0.47
V	−6.5	−3.2	14.9	0.05	0.837
R	−28.7	−14.4	14.9	0.93	0.379
D	−6	−3	16	0.03	0.859
P	8.4	4.2	14.9	0.08	0.789
GAD-1/CAN					
Constant		−7.34	6.11	3.17	0.113
T	22.11	11.06	6.11	3.27	0.113
C	2.04	1.02	6.11	0.03	0.13
V	−3.96	−1.98	6.11	0.1	0.874
R	−28.56	−14.28	6.11	5.46	0.759
D	−45.6	−22.8	6.59	11.98	0.067
P	6.67	3.33	6.11	0.3	0.018
GCD-2/CIT					
Constant		5.41	6.93	3.79	0.083
T	−28.8	−14.4	6.93	3.79	0.083
C	−31.9	−15.95	6.93	4.32	0.092
V	23.7	11.85	6.93	5.3	0.07
R	−37.88	−18.94	6.93	2.93	0.148
D	−20.03	−10.02	7.47	7.48	0.041
P	−5.4	−2.7	6.93	1.8	0.237

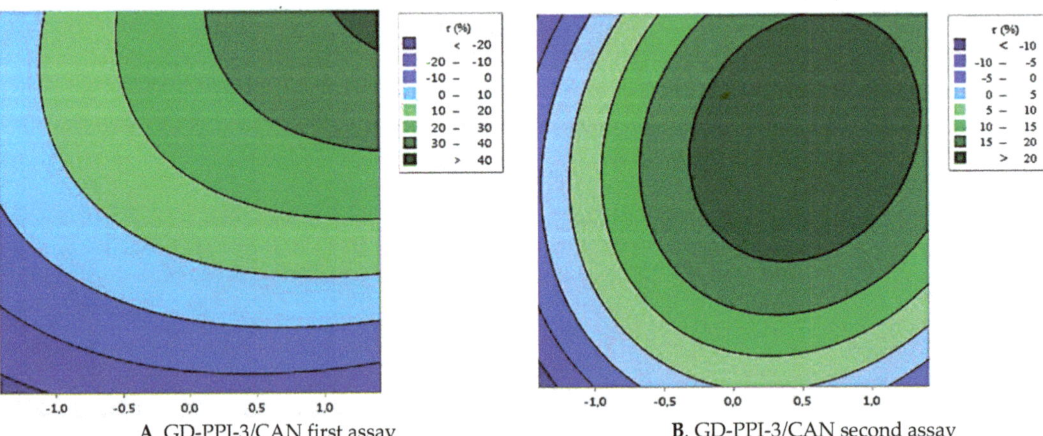

Figure 4. Contour plots showing the crossed effect of duration (D) and rate of stirring (R) on the retention rate (r) of cinnamon essential oil by GD-PPI-3 with the first sets of parameters (**A**) and the second one (**B**).

Table 7. Factors and their levels selected for the second assay of Box–Behnken design (response surface methodology) for the GD-PPI-3/CAN EO encapsulation.

Factors	Symbol	Levels		
		−1	0	+1
Stirring time (min)	D	60	240	420
Rate of stirring (rpm)	R	100	1500	2000

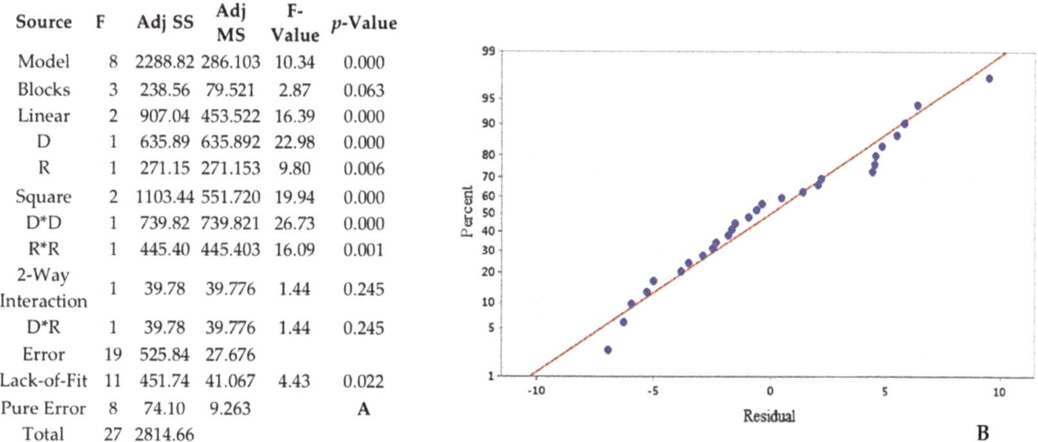

Source	F	Adj SS	Adj MS	F-Value	*p*-Value
Model	8	2288.82	286.103	10.34	0.000
Blocks	3	238.56	79.521	2.87	0.063
Linear	2	907.04	453.522	16.39	0.000
D	1	635.89	635.892	22.98	0.000
R	1	271.15	271.153	9.80	0.006
Square	2	1103.44	551.720	19.94	0.000
D*D	1	739.82	739.821	26.73	0.000
R*R	1	445.40	445.403	16.09	0.001
2-Way Interaction	1	39.78	39.776	1.44	0.245
D*R	1	39.78	39.776	1.44	0.245
Error	19	525.84	27.676		
Lack-of-Fit	11	451.74	41.067	4.43	0.022
Pure Error	8	74.10	9.263		
Total	27	2814.66			

A

B

Figure 5. Analysis of variance (ANOVA) for the response surface methodology (RSM) (**A**) and normal probability plot of the residuals of GD-PPI−3/CAN EO (2) (**B**).

As the model is trustworthy, we can focus on the influence and optimization of factors. Linear and square of each parameter were significant (p-value < 0.05), so they were both influencing the encapsulation rate following the curves independently because their interaction (D*R) was not significant (p-value = 0.245). The regression equation describing these mathematical relationships is:

$$(r) = 22.6 + 6.30\,D + 4.12\,R - 7.08\,D^2 - 5.49\,R^2 + 2.23\,D \times R \qquad (2)$$

Contour plot present in Figure 4B illustrates the level of parameters that allowed to reach the maximum of retention (>20%) which can be found with a stirring time between 240 and 420 min at a rate between 1500 and 2000 rpm.

3.3.2. GD-PAMAM-2/CIT

Second studied combination of dendrimer/EO showed that the model was significant with an F-value of 6.07 and p-value is 0.001 (Figure 6A). In addition, Figure 6B revealed a good correspondence between the linear regression model of RSM and the experimental data despite a slight rejection of the lack-of-lit test (p-value = 0.011). As for the first combination, linear and square of each parameter were significant but not their respective interaction. The regression equation describing these mathematical relationships is:

$$(r) = 13.03 - 3.54\,D - 6.43\,R - 3.69\,D^2 - 3.45\,R^2 + 3.40\,D \times R \qquad (3)$$

Source	DF	Adj SS	Adj MS	F-Value	*p*-Value
Model	8	1380.02	172.503	6.07	0.001
Blocks	3	75.54	25.181	0.89	0.466
Linear	2	861.85	430.927	15.16	0.000
D	1	200.55	200.549	7.06	0.016
R	1	661.31	661.306	23.27	0.000
Square	2	350.30	175.150	6.16	0.009
D*D	1	201.16	201.157	7.08	0.015
R*R	1	176.02	176.020	6.19	0.022
2-Way Interaction	1	92.32	92.322	3.25	0.087
D*R	1	92.32	92.322	3.25	0.087
Error	19	539.91	28.416		
Lack-of-Fit	11	478.05	43.459	5.62	0.011
Pure Error	8	61.86	7.732		**A**
Total	27	1919.93			

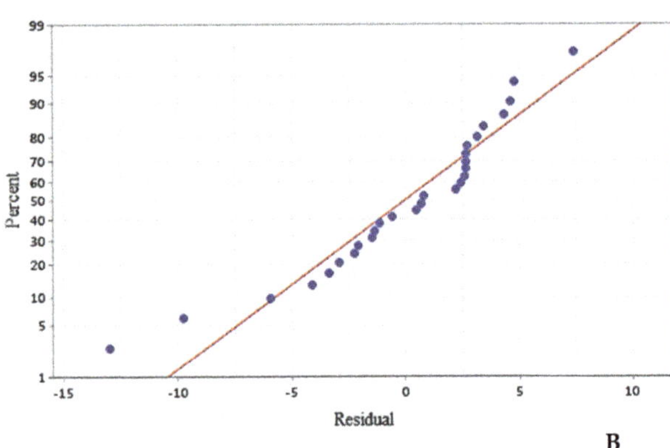

Figure 6. Analysis of variance (ANOVA) for the RSM (**A**) and normal probability plot of the residuals of GD-PAMAM-2/CIT EO (**B**).

Contour plot present in Figure 7. illustrates that a stirring during between 10 and 60 min at a rate between 150 and 1000 rpm allowed to reach the maximum of retention (>15%).

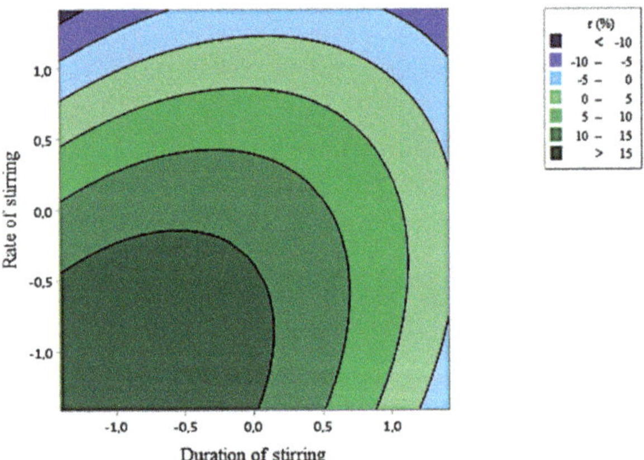

Figure 7. Contour plots showing the crossed effect of duration (D) and rate of stirring (R) on the retention rate (r) of citronella essential oil by GD-PAMAM-2.

3.3.3. GAD-1/CAN

Third studied combination of dendrimer/EO showed that the model is significant with an F-value of 7.06 and *p*-value < 0.001 (Figure 8A) and the lack-of-lit is non-significant (*p*-value = 0.645). In addition, Figure 8B reveals a good correspondence between the linear regression model of RSM and the experimental data. Linear and square of only the duration of stirring are significant (*p*-value of R is 0.175 and R2 is 0.258) and influence the encapsulation rate following the curves. The regression equation describing these mathematical relationships is:

$$(r) = 0.93 - 7.98\,D + 1.92\,R - 5.56\,D^2 - 1.66\,R^2 - 1.79\,D \times R \tag{4}$$

Source	DF	Adj SS	Adj MS	F-Value	*p*-Value
Model	8	1684.78	210.60	7.06	0.000
Blocks	3	102.11	34.04	1.14	0.358
Linear	2	1077.51	538.76	18.07	0.000
D	1	1018.30	1018.30	34.16	0.000
R	1	59.21	59.21	1.99	0.175
Square	2	479.60	239.80	8.04	0.003
D*D	1	457.14	457.14	15.33	0.001
R*R	1	40.57	40.57	1.36	0.258
2-Way Interaction	1	25.56	25.56	0.86	0.366
D*R	1	25.56	25.56	0.86	0.366
Error	19	566.46	29.81		
Lack-of-Fit	11	296.33	26.94	0.80	0.645
Pure Error	8	270.13	33.77		**A**
Total	27	2251.23			

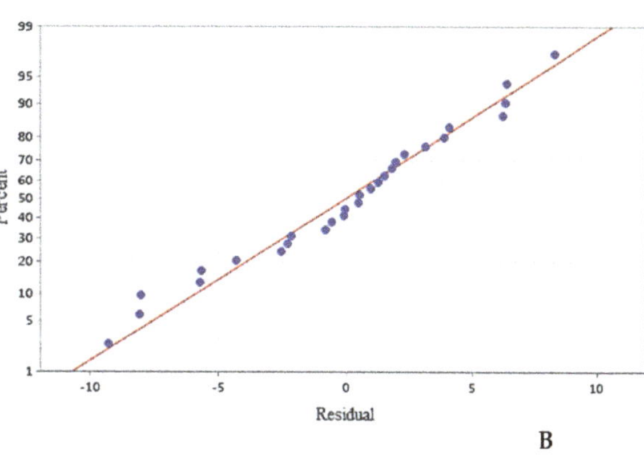

Figure 8. Analysis of variance (ANOVA) for the RSM (**A**) and normal probability plot of the residuals of GAD-1/CAN EO (**B**).

Contour plot present in Figure 9 illustrates the level of parameters that allow to reach the maximum of retention even if this one is very low (>5%). The best results can be found with a stirring time between 10 and 30 min at a rate between 1500 and 2000 rpm.

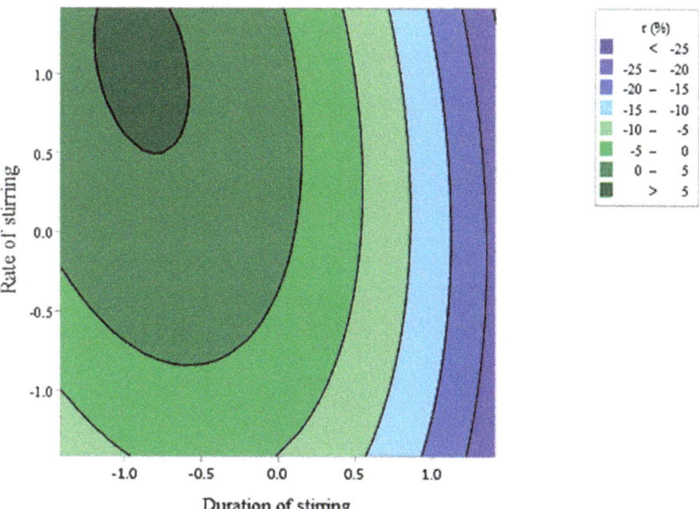

Figure 9. Contour plots showing the crossed effect of duration (D) and rate of stirring (R) on the retention rate (r) of cinnamon essential oil by GAD-1.

3.3.4. GCD-2/CIT

The last studied combination of dendrimer/EO showed that the model is significant with an F-value of 4.17 and *p*-value = 0.005 (Figure 10A) however, lack-of-lit is rejected with a *p*-value equal to 0.003 so results have to be discussed. Nevertheless, Figure 10B reveals a good correspondence between the linear regression model of RSM and experimental data which confirms the global correctness of the model. Only the linear effect rate of stirring was significant (*p*-value = 0.240) and the square effect of both parameters were significant. The regression equation describing these mathematical relationships is:

$$(r) = 8.62 + 3.55\,D + 7.41\,R - 11.65\,D^2 - 6.77\,R^2 - 2.04\,D \times R \qquad (5)$$

Source	DF	Adj SS	Adj MS	F-Value	*p*-Value
Model	8	4579.65	572.46	4.17	0.005
Blocks	3	946.91	315.64	2.30	0.110
Linear	2	1081.01	540.51	3.94	0.037
D	1	201.66	201.66	1.47	0.240
R	1	879.35	879.35	6.41	0.020
Square	2	2518.59	1259.30	9.17	0.002
D*D	1	2006.17	2006.17	14.61	0.001
R*R	1	676.79	676.79	4.93	0.039
2-Way Interaction	1	33.13	33.13	0.24	0.629
D*R	1	33.13	33.13	0.24	0.629
Error	19	2608.53	137.29		
Lack-of-Fit	11	2405.44	218.68	8.61	0.003
Pure Error	8	203.09	25.39		
Total	27	7188.18			

A

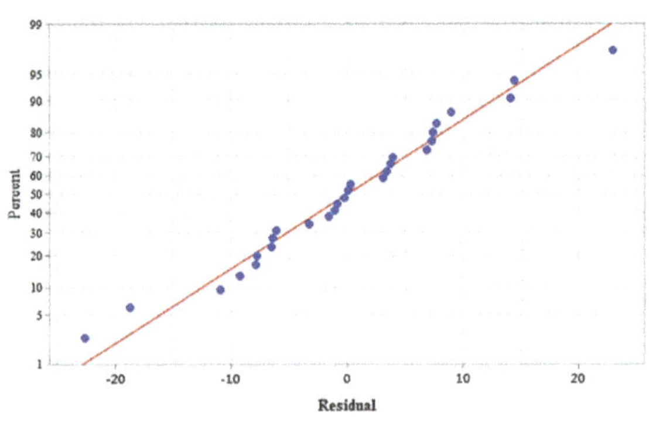

B

Figure 10. Analysis of variance (ANOVA) for the RSM (**A**) and normal probability plot of the residuals of GCD-2/CIT EO (**B**).

Contour plot present in Figure 11 illustrates the level of parameters that allowed to reach the maximum of retention (>10%) which was found with a stirring during around 60 min at a rate of 1500 rpm.

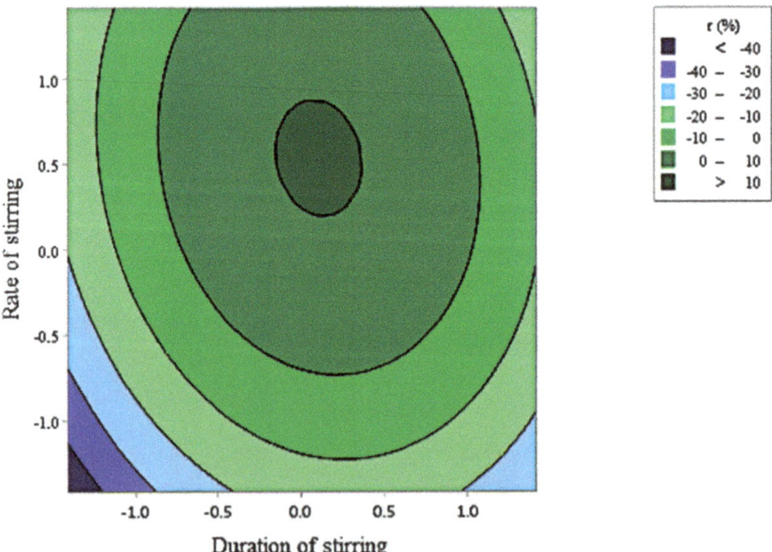

Figure 11. Contour plots showing the crossed effect of duration (D) and rate of stirring (R) on the retention rate (r) of citronella essential oil by GCD-2.

4. Conclusions

For the first time, essential oils encapsulation by bio-sourced dendrimers was successfully carried out, and this reaction was optimized using PBD and RSM. The first part proved that only the rate and the time of stirring influenced the retention rate among the six factors analyzed. The second part optimizes both factors for each couple dendrimer/EO and resulted in very different results. This is quite understandable considering the apolar nature of the EOs' constituents and the differences of structure between dendrimers. Indeed, we can see in Figure 1 that even if all dendrimers contain glycerol or glycerol derivatives in the intern structure or on the periphery of the dendrimer, and a polar surface, the properties of the cores are different. On one point, the core of GD-PAMAM-2 is more polar than the GD-PPI-3's one; on another point, some have strong steric hindrance and important electronic charge (GCD-2) while others are less energy-intensive (GAD-1). Previous study about encapsulation by dendrimers showed that the hydrodynamic radius of GD-PPI and GD-PAMAM influenced the encapsulation and that one occurred at the core level of dendrimers rather than at its periphery. Metal complexes were successfully encapsulated in the fourth and fifth generation of GD-PPI (around 25% of encapsulation rate), but not in the third probably because this one had a smaller hydrodynamic radius (2.81 nm) [20]. Organic compounds as β-estradiol, atrazine, diclofenac salt, or diuron have been also encapsulated in GD-PPI-4 and GD-PAMAM-3 up to 95% [18]. As the trans-cinnamaldehyde (Table 5), one of the major compounds responsible of herbicidal activity, is a smaller molecule than the previous encapsulated ones, it seems obvious that smaller dendrimer generations give here the best results for its encapsulation. Furthermore, this α,β-unsaturated aldehyde presents an important electronic density as the previous organic compounds used. It must be pointed out that chromatographic profiles were similar, for EOs encapsulated in dendrimers or not (control) which suggests that all compounds of each EOs were encapsulated in the same way (Figures 2 and 3). It can be concluded that first the size of molecules encapsulated in comparison with size of intern cavities of dendrimers, and secondly the amount of free electron in the EOs (aromatic circle and double bonds promotes electrostatic interactions) appear to be principal factors influencing the EOs encapsulation within dendrimers [28].

In the optimized conditions, the best encapsulation rates varied from 5 to 40% depending on the dendrimer-EO combination (Table 8). The combination of GD-PPI-3 with cinnamon EO leads to the most promising results with an r = 40% when the stirring is long (6 h) and strong (1735 rpm). As there is no other study on encapsulation of EOs within GDs yet, comparing these results with previous results is not possible. However a comparison with other encapsulations techniques can be done: for example, dendrimers have a better encapsulation rate than the powder optimized by Huynh T. V. et al. who obtained 18% as optimum EO concentration [29]. On the opposite, the rate of encapsulation is quite lower than encapsulation by coacervation in gelatin optimized by Sutaphanit P. and Chitprasert P. (66.5 to 98.4%) but the release from these capsules is almost impossible (stable for 18 months storage) [30]. In another field of application, optimized encapsulation of gallic acid in calcium alginate microbeads was of the same order (42.8%) [31].

Table 8. Optimized values of stirring rate and time for all combinations obtained using RSMs.

Combination	Stirring Rate	Stirring Time	Encapsulation Rate
GD-PPI-3/CAN EO	1735 rpm	366 min	39.92%
GD-PAMAM-2/CIT EO	120 rpm	10 min	19.93%
GAD-1/CAN EO	2142 rpm	9 min	9.75%
GCD-2/CIT EO	1528 rpm	65 min	10.78%

In the context of the use of dendrimer-EOs formulations as biopesticide, it is essential to go further in the study of the encapsulation rate with a dynamic study of the release of EOs by the dendrimer. It is also worthwhile to determine the stability and biological effects of the new biosourced herbicide formulation. In addition, it would be relevant to study with a more fundamental point of view the encapsulation of the selected pure compounds from EO like trans-cinnamaldehyde within GD-PPI-3 to better understand the interactions between EO constituent and dendrimer particularly through NMR studies. This work is in progress.

This article shows for the first time that it is possible to effectively encapsulate essential oils in dendrimers. Given the numerous biocidal properties of essential oils, this technique opens the road to numerous applications in agronomy but also in other sectors where a slow release of essential oils is being researched, such as in pharmaceuticals or in the food industry with the design of innovative packaging.

Author Contributions: Conceptualization, M.-L.F. and S.B.; methodology, C.M., M.-L.F., S.B., and Y.B.; formal analysis, C.M.; data curation, C.M.; writing—original draft preparation, C.M.; writing—review and editing, C.M., M.-L.F., S.B., and Y.B.; supervision, Y.B., M.-L.F., and S.B. All authors have read and agreed to the published version of the manuscript.

Funding: We are grateful to the Universities of Reims Champagne Ardenne (France) and Liège (Belgium) for material funds and the doctoral position to Chloë Maes. This research was supported by the Education, Audiovisual and Culture Executive Agency (EACEA), through EOHUB project 600873EPP-1-2018-1ES-EPPKA2-KA. Published with the assistance of the University Foundation of Belgium.

Institutional Review Board Statement: Not applicable.

Informed Consent Statement: Not applicable.

Data Availability Statement: The data presented in this study are available on request from the corresponding author.

Acknowledgments: We thank Franck Michiels and Pierre Jacquet for their technical help, and Chun Yu Yang for his proofreading.

Conflicts of Interest: The authors declare no conflict of interest.

References

1. Bruggen, V.J., Jr. Environmental and health effects of the herbicide glyphosate. *Sci. Total Environ.* **2017**, *616–617*, 255–268. [CrossRef] [PubMed]
2. Werrie, P.-Y.; Durenne, B.; Delaplace, P.; Fauconnier, M.-L. Phytotoxicity of Essential Oils: Opportunities and Constraints for the Development of Biopesticides. A Review. *Foods* **2020**, *9*, 1291. [CrossRef] [PubMed]
3. De Clerck, C.; Dal Maso, S.; Parisi, O.; Dresen, F.; Zhiri, A.; Jijakli, M.H. Screening of Antifungal and Antibacterial Activity of 90 Commercial Essential Oils against 10 Pathogens of Agronomical Importance. *Foods* **2020**, *9*, 1418. [CrossRef]
4. Bakkali, F.; Averbeck, S.; Averbeck, D.; Idaomar, M. Biological effects of essential oils—A review. *Food Chem. Toxicol.* **2008**, *46*, 446–475. [CrossRef] [PubMed]
5. Nea, F.; Tanoh, E.A.; Wognin, E.L.; Kenne Kemene, T.; Genva, M.; Saive, M.; Tonzibo, Z.F.; Fauconnier, M.-L. A new chemotype of Lantana rhodesiensis Moldenke essential oil from Côte d'Ivoire: Chemical composition and biological activities. *Ind. Crops Prod.* **2019**, *141*, 111766. [CrossRef]
6. Tanoh, E.A.; Boué, G.B.; Nea, F.; Genva, M.; Wognin, E.L.; Ledoux, A.; Martin, H.; Tonzibo, Z.F.; Frederich, M.; Fauconnier, M.-L. Seasonal Effect on the Chemical Composition, Insecticidal Properties and Other Biological Activities of Zanthoxylum leprieurii Guill. & Perr. Essential Oils. *Foods* **2020**, *9*, 550.
7. Benali, T.; Habbadi, K.; Khabbach, A.; Marmouzi, I.; Zengin, G.; Bouyahya, A.; Chamkhi, I.; Chtibi, H.; Aanniz, T.; Achbani, E.H.; et al. GC–MS Analysis, Antioxidant and Antimicrobial Activities of Achillea Odorata Subsp. Pectinata and Ruta Montana Essential Oils and Their Potential Use as Food Preservatives. *Foods* **2020**, *9*, 668. [CrossRef]
8. Koul, O.; Walia, S.; Dhaliwal, G.S. Essential Oils as Green Pesticides: Potential and Constraints. *Biopestic. Int.* **2008**, *4*, 63–84.
9. Maes, C.; Bouquillon, S.; Fauconnier, M.-L. Encapsulation of Essential Oils for the Development of Biosourced Pesticides with Controlled Release: A Review. *Molecules* **2019**, *24*, 2539. [CrossRef]
10. Tworkoski, T. Herbicide effects of essential oils. *Weed Sci.* **2002**, *50*, 425–431. [CrossRef]
11. Cavalieri, A.; Caporali, F. Effects of essential oils of cinnamon, lavender and peppermint on germination of Mediterranean weeds. *Allelopath. J.* **2010**, *25*, 441–452.
12. Lins, L.; Dal Maso, S.; Foncoux, B.; Kamili, A.; Laurin, Y.; Genva, M.; Jijakli, M.H.; De Clerck, C.; Fauconnier, M.L.; Deleu, M. Insights into the Relationships Between Herbicide Activities, Molecular Structure and Membrane Interaction of Cinnamon and Citronella Essential Oils Components. *Int. J. Mol. Sci.* **2019**, *20*, 4007. [CrossRef] [PubMed]

13. Chaimovitsh, D.; Shachter, A.; Abu-Abied, M.; Rubin, B.; Sadot, E.; Dudai, N. Herbicidal Activity of Monoterpenes Is Associated with Disruption of Microtubule Functionality and Membrane Integrity. *Weed Sci.* **2017**, *65*, 19–30. [CrossRef]

14. Díaz-Galindo, E.P.; Nesic, A.; Bautista-Baños, S.; Dublan García, O.; Cabrera-Barjas, G. Corn-Starch-Based Materials Incorporated with Cinnamon Oil Emulsion: Physico-Chemical Characterization and Biological Activity. *Foods* **2020**, *9*, 475. [CrossRef]

15. Espín, J.C.; García-Ruiz, P.A.; Tudela, J.; García-Cánovas, F. Study of stereospecificity in mushroom tyrosinase. *Biochem. J.* **1998**, *331*, 547–551. [CrossRef]

16. Baser, K.H.C.; Buchbauer, G. *Handbook of Essential Oils*, 3rd ed.; CRC Press: Boca Raton, FL, USA, 2020; ISBN 9781351246460.

17. Balieu, S.; El Zein, A.; De Sousa, R.; Jérôme, F.; Tatibouët, A.; Gatard, S.; Pouilloux, Y.; Barrault, J.; Rollin, P.; Bouquillon, S. One-Step Surface Decoration of Poly(propyleneimines) (PPIs) with the Glyceryl Moiety: New Way for Recycling Homogeneous Dendrimer-Based Catalysts. *Adv. Synth. Catal.* **2010**, *352*, 1826–1833. [CrossRef]

18. Menot, B.; Stopinski, J.; Martinez, A.; Oudart, J.-B.; Maquart, F.-X.; Bouquillon, S. Synthesis of surface-modified PAMAMs and PPIs for encapsulation purposes: Influence of the decoration on their sizes and toxicity. *Tetrahedron* **2015**, *71*, 3439–3446. [CrossRef]

19. Menot, B.; Salmon, L.; Bouquillon, S. Platinum Nanoparticles Stabilized by Glycerodendrimers: Synthesis and Application to the Hydrogenation of α,β-Unsaturated Ketones under Mild Conditions. *Eur. J. Inorg. Chem.* **2015**, *2015*, 4518–4523. [CrossRef]

20. Balieu, S.; Cadiou, C.; Martinez, A.; Nuzillard, J.M.; Oudart, J.B.; Maquart, F.X.; Chuburu, F.; Bouquillon, S. Encapsulation of contrast imaging agents by polypropyleneimine-based dendrimers. *J. Biomed. Mater. Res. Part A* **2013**, *101*, 613–621. [CrossRef]

21. Hayouni, S.; Menot, B.; Bouquillon, S. Dendrimer-Type Compounds, Methods for Producing Same and Uses Thereof. WO/2020/021034, 30 January 2020.

22. Hayouni, S.; Menot, B.; Bouquillon, S. PCT/EP2019/070092: Dendrimer-Type Compounds, Methods for Producing Same and Uses Thereof. WO/2020/021032A1, 30 January 2020.

23. Mounesan, M.; Akbari, S.; Brycki, B.E. Extended-release essential oils from poly(acrylonitrile) electrospun mats with dendritic materials. *Ind. Crops Prod.* **2020**, *160*, 113094. [CrossRef]

24. Yousefi, M.; Narmani, A.; Jafari, S.M. Dendrimers as efficient nanocarriers for the protection and delivery of bioactive phytochemicals. *Adv. Colloid Interface Sci.* **2020**, *278*, 102125. [CrossRef] [PubMed]

25. Ferreira, S.; Bruns, R.; Ferreira, H.; Matos, G.; David, J.; Brandão, G.; da Silva, E.; Portugal, L.; dos Reis, P.; Souza, A.; et al. Box-Behnken design: An alternative for the optimization of analytical methods. *Anal. Chim. Acta* **2007**, *597*, 179–186. [CrossRef] [PubMed]

26. Ghaderi-Ghahfarokhi, M.; Barzegar, M.; Sahari, M.A.; Ahmadi Gavlighi, H.; Gardini, F. Chitosan-cinnamon essential oil nanoformulation: Application as a novel additive for controlled release and shelf life extension of beef patties. *Int. J. Biol. Macromol.* **2017**, *102*, 19–28. [CrossRef] [PubMed]

27. Kfoury, M.; Auezova, L.; Greige-Gerges, H.; Fourmentin, S. Promising applications of cyclodextrins in food: Improvement of essential oils retention, controlled release and antiradical activity. *Carbohydr. Polym.* **2015**, *131*, 264–272. [CrossRef] [PubMed]

28. Amariei, G.; Boltes, K.; Letón, P.; Iriepa, I.; Moraleda, I.; Rosal, R. Poly(amidoamine) dendrimers grafted on electrospun poly(acrylic acid)/poly(vinyl alcohol) membranes for host–guest encapsulation of antioxidant thymol. *J. Mater. Chem. B* **2017**, *5*, 6776–6785. [CrossRef]

29. Huynh, T.V.; Caffin, N.; Dykes, G.A.; Bhandari, B. Optimization of the Microencapsulation of Lemon Myrtle Oil Using Response Surface Methodology. *Dry. Technol.* **2008**, *26*, 357–368. [CrossRef]

30. Chitprasert, P.; Sutaphanit, P. Holy basil (ocimum sanctum linn.) Essential oil delivery to swine gastrointestinal tract using gelatin microcapsules coated with aluminum carboxymethyl cellulose and beeswax. *J. Agric. Food Chem.* **2014**, *62*, 12641–12648. [CrossRef]

31. Essifi, K.; Lakrat, M.; Berraaouan, D.; Fauconnier, M.-L.; El Bachiri, A.; Tahani, A. Optimization of gallic acid encapsulation in calcium alginate microbeads using Box-Behnken Experimental Design. *Polym. Bull.* **2020**, 1–26. [CrossRef]

Article

Insecticidal Activity of 25 Essential Oils on the Stored Product Pest, *Sitophilus granarius*

Sébastien Demeter [1,*], Olivier Lebbe [1], Florence Hecq [1], Stamatios C. Nicolis [2], Tierry Kenne Kemene [3], Henri Martin [3], Marie-Laure Fauconnier [3] and Thierry Hance [1]

[1] Biodiversity Research Center, Earth and Life Institute, Université Catholique de Louvain, 4-5 Place Croix du Sud, 1348 Louvain-la-Neuve, Belgium; olivier.lebbe@uclouvain.be (O.L.); florence.hecq@gmail.com (F.H.); thierry.hance@uclouvain.be (T.H.)
[2] Interdisciplinary Center for Nonlinear Phenomena and Complex System, Université Libre de Bruxelles, Campus Plaine, CP 231 bd du Triomphe, 1050 Brussels, Belgium; stamatios.nicolis@ulb.be
[3] Laboratory of Chemistry of Natural Molecules, Gembloux Agro-Bio Tech, Université de Liège, 2 Passage des Déportés, 5030 Gembloux, Belgium; kenne@gmx.com (T.K.K.); henri.martin@uliege.be (H.M.); marie-laure.fauconnier@uliege.be (M.-L.F.)
* Correspondence: sebastien.demeter@uclouvain.be

Abstract: The granary weevil *Sitophilus granarius* is a stored product pest found worldwide. Environmental damages, human health issues and the emergence of resistance are driving scientists to seeks alternatives to synthetic insecticides for its control. With low mammal toxicity and low persistence, essential oils are more and more being considered a potential alternative. In this study, we compare the toxicity of 25 essential oils, representing a large array of chemical compositions, on adult granary weevils. Bioassays indicated that *Allium sativum* was the most toxic essential oil, with the lowest calculated lethal concentration 90 (LC90) both after 24 h and 7 days. *Gaultheria procumbens*, *Mentha arvensis* and *Eucalyptus dives* oils appeared to have a good potential in terms of toxicity/cost ratio for further development of a plant-derived biocide. Low influence of exposure time was observed for most of essential oils. The methodology developed here offers the possibility to test a large array of essential oils in the same experimental bioassay and in a standardized way. It is a first step to the development of new biocide for alternative management strategies of stored product pests.

Keywords: essential oil; insecticide; eco-friendly; stored product pest; *Sitophilus granarius*; *Allium sativum*; *Gaultheria procumbens*; *Mentha arvensis*; *Eucalyptus dives*

Citation: Demeter, S.; Lebbe, O.; Hecq, F.; Nicolis, S.C.; Kenne Kemene, T.; Martin, H.; Fauconnier, M.-L.; Hance, T. Insecticidal Activity of 25 Essential Oils on the Stored Product Pest, *Sitophilus granarius*. Foods **2021**, 10, 200. https://doi.org/10.3390/foods10020200

Academic Editor: Francesco Visioli
Received: 4 December 2020
Accepted: 15 January 2021
Published: 20 January 2021

Publisher's Note: MDPI stays neutral with regard to jurisdictional claims in published maps and institutional affiliations.

1. Introduction

Loss of food during storage by pest infestation is a major problem in our societies in both developed and developing countries, causing significant financial losses [1–4]. Stored cereals are, indeed, a source of food for many insects, mites and fungi which degrade the product quality and can cause from 9 to 20% of net losses [5]. Around 1660 insect species worldwide are known to affect the quality of stored food products [6]. Despite this worrying situation, few research funds are allocated to offset these losses [7].

Since 1960, stored product pests have been mainly controlled by synthetic contact pesticides [8,9]. The utilization of those pesticides is being criticized more and more. Appearance of resistance in addition to the increased risks of residues dangerous to the environment and human health have led to an increasingly restricted use of those compounds [9,10]. These environmental concerns and demand for food safety have underlined the need for alternative research [10,11]. In the last decades, plant essential oils have been reported to be a potential alternative for many applications such as anti-microbial, antifungal or herbicide uses [12]. More particularly, essential oils also have interesting properties to replace synthetic insecticides [13,14]. Isman and Grieneisen [15] showed that from 1980 to 2012 the proportion of papers on botanicals among all papers on insecticides raised from

1.43% to 21.38%. Increasing interest in essential oils as an alternative to synthetic pesticides comes from their characteristics [16]. Due to their high volatility, temperature and UV light degradation sensitivity, essential oils are less persistent in the environment than traditional pesticides [17]. In addition, most essential oils have low mammalian toxicity in comparison with synthetic insecticides and are considered as eco-friendly [18]. For instance, Stroh et al. [19] showed that eugenol was 1500 times less toxic than pyrethrum and 15,000 times less toxic than the organophosphate azinphosmethyl for juvenile rainbow trout based on 96 h-LC$_{50}$ values.

In temperate regions, the granary weevil is considered as one of the major pests of stored grain [9,20–22]. Many authors [23–29] have investigated the use of essential oils as alternative insecticides against *S. granarius*. Yildirim et al. [30] demonstrated the high fumigation toxicity of *Satureja hortensis* among eleven essential oils from Lamiaceae family on *S. granarius*. Others have highlighted the contact toxicity by topical application of essential oils, such as Conti et al. [28] with *Hyptis* genera plants. A few less have worked on treated grains taking into account contact, fumigation and ingestion intoxication paths together [31]. The repellency potential of essential oils was also analysed [32,33] for *S. granarius*. In addition, essential oils were reported as a good food deterrent, as in the case of *H. spicigera* essential oils against *Sitophilus zeamais*, preventing grain degradation [34].

Nevertheless, few actual applications have emerged for the protection of stored foodstuffs and we still lack a systematic screening of potently active oils under conditions mimicking storage reality and with a standardized strain of insects. The aim of our study was precisely to test and rank 25 essential oils commonly used and available on the market against *S. granarius*. Special care has been taken for the selection of essential oil based on a large array of chemical composition (different major compounds or groups of major compounds, Table 1). In order to remain under realistic conditions for industrial large-scale application, data as price, availability on the market or health implications has been taken into account in our discussion. To allow comparison, a standardized strain of *S. granarius* was used for all the test performed under the same experimental conditions. Determination of essential oils toxicity has been done by treating the wheat grains directly, considering that the presence of wheat may influence results [35] and mostly because, in practice, it is the grain itself that will be treated in storage facilities.

Table 1. List of the essential oils tested and their composition for compounds (main compounds representing more than 10% of the total composition on peak area basis).

Essential Oils	Major Compounds (>10%)	Essential Oils	Major Compounds (>10%)
Abies sibirica Ledeb.	Bornyl acetate (20.41%), camphene (19.51%), limonene (18.04%), α-pinene (15.71%)	*Melaleuca alternifolia* (Maiden & Betche) Cheel	Terpinene-4-ol (40.14%), γ-terpinene (18.75%)
Allium sativum L.	Diallyl disulfide (36.60%), diallyl trisulfide (32.33%)	*Mentha arvensis* L.	Menthol (73.72%)
Cinnamomum camphora (L.) J. Presl ct cineole	1,8-cineole (53.11%), sabinene (14.50%)	*Myristica fragrans* Houtt.	α-Pinene + α-thujene (21.78%), sabinene (17.91%), β-pinene (14.68%)

Table 1. *Cont.*

Essential Oils	Major Compounds (>10%)	Essential Oils	Major Compounds (>10%)
Citrus limon (L.) Burm. F.	Limonene (68.13%), β-pinene (12.04%)	*Myrtus communis* L.	α-Pinene (54.71%), 1,8-cineole (24.31%)
Copaifera officinalis L.	β-Caryophyllene (64.25%)	*Ocimum basilicum* L.	Estragol (73.43%), linalool (18.85%)
Cuminum cyminum L.	Cuminaldehyde (28.11%), γ-terpinene (20.88%), *p*-cymene (18.26%), β-pinene (14.18%)	*Ocimum sanctum* L.	Eugenol (33.7%), β-caryophyllene (21.8%), methyleugenol (20.5%)
Eucalyptus citriodora Hook	Citronellal (71.09%)	*Origanum majorana* L.	Terpinen-4-ol (21.67%), *cis*-thujanol (15.69%), γ-terpinene (14.14%)
Eucalyptus dives Schauer	Piperitone (47.87%), α-phellandrene (23.33%)	*Rosmarinus officinalis* L. CT camphor	α-pinene (24.62%), 1,8-cineole (16.43%), camphor (16%), camphene (10.90%)
Eucalyptus globulus Labill.	1,8-Cineole (66.10%)	*Rosmarinus officinalis* L. CT verbenone	α-Pinene + α-thujene (31.84%), camphor (10.65%)
Gaultheria procumbens L.	Methyl salicylate (99%)	*Thymus vulgaris* L. CT geraniol	Geraniol (58.25%), geranyl acetate (14.03%)
Illicium verum Hook. F.	*trans*-Anethole (77.71%)	*Vetiveria zizanioides* (L.) Stapf	Khuenic acid (10.48%)
Lavandula hybrida super	Linalool (33.90%), linalyl acetate (33.20%)	*Zingiber officinale* Roscoe	α-Zingiberene (19.87%), β-sesquiphellandrene (14.64%), camphene (12.18%)
Matricaria recutita (L.) Rauschert	E-(*trans*)-β-farnesene (41.17%)	-	-

2. Materials and Methods

2.1. Biological Material

The granary weevil, *S. granarius*, was collected in Belgium from infested wheat grain stock in 2016. They were reared at the Biodiversity Section of the Earth and Life Institute, under controlled conditions in a climatic chamber (28 °C ± 1, 75 ± 5% RH, in the dark) on organic wheat (*Triticum aestivum*).

2.2. Selected Essential Oils (EO) and Their Composition

EOs were selected based on their availability on the market and their composition. Selected essential oils have all a distinctive major compound or a combination of major compounds to make sure to test a large range of composition.

Essential oils have been mainly obtained from Pranarom S.A. (7822—Ghislenghien, Belgium) as well as their composition. Only *Ocimum sanctum* essential oil has been purchased from "Herb and tradition" S.A company (CP59560—Comines, France) and was analyzed by GC-MS. List of the essential oils tested and their composition is indicated in Table 1. The GC-MS used for EOs characterization was a Hewlett Packard system (HP Inc., Palo Alto, CA, USA) in splitless injection mode system, with a HP INNOWAX column of 60 m, 0.25 mm of diameter and 0.5 μm of film thickness. The initial temperature of 50 °C was maintained for 6 min before a progressive warming of 2 °C per minute up to 250 °C. Once the temperature peak of 250 °C was reached, it has been maintained for 20 min. The injector and interface temperature were 250 °C whereas that of the source was 230 °C. The gas vector was helium at a pressure of 23 psi and the total ion chromatogram was recorded by using an electron-impact source at 70 eV of ion kinetic energy. The compound identification was made by comparison of the spectra to National Institute of Standards and Technology (NIST, Gaithersburg, MD, USA) spectral library and pure commercial standards injection in the same chromatographic conditions.

2.3. Toxicity Test in Treated Grain

To be as close as possible to realistic application conditions, we have chosen to treat the grains directly with a standardized quantity of oils. A determined quantity of insects of the same age group was then directly placed on the grains. Consequently, the observed mortality is a result of contact with the treated grain, attempts at nutrition and a fumigation effect.

Toxicity tests were performed in 15 mL plastic Falcon tubes containing 8 g of treated wheat. One mL of essential oil diluted in acetone at concentrations of respectively 1; 2; 3; 4 and 5% (*v/v*) were applied on the wheat except for *Gaultheria procumbens* for which concentrations of 5; 3; 2; 1 and 0.75% were used. Moreover, because of its efficiency, the same tests of mortality have been realized for *Allium sativum* at lower concentrations of 0.75%; 0.5%, 0.25% and 0.125%.

After EO application, samples were mixed by a vortex for 1 min to homogenize the treatment. The control treatment consisted of five Falcons with 8 g of wheat treated with 1 mL of acetone only. Treated wheat dried for 15 min under hood to eliminate the acetone. Then, twenty insects per falcon were added to the wheat and Falcon were closed by a tulle to allow air circulation. Tubes were placed under controlled condition (28 °C ± 1 °C; 75 ± 5% RH). Temperature and humidity were chosen as the optimum for *S. granarius* [36] and to be representative of the conditions at the harvest period. Five repetitions were performed for each concentration.

The mortality was recorded after 24 h and 7 days of exposure. Light is repellent to *S. granarius* [37]. This particularity was used to identify dead individuals by placing a cold lamp of 100 watt in front of eyes of insects for 5 s. Individuals unable to move were considered dead.

2.4. Data Analysis

In the control treatment, in one case the average mortality reached 5 percent and consequently, the Abbott formula [38] has been used to correct mortality.

The relationship between the mortality rate and the concentrations of the different oils tested was fitted with a Hill function using Scipy module of Python v.3.8.2 (Beaverton, OR, USA). This allowed us to estimate the LC50 (lethal concentration that produces 50% of mortality) and LC90 (lethal concentration that produces 90% of mortality). The Hill function is frequently used in different disciplines, from biochemistry and cellular biology to Physics [39] with the following Equation (1):

$$M = \frac{C^n}{C^n + K^n} \tag{1}$$

where M is the mortality proportion; C is the concentration of oil used; K a threshold concentration value beyond which the mortality exceeds 50% (which corresponds to the LC50) and n a cooperativity exponent. A value of n that is larger than 1 signals the presence of cooperative processes between the concentration level and the propagation of the mortality inside the population. In order to calculate the resulting LC for an arbitrary proportion of the population by rearranging the previous Equation (1):

$$C_x = \left(\frac{M_x}{1 - M_x}\right)^{1/n} K \tag{2}$$

which as in Equation (3) gives for the *LC*90

$$C_{90} = LC90 = 9^{1/n} K \tag{3}$$

LC90 has been used to compare essential oils' toxicity. Toxicities are considered significantly different if its standard deviation does not overlap.

3. Results

Mortality Analyses

Mortality levels clearly varied among oils. When tested at the highest concentration of 5%, nine out of 25 essential oils provoked a mortality of less than 60% of the individuals after 24 h (ranging from 0 to 59%). We considered that this threshold must be exceeded to give sufficient efficiency in practice. In consequence, for these oils lower concentrations were not further tested. Looking at the results, it appeared that EOs listed in Table 2 are not effective at this concentration on *S. granarius*.

Table 2. Essential oils tested at a concentration of 5% for which the mortality was not satisfactory.

Essential Oil	Major Compounds	Mortality (24 h)	Control Mortality (24 h)
Cinnamomum camphora CT cinéole	1,8 Cineole (53.11%), sabinene (14.50%)	59% ± 10.2	0%
Zingiber officinale	α-Zingiberene (19.87%), β-sesquiphellandrene (14.64%), camphene (12.18%)	45% ± −5.5	0%
Eucalyptus globulus	1,8-Cineole (66.10%)	33% ± 9.1	0%
Abies sibirica	Bornyl acetate (20.41%), camphene (19.51%), limonene (18.04%), α-pinene (15.71%)	9% ± 10.2	0%
Matricaria recutita	E-(trans)-β-Farnesene (41.17%)	7% ± 1.1	0%
Copaifera officinalis	β-Caryophyllene (64.25%)	5% ± 6.3	0%
Vetiveria zizanoides	Khuenic Acid (10.48%)	2% ± 0.5	0%
Citrus limon	Limonene (68.13%), β-pinene (12.04%)	0%	0%
Myrtus communis	α-Pinene (54.71%), 1,8 cineole (24.31%)	0%	1% ± 0.45

For the 16 remaining EOs, a positive relation was observed between mortality and concentration. Most of them showed a zero or almost zero mortality at a concentration of 1% except *A. sativum* which still provoked 75% of mortality after 24 h at that concentration and represents therefore the most toxic oil tested. Among the remaining oil, *G. procumbens*, *O. sanctum* and *Eucalyptus* dives reached respectively 81%, 68% and 51% of mortality (24 h) for a 2% concentration (Table 3).

For most of EOs tested, time of exposure did not have a significant effect on percentage of mortality, indicating that a knock down effect is rapidly observed (Table 4). However, this observation does not hold for three EOs after 24 h and 7 days, *Thymus vulgaris* CT geraniol, *Myristica fragrans* and *O. sanctum*, indicating a cumulative toxic effect probably linked to physiological or neurological disorders.

With a LC90 of 1.04% after 7 days of exposure, *A. sativum* is the most toxic essential oil tested. It is followed by *G. procumbens* and *O. sanctum* that showed similar results with LC90 of 2.10 and 2.11% (7 days). The third position in the list of the most toxic essential oils is shared by *Mentha arvensis*, *T. vulgaris* CT geraniol and *E. dives* which present respectively a LC90 of 3.08; 3.08 and 3.11% after 7 days of exposure.

Calculation of mortality curves was realized for 24 h and 7 days treatment (Figure 1). Table 4 indicates the LC90 after 24 h and 7 days for these 16 essential oils tested.

Table 3. Summary of mortality percentages after 24 h hours of exposure for the concentrations tested ($n = 5$).

Essential Oils	5%	4%	3%	2%	1%	Control
A. sativum	99 ± 2.2%	100%	100%	98 ± 2.8%	75 ± 7.9%	1 ± 0.45%
C. cyminum	90 ± 7.1%	68 ± 12.5%	55 ± 25%	11 ± 6.5%	0%	0%
E. citriodora	96 ± 5.5%	79 ± 16.3%	56 ± 18.5%	3 ± 4.5%	0%	0%
E. dives	100%	96 ± 4.2%	80 ± 6.1%	51 ± 7.4%	0%	0%
G. procumbens	100%	-	96 ± 4.2%	81 ± 6.5%	5 ± 6.1%	0%
I. verum	100%	87 ± 5.7%	50 ± 12.7%	7 ± 4.5%	0%	0%
L. intermedia super	88.89 ± 5.5%	80.81 ± 16.7%	30.3 ± 12.9%	5 ± 5%	0%	0%
M. alternifolia	98 ± 4.5%	86.87 ± 9.2%	61 ± 15.6%	3 ± 4.5%	2 ± 4%	0%
M. arvensis	100%	93 ± 8.4%	73 ± 13.0%	41 ± 10.8%	0%	0%
M. fragrans	75 ± 17.3%	52 ± 14.8%	46 ± 9.6%	25 ± 11.7%	0%	0%
O. basilicum spp basilicum	97 ± 2.7%	80 ± 19.7%	41 ± 14.7%	12 ± 5.7%	0%	0%
O. sanctum	99 ± 2.3%	98 ± 2.8%	75.75 ± 11.5%	68 ± 6.7%	6 ± 4.2%	0%
O. majorana	97 ± 4.5%	81 ± 6.5%	50 ± 16.6%	4 ± 4.2%	0%	0%
R. officinalis CT camphor	93 ± 7.6%	69 ± 10.8%	8 ± 9.7%	0%	0%	0%
R. officinalis CT verbenone	90 ± 7.9%	12 ± 5.7%	2 ± 2.7%	0%	0%	0%
T. vulgaris CT geraniol	74 ± 14.7%	89 ± 8.2%	50 ± 11.2%	20 ± 7.1%	0%	0%

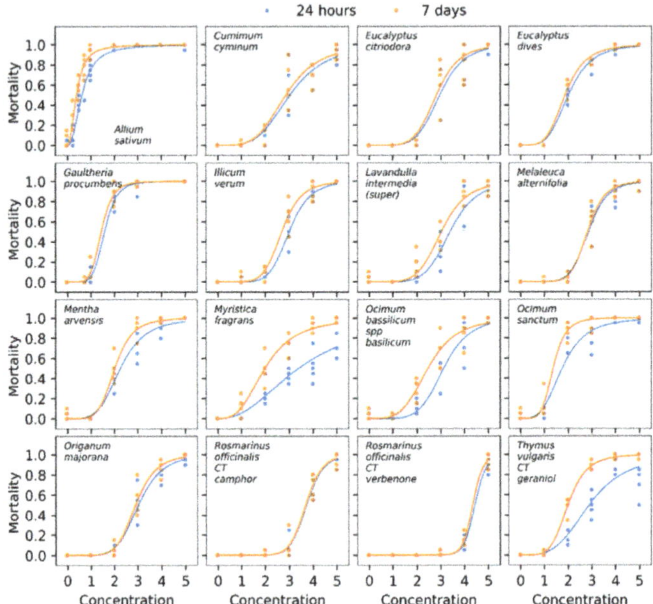

Figure 1. Mortality curves of EOs tested on *S. granarius* 24 h after treatment (blue) and 7 days after treatment (orange).

Table 4. Summary of mortality data presented at the Figure 1 for the 16 essential oils tested. Lethal concentrations are expressed in percent.

Essential Oil	Exposure Time	LC50	LC90	R^2	n
Allium sativum	24 h	0.64 ± 0.02	$1.43 \leq 1.58 \leq 1.75$	0.983	2.4 ± 0.19
	7 days	0.42 ± 0.02	$0.93 \leq 1.04 \leq 1.17$	0.976	2.4 ± 0.20
Cumimum cyminum	24 h	3.05 ± 0.12	$4.72 \leq 5.27 \leq 6.02$	0.942	4.0 ± 0.59
	7 days	2.89 ± 0.10	$4.42 \leq 4.88 \leq 5.50$	0.952	4.2 ± 0.57
Eucalyptus citriodora	24 h	2.98 ± 0.08	$4.02 \leq 4.34 \leq 4.77$	0.956	5.8 ± 0.88
	7 days	2.84 ± 0.09	$3.76 \leq 4.11 \leq 4.58$	0.945	6 ± 1.03
Eucalyptus dives	24 h	2.03 ± 0.04	$3.21 \leq 3.4 \leq 3.61$	0.991	4.3 ± 0.33
	7 days	1.90 ± 0.038	$2.94 \leq 3.11 \leq 3.32$	0.991	4.4 ± 0.37
Gaultheria procumbens	24 h	1.59 ± 0.04	$2.15 \leq 2.26 \leq 2.4$	0.993	6.2 ± 0.55
	7 days	1.46 ± 0.04	$1.99 \leq 2.10 \leq 2.23$	0.99	6.1 ± 0.48
Illicum verum	24 h	3.02 ± 0.04	$3.97 \leq 4.14 \leq 4.35$	0.986	6.9 ± 0.68
	7 days	2.72 ± 0.05	$3.58 \leq 3.78 \leq 4.01$	0.98	6.7 ± 0.74
Lavandulla intermedia (super)	24 h	3.41 ± 0.08	$4.57 \leq 4.89 \leq 5.29$	0.958	6.1 ± 0.81
	7 days	3.05 ± 0.08	$4.16 \leq 4.48 \leq 4.90$	0.96	5.7 ± 0.82
Melaleuca alternifolia	24 h	2.86 ± 0.06	$3.67 \leq 3.89 \leq 4.16$	0.976	7.2 ± 0.98
	7 days	2.84 ± 0.05	$3.56 \leq 3.76 \leq 4.02$	0.979	7.8 ± 1.11
Mentha arvensis	24 h	2.27 ± 0.06	$3.55 \leq 3.83 \leq 4.16$	0.98	4.2 ± 0.42
	7 days	2.04 ± 0.05	$2.86 \leq 3.08 \leq 3.36$	0.98	5.3 ± 0.73
Myristica fragrans	24 h	3.40 ± 0.17	$7.31 \leq 8.68 \leq 10.72$	0.946	2.3 ± 0.35
	7 days	2.01 ± 0.07	$3.69 \leq 4.09 \leq 4.59$	0.975	3.1 ± 0.31
Ocimum bassilicum spp basilicum	24 h	3.14 ± 0.08	$4.30 \leq 4.63 \leq 5.05$	0.961	5.7 ± 0.78
	7 days	2.49 ± 0.08	$3.87 \leq 4.24 \leq 4.73$	0.964	4.1 ± 0.50
Ocimum sanctum	24 h	1.77 ± 0.07	$2.94 \leq 3.26 \leq 3.66$	0.973	3.6 ± 0.40
	7 days	1.40 ± 0.05	$1.96 \leq 2.11 \leq 2.27$	0.981	5.4 ± 0.51
Origanum majorana	24 h	3.04 ± 0.06	$4.13 \leq 4.36 \leq 4.65$	0.978	6.1 ± 0.67
	7 days	2.94 ± 0.05	$3.85 \leq 4.06 \leq 4.32$	0.98	6.8 ± 0.81
Rosmarinus officinalis CT camphor	24 h	3.72 ± 0.04	$4.43 \leq 4.58 \leq 4.75$	0.981	10.6 ± 1.16
	7 days	3.70 ± 0.05	$4.39 \leq 4.54 \leq 4.73$	0.978	10.7 ± 1.26
Rosmarinus officinalis CT verbenone	24 h	4.45 ± 0.03	$4.93 \leq 5.00 \leq 5.07$	0.99	18.8 ± 1.14
	7 days	4.36 ± 0.03	$4.80 \leq 4.86 \leq 4.94$	0.992	20.1 ± 1.41
Thymus vulgaris CT geraniol	24 h	2.90 ± 0.11	$4.75 \leq 5.33 \leq 6.10$	0.948	3.6 ± 0.5
	7 days	2.02 ± 0.03	$2.95 \leq 3.08 \leq 3.23$	0.994	5.2 ± 0.38

4. Discussion

4.1. Insecticidal Potential

This study compares the toxicity of 25 essential oils on the granary weevil. Sixteen of these were found to have an interesting insecticidal activity on *S. granarius*. Our results show that *A. sativum*, *G. procumbens*, *O. sanctum*, *M. arvensis*, *T. vulgaris* (geraniol) and *E. dives* present a potential to control *S. granarius* population directly in the grain.

Garlic essential oil has been identified as the most toxic oil with a LC90 two to four times lower than other EOs, probably because of its content in sulfur compounds. Its toxicity on other insect pests of stored products like *Tenebrio molitor* [40], *Sitotroga cerealella* [41], *Tribolium castaneum* and *Sitophilus zeamais* [42,43] has already been described. The efficiency of garlic essential oils and his constituents may vary with the target species, the stage of life and the exposure mode (fumigation or contact). For example, Ho et al. [42] calculated a KD50 (knock down) of 1.32 mg/cm^2 and 7.65 mg/cm^2 of garlic essential oil against *T. castaneum* and *S. zeamais* respectively. In addition, Plata-Rueda et al. [40] have identified diallyl disulfide as the most toxic compounds present in the garlic essential oil explaining its efficiency on *Tenebrio molitor*. Contact and fumigation toxicities of diallyl trisulfide has been highlighted by Huang et al. [43] on *T. castaneum* and *S. zeamais*. Contrary to most other essential oils, these molecules are not present in the garlic clove itself, but arise from the conversion of thiosulfinates (water-soluble) to sulfides (oil-soluble) during the hydrodistillation process [44]. In short, the main sulfur compounds in the whole garlic clove are cysteine sulfoxides like allylcysteine sulfoxide (alliine) and methylcysteine sulfoxide (methiine) which are located in the clove mesophyll storage cells. After crushing the clove, those compounds come in contact with the enzyme *alliinase* that is normally localized in the vascular bundle sheath cells. The vast majority of cysteine sulfoxides are then converted in sulfenic acids which self-condense to thiosulfinates like allicin which is the most abundant compound (60–90% of total thiosulfinate). Allicin is quite unstable depending on the medium and temperature. Upon hydrodistillation, thiosulfinates are transformed into diallyl trisulfide, diallyl disulfide and allyl methyl trisulfide as major products [44].

Essential oils toxicity of *M. avensis* [45], *G. procumbens* [46] and *E. dives* [47] as well as geraniol (main compound of *T. vulgaris* essential oil) [48] was also been highlighted for their activities against various stored product pests. In addition, Yazdgerdian et al. [46] identified *G. procumbens* as the most toxic oil, both by fumigation (6.8 µL/L air) and contact on treated wheat (0.235µL/g), among five essential oils tested on *S. oryzae*. These results confirm the toxicity of *G. procumbens* observed in our study. However, although many studies highlighted toxicities of essential oils, lack of a common protocol or of major compounds description often prevent from reliable and univocal comparison. For example, in the study of Teke et al. [49] the fennel essential oils applied on *S. granarius* contains 71.64% of estragol, which closely resembles the composition of the basil oil in our study (73.43% estragol). However, in their case they realized topical application without grain presence which is quite different that in our case.

At the opposite, Zohry et al. [50] tested toxicity of 10 essential oils on *S. granarius* by exposure to treated wheat in a protocol close to that of this study. Garlic oil was identified as the most toxic one with a concentration of oil per grams of grain similar to ours. However, no precision on composition of EOs are available in their publication, which do not allow a deeper comparison. Further studies on the evaluation of the industrial potential of essential oils need to be based on a common protocol taking into account the influence of the media [35] and a full description of the composition of essential oils.

Despite numerous studies on the toxicity of essential oil on stored product pests, little data is available on the mechanism of action of the insecticidal effect of these essential oils as a mixture of molecules. However, some studies highlight some mechanisms. For instance, Jankowska et al. [51] showed that menthol acts on octopamine receptors and trigger protein kinase A phosphorylation pathway on cockroach DUM neurons. Hong et al. [52] indicate a potential interference of methyl salicylate and eugenol with octopamynergic system as well. Action on octopamine receptors is an advantage in the elaboration of an insecticide due to absence of key role in vertebrates involving a relative security for human health. However, methyl salicylate is known to have a LD50 oral (rat) of 887 mg/kg indicating that it should have another mechanism of action on mammals. Therefore, the mere fact that octopamynergic system is targeted by an essential oil cannot guarantee safety for human health. β-caryophyllene was identified as an inhibitor of the activities of acetylcholine esterase, polyphenol oxidase and carboxylesterase on *Aphis gossypii* [53]. α-phellandrene is believed to have a neurotoxic effect on *Lucinia cuprina* [54]. Diallyl disulfide is known to impact digestion of *Ephestia kuehniella* by decreasing activity of digestive enzymes [55]. Diallyl trisulfide, another major compound of garlic EO, has been recently described as a regulator of the expression of the chitin synthase A gene which generates alteration of the morphology and inhibition of the oviposition of *Sitotroga cerealella* [56]. Finally, essential oils are complex mixture of molecules, possibly interacting and entering in synergy for their mechanism of action. Therefore, it is important to analyse their impact on insect as a whole.

For example, a recent study shows that *M. arvensis* EO is associated with a systemic mode of action on *S. granarius* since it is capable of altering the nervous and muscular systems, cellular respiration processes and the cuticle, the first protective barrier of insects [57].

4.2. Human Health Risk

Toxicity on the target pests is a first step for any kind of new pesticide elaboration. However, in the perspective of a potential utilization of essential oils in an industrial context, it is also essential to focus on some other aspects, such as the price, the wheat deterioration or the mammal toxicity to determine their actual industrial potential. Concerning mammal toxicity, the WHO classification ranked compounds from "extremely hazardous" to "unlikely to present acute hazard" based on the concentration in mg/kg that provoke 50% of mortality in rat (WHO, 2009). Concerning *A. sativum*, diallyl trisulfide is ranked as "unlikely to present acute hazard" while diallyl disulfide is considered as moderately hazardous with an oral LD50 (rat) of 260 mg/kg. Even if this toxicity is two to four times lower than deltamethrin currently used in granaries, it remains to be carefully considered in the case of a conception of healthy and ecofriendly alternatives to insecticide.

Gaultheria procumbens which showed the second highest acute toxicity to *S. granarius* is constituted at 99% of methyl salicylate, a molecule classified as moderately hazardous for human health. Because of this specific composition, this essential oil should be use in association to avoid a rapid development of resistance. Further analyses have also to be done on the persistence of methyl salicylate, on its environmental and mammal toxicity to estimate the potential of this EO as a stand-alone or mix product. Two molecules of *O. sanctum* (eugenol and methyl eugenol) as well are classified as moderately hazardous to mammals and need to be considered with the same caution.

For the three other oils identified, major compounds are all classified as "slightly hazardous" to "unlikely to present acute hazard" and their use should not be a problem to treat food product.

4.3. Prices

If we considered prices (Table 5), essential oils are quite expensive, particularly garlic oil probably because its low availability and its use mainly as an aroma in food industry. Moreover, sulfides are also well known for their unpleasant odor complicating its practical application. These two points explained its low practical applications. *O. sanctum* also seems too expensive to be used at an industrial scale.

Table 5. Price of the most lethal oils tested and the mammal's toxicity of their major compounds.

Essential Oil	Price ($/kg)	Major Compounds	DL50 Oral Rat Toxicity (Mg/Kg)	WHO Classification
A. sativum	130–250	Diallyl disulfide (36.6%)	260 *	II
		Diallyl trisulfide (32.33%)	5800 *	U
G. procumbens	55	Methyl salicylate (99%)	887 **	II
E. dives	34	Piperitone (47.87%)	3350 ***	III
		α-phellandrene (23.33%)	5700 *	U
M. arvensis	22	Menthol (73.72%)	3300 **	III
O. sanctum	200	Eugenol (33.7%)	1930 **	II
		β-caryophyllene (21.8%)	>5000 ****	U
		Methyl eugenol (20.5%)	810 *****	II
T. vulgaris CT geraniol	-	Geraniol (58.25%)	3600 **	III
		Geranyl acetate (14.03%)	6330 *	U

Data obtained from safety data sheet from: * Cayman (Ann Arbor, MI, USA); ** Fisher Science education (Rochester, NY, USA); *** Echemi (Qingdao, China); **** Carl Roth (D-76185 Karlsruhe, Germany); ***** CDH Fine Chemicals (New Delhi, India). Prices have been obtained from Ultra Internationnal B.V. (Spijkenisse, The Netherlands). WHO Classification: II: Moderately hazardous; III: Slightly hazardous; U: Unlikely to present acute hazard.

Gaultheria procumbens, *M. arvensis* and *E. dives* are among the less costly essential oils on the market. Moreover, these three oils are easily available on the market. Based on our results, their toxicity and their price, these three essential oils could represent good opportunity to develop a botanical insecticide to control insect pest in stored product. We did not obtain a commercial price for *T. vulgaris* CT geraniol at an industrial scale.

4.4. Duration of Exposure

Only three essential oils (*M. fragrans*, *O. sanctum* and *T. vulgaris* CT geraniol) showed an increase in mortality 7 days after the treatment (Figure 1). This could be the consequence of a cumulative contamination during all the period, including by feeding. It is also possible that physiological disorders took times and was linked to an arrest of feeding and water losses.

For the other essential oils, little differences of mortality were observed after 1 and 7 days of exposure. Several hypotheses could explain that observation. First as mortality arise soon after the insect introduction, we may expect a strong selection effect on susceptible individuals, leaving alive after one day only more resistant individuals. Secondly, the absorption of essential oils by the grains (by fumigation or contact) could reduce the biodisponibility of the active compounds and thus the lack of efficiency on long terms period. Indeed, Lee et al. [35] put into light that fumigation toxicity of certain essential oils is three to nine times lesser in presence of wheat due to the absorption phenomena.

Thirdly, our experiment has been conducted at 28 °C. The evaporation rate of essential oils is rapid at this temperature and a substantial part of the essential oil may have vanished after 24 h. Heydarzade et al. [58] highlighted the low persistence of essential oils of *Teucrium polium* and *Foeniculum vulgare*. Treated filter paper induced a 99% mortality at time zero and 0% 30 h after application on *Callosobruchus maculatus* adults. This downgrade of activity is supposed to be caused by high volatility and/or quick degradation of active compounds.

Studies must be carried out on the combined influence of evaporation and absorption by grains of essential oils in order to demonstrate their toxicity persistence over time. In further studies, it is a priority to include GC-MS analyses of treated wheat that allowed scientists to determine the behavior of essential oils and its remanence at the surface and inside the treated wheat until the end of experiment. This factor is essential to control insect pests that lay eggs into the grain, which causes a delay between treatment and the potential contact with the insecticide product by emerging individuals.

Finally, we cannot exclude that the low difference between mortalities for both exposure times could be explained by the absence of accumulation of toxic compounds in the insect and its capacity to metabolize them. The few cases where a difference was identified between both exposure times could be explain by a more physiologic mode of action inducing drying or no feeding effect which induces slower death pattern.

5. Perspectives

Moreover, to precise if these essential oils could be a viable alternative to pesticide in an industrial point of view, further studies has to be conduct on the comparison of their efficiency with the one of actual synthetic insecticides and/or natural substances well known for their insecticidal properties in a protocol mimicking the actual mode of treatment. To answer eventually the question: "Are these essential oils actually a good alternative to the current standards", future studies should include a positive control with a treatment protocol based on pulverization.

Experiments should also be carried out at a larger scale, such as experimental granaries, with the purpose of estimating the quantity of oil per ton of wheat needed and thus the practical applicability of these treatments. Indeed, under mass storage conditions, the application of essential oils during the grain filing process in the silo is based on nano-drop pulverization which could greatly increase the evaporation of the product. Moreover, the formulation of the essential oil is also of tremendous importance as discussed by Maes et al. [59]. In our cases, dilutions were made using acetone which is also quite different from actual industrial application. These points should be further analyzed in details.

6. Conclusions

Considering insecticidal effects, prices, availability and mammal toxicity of essential oils tested, *M. arvensis*, *E. dives* and *G. procumbens* can be considered as good potential alternatives to the synthetic pesticides presently used to control grain weevils. As essential oils are products of very variable composition, studies must be performed to clearly identify the compound(s) responsible of the insecticidal toxicity of these three essential oils to avoid variable responses to future treatments. More investigations need to be done on the mechanism of action of these oils, including the role of minor components, both on insects and mammals, to secure their industrial use.

Author Contributions: Conceptualization, S.D.; M.-L.F. and T.H.; writing—original draft, S.D.; writing—review & editing, S.D.; M.-L.F.; S.C.N. and T.H.; data curation, S.D., O.L., F.H., S.C.N., T.K.K., H.M.; formal analysis: S.D.; S.C.N.; supervision: M.-L.F.; T.H.; funding acquisition: T.H. and M.-L.F. All authors have read and agreed to the published version of the manuscript.

Funding: This study is part of the Walinnov project OILPROTECT (1610128) granted by Wallonia via the "SPF-Economie Emploi Recherche", Win^{2}Wal program. This research was funded by the Education, Audio-visual and Culture Executive Agency (EACEA) through the EOHUB project 600873-EPP1-2018-1ES-EPPKA2-KA.

Institutional Review Board Statement: Not applicable.

Informed Consent Statement: Not applicable.

Data Availability Statement: The data presented in this study are openly available in 10.6084/m9.figshare.13603526.

Acknowledgments: This paper is publication BRC275 of the Biodiversity Research Center, Université catholique de Louvain.

Conflicts of Interest: The authors declare no conflict of interest. The funders had no role in the design of the study; in the collection, analyses, or interpretation of data; in the writing of the manuscript, or in the decision to publish the results.

References

1. Haff, R.P.; Slaughter, D.C. Real-time x-ray inspection of wheat for infestation by the granary weevil, *Sitophilus granarius* (L.). *Trans. ASAE* **2004**, *47*, 531–537. [CrossRef]
2. Fornal, J.; Jeliński, T.; Sadowska, J.; Grundas, S.; Nawrot, J.; Niewiada, A.; Warchalewski, J.R.; Błaszczak, W. Detection of granary weevil *Sitophilus granarius* (L.) eggs and internal stages in wheat grain using soft X-ray and image analysis. *J. Stored Prod. Res.* **2007**, *43*, 142–148. [CrossRef]
3. Gustavsson, J.; Cederberg, C.; Sonesson, U.; Van Otterdijk, R.; Meybeck, A. Global Food Losses and Food Waste—Extent, Causes and Prevention. Available online: http://www.fao.org/3/a-i2697e.pdf (accessed on 18 January 2021).
4. Gustavsson, J.; Cederberg, C.; Sonesson, U.; Van Otterdijk, R.; Meybeck, A. Global Food Losses and Food Waste. Available online: https://www.madr.ro/docs/ind-alimentara/risipa_alimentara/presentation_food_waste.pdf (accessed on 18 January 2021).
5. Phillips, T.W.; Throne, J.E. Biorational approaches to managing stored-product insects. *Annu. Rev. Èntomol.* **2010**, *55*, 375–397. [CrossRef] [PubMed]
6. Semeão, A.A.; Campbell, J.F.; Hutchinson, J.M.S.; Whitworth, R.J.; Sloderbeck, P.E. Spatio-temporal distribution of stored-product insects around food processing and storage facilities. *Agric. Ecosyst. Environ.* **2013**, *165*, 151–162. [CrossRef]
7. Kumar, D.; Kalita, P.K. Reducing Postharvest Losses during Storage of Grain Crops to Strengthen Food Security in Developing Countries. *Foods* **2017**, *6*, 8. [CrossRef] [PubMed]
8. Zettler, J.; Arthur, F.H. Chemical control of stored product insects with fumigants and residual treatments. *Crop. Prot.* **2000**, *19*, 577–582. [CrossRef]
9. Kljajić, P.; Perić, I. Susceptibility to contact insecticides of granary weevil *Sitophilus granarius* (L.) (Coleoptera: Curcu-lio-nidae) originating from different locations in the former Yugoslavia. *J. Stored Prod. Res.* **2006**, *42*, 149–161. [CrossRef]
10. Kaan, P.; Ömer, C.K.; Yasemin, Y.Y.; Salih, G.; Betül, D.; Kemal, H.C.B.; Fatih, D. Insecticidal activity of edible *Crithmum maritimum* L. essential oil against Coleop-teran and Lepidopteran insects. *Ind. Crops Prod.* **2016**, *89*, 383–389.
11. Kostyukovsky, M.; Trostanetsky, A.; Quinn, E. Novel approaches for integrated grain storage management. *Isr. J. Plant Sci.* **2016**, *63*, 7–16. [CrossRef]
12. Lins, L.; Maso, S.D.; Foncoux, B.; Kamili, A.; Laurin, Y.; Genva, M.; Jijakli, M.H.; Fauconnier, M.-L.; Fauconnier, M.-L.; Deleu, M. Insights into the Relationships Between Herbicide Activities, Molecular Structure and Membrane Interaction of Cinnamon and Citronella Essential Oils Components. *Int. J. Mol. Sci.* **2019**, *20*, 4007. [CrossRef] [PubMed]

13. Huang, Y.; Liao, M.; Yang, Q.; Shi, S.; Xiao, J.; Cao, H. Knockdown of NADPH-cytochrome P450 reductase and CYP6MS1 in-creases the susceptibility of *Sitophilus zeamais* to terpinen-4-ol. *Pest. Biochem. Phys.* **2020**, *162*, 15–22. [CrossRef] [PubMed]

14. Oliveira, A.P.; Santana, A.S.; Santana, E.D.; Lima, A.P.S.; Faro, R.R.; Nunes, R.S.; Lima, A.D.; Blank, A.F.; Araújo, A.P.A.; Cristaldo, P.F.; et al. Nanoformulation prototype of the essential oil of Lippia sidoides and thymol to population management of *Sitophilus zeamais* (Coleoptera: Curculionidae). *Ind. Crop. Prod.* **2017**, *107*, 198–205. [CrossRef]

15. Isman, M.B.; Gieneisen, M.L. Botanical insecticide research: Many publications, limited useful data. *Trends Plant. Sci.* **2013**, *19*, 1–6. [CrossRef] [PubMed]

16. Campolo, O.; Giunti, G.; Russo, A.; Palmeri, V.; Zappalà, L. Essential Oils in Stored Product Insect Pest Control. *J. Food Qual.* **2018**, *2018*, 6906105. [CrossRef]

17. Koul, O.; Walia, S.; Dhaliwal, G.S. Essential oils as green pesticides: Potential and constraints. *Biopestic. Int.* **2008**, *4*, 63–84.

18. Mossa, A.-T.H. Green Pesticides: Essential Oils as Biopesticides in Insect-pest Management. *J. Environ. Sci. Technol.* **2016**, *9*, 354–378. [CrossRef]

19. Stroh, J.; Wan, M.T.; Isman, M.B.; Moul, D.J. Evaluation of the Acute Toxicity to Juvenile Pacific Coho Salmon and Rainbow Trout of Some Plant Essential Oils, a Formulated Product, and the Carrier. *Bull. Environ. Contam. Toxicol.* **1998**, *60*, 923–930. [CrossRef]

20. Niewiada, A.; Nawrot, J.; Szafranek, J.; Szafranek, B.; Synak, E.; Jeleń, H.; Wąsowicz, E. Some factors affecting egg-laying of the granary weevil (*Sitophilus granarius* L.). *J. Stored Prod. Res.* **2005**, *41*, 544–555. [CrossRef]

21. Nea, F.; Kambiré, D.A.; Genva, M.; Tanoh, E.A.; Wognin, E.L.; Martin, H.; Brostaux, Y.; Tomi, F.; Lognay, G.C.; Tonzibo, Z.F.; et al. Composition, seasonal variation, and biological activities of Lantana camara es-sential oils from Côte d'Ivoire. *Molecules* **2020**, *25*, 2400. [CrossRef]

22. Tanoh, E.A.; Boué, G.B.; Nea, F.; Genva, M.; Wognin, E.L.; LeDoux, A.; Martin, H.; Tonzibo, F.Z.; Frederich, M.; Fauconnier, M.-L. Seasonal Effect on the Chemical Composition, Insecticidal Properties and Other Biological Activities of Zanthoxylum leprieurii Guill. & Perr. Essential Oils. *Foods* **2020**, *9*, 550. [CrossRef]

23. Polatoğlu, K.; Karakoç, O.C.; Gören, N. Phytotoxic, DPPH scavenging, insecticidal activities and essential oi lcomposition of Achillea vermicularis, A. teretifolia and proposed chemotypes of *A. biebersteinii* (Asteraceae). *Ind. Crops Prod.* **2013**, *51*, 35–45. [CrossRef]

24. Saban, K.; Aslan, O.; Çalmaşur, A.; Cakir, A.C. Toxicity of essential oils isolated from three Ar-temisia species and some of their major components to granary weevil *Sitophilus granarius* (L.) (Coleoptera: Curculinon-idae). *Ind. Crops Prod.* **2016**, *23*, 162–170.

25. Tapondjou, L.; Adler, C.; Bouda, H.; Fontem, D. Efficacy of powder and essential oil from Chenopodium ambrosioides leaves as post-harvest grain protectants against six-stored product beetles. *J. Stored Prod. Res.* **2002**, *38*, 395–402. [CrossRef]

26. Hamza, A.F.; El-Orabi, M.N.; Gharieb, O.H.; El-Saeady, A.-H.A.; Hussein, A.-R.E. Response of *Sitophilus granarius* L. to fumigant toxicity of some plant volatile oils. *J. Radiat. Res. Appl. Sci.* **2016**, *9*, 8–14. [CrossRef]

27. Zoubiri, S.; Baaliouamer, A. Chemical composition and insecticidal properties of some aromatic herbs essential oils from Algeria. *Food Chem.* **2011**, *129*, 179–182. [CrossRef]

28. Conti, B.; Canale, A.; Cioni, P.L.; Flamini, G.; Rifici, A. Hyptis suaveolens and Hyptis spic-igera (Lamiaceae) essential oils: Qualitative analysis, contact toxicity and repellent activity against *Sitophilus granarius* (L.) (Cole-optera: Dryophthoridae). *J. Pest. Sci.* **2011**, *84*, 219–228. [CrossRef]

29. Abdelli, M.; Moghrani, H.; Aboun, A.; Maachi, R. Algerian Mentha pulegium L. leaves essential oil: Chemical composition, anti-microbial, insecticidal and antioxidant activities. *Ind. Crop. Prod.* **2016**, *94*, 197–205. [CrossRef]

30. Yildirim, E.; Kordali, S.; Yazici, G. Insecticidal effects of essential oils of eleven plant species from Lamiaceae on *Sitophilus granarius* (L.) (Coleoptera:Curculionidae). *Rom. Biotechnol. Lett.* **2011**, *16*, 6702–6709.

31. Jembere, B.; Obeng-Ofori, D.; Hassanali, A.; Nyamasyo, G.N.N. Products derived from the leaves of Ocimum kili-mand-scharicum (Labiatae) as post-harvest grain protectants against the infestation of three major stored product insect pests. *Bull. Entomol. Res.* **1995**, *85*, 361–367. [CrossRef]

32. Germinara, G.S.; Stefano, M.G.; De Acutis, L.; Pati, S.; Delfine, S.; De Cristofaro, A.; Rotundo, G. Bioac-tivities of *Lavandula angustifolia* essential oil against the stored grain pest *Sitophilus granarius*. *Bull. Insectol.* **2017**, *70*, 129–138.

33. Plata-Rueda, A.; Campos, J.M.; da Silva, R.G.; Martínez, L.C.; Dos Santos, M.H.; Fernandes, F.L.; Serrão, J.E.; Zanuncio, J.C. Ter-penoid constituents of cinnamon and clove essential oils cause toxic effects and behavior repellency response on granary weevil, *Sitophilus granarius*. *Ecotoxicol. Environ. Saf.* **2018**, *156*, 263–270. [CrossRef] [PubMed]

34. Ngamo, T.L.; Goudoum, A.; Ngassoum, M.B.; Mapongmetsem, M.; Lognay, G.; Hance, T. Chronic toxicity of essential oils of 3 local aromatic plants towards *Sitophilus zeamais* Motsch (Coleoptera: Curculionidae). *Afr. J. Agric. Res.* **2007**, *2*, 164–167.

35. Lee, B.-H.; Annis, P.C.; Tumaalii, F.; Choi, W.-S. Fumigant toxicity of essential oils from the Myrtaceae family and 1,8-cineole against 3 major stored-grain insects. *J. Stored Prod. Res.* **2004**, *40*, 553–564. [CrossRef]

36. Longstaff, B.C. Biology of the grain pest species of the genus *Sitophilus* (Coleoptera: Curculionidae): A critical review. *Prot. Ecol.* **1981**, *2*, 83–130.

37. Gopal, A.; Benny, P. Neo-simple methodology for the evaluation of potential botanical insect repellents and the rapid com-par-ative study on specific chemical and photo sensitivity of selected insects. *Int. J. Life Sci.* **2018**, *6*, 87–104.

38. Abbott, W.S. A method of computing the effectiveness of an insecticide. 1925. *J. Am. Mosq. Control. Assoc.* **1987**, *3*, 302–303. [PubMed]

39. Likhoshvai, V.A.; Ratushny, A. Generalized hill function method for modeling molecular processes. *J. Bioinform. Comput. Biol.* **2007**, *5*, 521–531. [CrossRef] [PubMed]

40. Plata-Rueda, A.; Martínez, L.C.; Dos Santos, M.H.; Fernandes, F.L.; Wilcken, C.F.; Soares, M.A.; Serrão, J.E.; Zanuncio, J.C. Insecticidal activity of garlic essential oil and their constituents against the mealworm beetle, *Tenebrio molitor* Linnaeus (Coleoptera: Tenebrionidae). *Sci. Rep.* **2017**, *7*, srep46406. [CrossRef]

41. Yang, F.-L.; Zhu, F.; Lei, C.-L. Insecticidal activities of garlic substances against adults of grain moth, *Sitotroga cerealella* (Lepidoptera: Gelechiidae). *Insect. Sci.* **2011**, *19*, 205–212. [CrossRef]

42. Ho, S.; Koh, L.; Ma, Y.; Huang, Y.; Sim, K. The oil of garlic, *Allium sativum* L. (Amaryllidaceae), as a potential grain protectant against *Tribolium castaneum* (Herbst) and *Sitophilus zeamais* Motsch. *Postharvest Biol. Technol.* **1996**, *9*, 41–48. [CrossRef]

43. Huang, Y.; Chen, S.X.; Ho, S.H. Bioactivities of methyl allyl disulfide and diallyl trisulfide from essential oil of garlic to two species of stored-product pests, *Sitophilus zeamais* (Coleoptera: Curculionidae) and *Tribolium castaneum* (Coleoptera: Tenebrionidae). *J. Econ. Èntomol.* **2000**, *93*, 537–543. [CrossRef] [PubMed]

44. Lawson, L.D. Garlic: A review of its medicinal effects and indicated active compounds. *ACS Symp. Ser.* **1998**, *691*, 176–209. [CrossRef]

45. Kumar, A.; Shukla, R.; Singh, P.; Singh, A.K.; Dubey, N.K. Use of essential oil from *Mentha arvensis* L. to control storage moulds and insects in stored chickpea. *J. Sci. Food Agric.* **2009**, *89*, 2643–2649. [CrossRef]

46. Yazdgerdian, A.R.; Akhtar, Y.; Isman, M.B. Insecticidal effects of essential oils against woolly beech aphid, *Phyllaphis fagi* (Hemiptera:Aphididae) and rice weevil, *Sitophilus oryzae* (Coleoptera:Curculionidae). *J. Entomol. Zool.* **2015**, *3*, 265–271.

47. Park, J.-H.; Lee, H.-S. Toxicities of Eucalyptus dives oil, 3-carvomenthone, and its analogues against sotred product insects. *J. Food Prot.* **2018**, *81*, 653–658. [CrossRef]

48. Jeon, J.H.; Lee, C.H.; Lee, H.-S. Food Protective Effect of Geraniol and Its Congeners against Stored Food Mites. *J. Food Prot.* **2009**, *72*, 1468–1471. [CrossRef]

49. Teke, M.A.; Mutlu, Ç. Insecticidal and behavioral effects of some plant essential oils against *Sitophilus granarius* L. and *Tribolium castaneum* (Herbst). *J. Plant Dis. Prot.* **2020**, 1–11. [CrossRef]

50. Zohry, N.M.H.; Ali, S.A.; Ibrahim, A.A. Toxycity of ten edible and essential plant oils against the granary weevil, *Sitophilus granarius* L. (Coleoptera:Curculionidae). *Egypt Acad. J. Biolog.* **2020**, *12*, 219–227.

51. Jankowska, M.; Wiśniewska, J.; Fałtynowicz, Ł.; Lapied, B.; Stankiewicz, M. Menthol Increases Bendiocarb Efficacy Through Activation of Octopamine Receptors and Protein Kinase A. *Molecules* **2019**, *24*, 3775. [CrossRef]

52. Hong, T.-K.; Perumalsamy, H.; Jang, K.-H.; Na, E.-S.; Ahn, Y.-J. Ovicidal and larvicidal activity and possible mode of action of phenylpropanoids and ketone identified in *Syzygium aromaticum* bud against *Bradysia procera*. *Pestic. Biochem. Physiol.* **2018**, *145*, 29–38. [CrossRef]

53. Yu-qing, L.; Ming, X.; Qing-chen, Z.; Fang-yuan, Z.; Jiqian, W. Toxicity of β-caryophyllene from *Vitex negundo* (Lamiales: Ver-benaceae) to *Aphis gossypii* Glover (Homoptera: Aphididae) and its action mechanism. *Acta Entomol. Sin.* **2010**, *53*, 396–404.

54. Chabaan, A.; Rhichardi, V.S.; Carrer, A.R.; Brum, J.S.; Cipriano, R.R.; Martins, C.E.; Silva, M.; Deschamps, C.; Molento, M. In-secticide activity of *Curcuma longa* (leaves) essential oil and its major compound α-phellandrene against *Lucilia cuprina* larvae (Diptera: Calliphoridae): Histological and ultrastructural bi-omarkers assessment. *Pestic. Biochem. Phys.* **2019**, *153*, 17–27. [CrossRef] [PubMed]

55. Shahriari, M.; Sahebzadeh, N. Effect of diallyl disulfide on physiological performance of *Ephestia kuehniella* Zeller (Lepi-doptera: Pyralidae). *Arch. Phytopathol. Pflanzenschutz.* **2017**, *50*, 33–46. [CrossRef]

56. Sakhawat, S.; Muhammad, H.; Meng-Ya, W.; Su-Su, Z.; Muhammad, I.; Gang, W.; Feng-Lian, Y. Down-regulation of chitin synthase A gene by diallyl trisulfide, an active substance from garlic essential oil, inhibits oviposition and alters the morphology of adult *Sitotroga cerealella*. *J. Pest. Sci.* **2020**, *93*, 1097–1106.

57. Renoz, F.; Demeter, S.; Degand, H.; Nicolis, S.C.; Lebbe, O.; Martin, H.; Deneubourg, J.-L.; Fauconnier, M.-L.; Morsomme, P.; Hance, T. Label-free quantitative proteomic analysis revealed dramatic physiological changes of *Sitophilus granarius* in response to essential oil from *Mentha arvensis*. *J. Pest Sci.* **2021**. submitted for publication.

58. Heydarzade, A.; Moravvej, G. Contact toxicity and persistence of essential oils from *Foeniculum vulgare*, *Teucrium polium* and *Satureja hortensis* against *Callosobruchus maculatus* (Fabricius) adults (Coleoptera:Bruchidae). *Türk. Entomol. Derg.* **2012**, *36*, 507–519.

59. Maes, C.; Bouquillon, S.; Fauconnier, M.-L. Encapsulation of Essential Oils for the Development of Biosourced Pesticides with Controlled Release: A review. *Molecules* **2019**, *24*, 2539. [CrossRef]

Article

Screening of Antifungal and Antibacterial Activity of 90 Commercial Essential Oils against 10 Pathogens of Agronomical Importance

Caroline De Clerck [1,*,†], Simon Dal Maso [1,†], Olivier Parisi [1], Frédéric Dresen [1], Abdesselam Zhiri [2] and M. Haissam Jijakli [1,*]

[1]   Integrated and Urban Plant Pathology Laboratory, Gembloux Agro-Bio Tech (Liege University), Passage des Déportés 2, 5030 Gembloux, Belgium; S.DalMaso@uliege.be (S.D.M.); oparisi@gmail.com (O.P.); Frederic.Dresen@uliege.be (F.D.)

[2]   Pranarom International, Avenue des Artisans 37, 7822 Ghislenghien, Belgium; azhiri@pranarom.com

*   Correspondence: Caroline.declerck@uliege.be (C.D.C.); MH.Jijakli@uliege.be (M.H.J.)

†   These authors contributed equally to this work.

Received: 2 September 2020; Accepted: 3 October 2020; Published: 7 October 2020

Abstract: Nowadays, the demand for a reduction of chemical pesticides use is growing. In parallel, the development of alternative methods to protect crops from pathogens and pests is also increasing. Essential oil (EO) properties against plant pathogens are well known, and they are recognized as having an interesting potential as alternative plant protection products. In this study, 90 commercially available essential oils have been screened in vitro for antifungal and antibacterial activity against 10 plant pathogens of agronomical importance. EOs have been tested at 500 and 1000 ppm, and measures have been made at three time points for fungi (24, 72 and 120 h of contact) and every two hours for 12 h for bacteria, using Elisa microplates. Among the EOs tested, the ones from *Allium sativum*, *Corydothymus capitatus*, *Cinnamomum cassia*, *Cinnamomum zeylanicum*, *Cymbopogon citratus*, *Cymbopogon flexuosus*, *Eugenia caryophyllus*, and *Litsea citrata* were particularly efficient and showed activity on a large panel of pathogens. Among the pathogens tested, *Botrytis cinerea*, *Fusarium culmorum*, and *Fusarium graminearum* were the most sensitive, while *Colletotrichum lindemuthianum* and *Phytophthora infestans* were the less sensitive. Some EOs, such as the ones from *A. sativum*, *C. capitatus*, *C. cassia*, *C. zeylanicum*, *C. citratus*, *C. flexuosus*, *E. caryophyllus*, and *L. citrata*, have a generalist effect, and are active on several pathogens (7 to 10). These oils are rich in phenols, phenylpropanoids, organosulfur compounds, and/or aldehydes. Others, such as EOs from *Citrus sinensis*, *Melaleuca cajputii*, and *Vanilla fragrans*, seem more specific, and are only active on one to three pathogens. These oils are rich in terpenes and aldehydes.

Keywords: essential oil; biocontrol; antifungal; antibacterial; biopesticide

1. Introduction

Fruits, vegetables, and cereals are important components of the human diet at every age [1]. The increased demand for these commodities exert significant pressure on the environment, leading to intensive agriculture and the use of chemical pesticides. However, the use of these chemicals, and the resulting presence of their residues in food and water, are leading to several health safety breakdowns. Moreover, the use of chemical pesticides affects the environment and the biodiversity. The constant (and sometimes inadequate) use of pesticides is also responsible of the development of pathogen resistances leading to possible food safety issues [2].

Today, the demand for a reduction of chemical pesticides, and for the development of alternative ways to protect crops from pathogens and pests, is growing [3]. In response, research and development in the field of biopesticides has grown exponentially in the last 20 years.

Among the natural alternatives to chemical pesticides, products based on plant extracts and/or plant essential oils (EOs) have received increasing attention because of their generally recognized as safe (GRAS) compounds, due to their very low human toxicity, high volatility, and rapid degradation [4].

Essential oils possess a strong odor and are produced by aromatic plants as secondary metabolites [5]. They are usually obtained from several plant parts by steam hydrodistillation [6]. They are made of a mixture of volatile compounds (between 20 and 100), even if they are, in most cases, characterized by two or three main compounds, representing the major part of the EO (20–70%). As an example, EO of *Citrus limon* is composed, in majority, of limonene and β-pinene [7,8]. Two kinds of molecules can enter in the composition of essential oils: terpenes and terpenoids (e.g., limonene, linalool); and aromatic and aliphatic molecules (e.g., cinnamaldehyde, safrole) [9]. All of these components are characterized by a low molecular weight [10].

Essential oils were known, for a long time, for their antimicrobial and medicinal properties. The latter have, among others, led to the development of aromatherapy, where they are used as bactericide (e.g., tea tree and cinnamon EOs), fungicide (*Lavandula spica* EO), or virucid (*Cinnamomum camphora*) [5,11].

In the last 20 years, the antibacterial and antifungal properties of essential oils have been assessed against a large variety of plant pathogens in order to determine their potential as alternative plant protection products [6,12]. The complex composition of essential oils is interesting, as they could act as multisite chemicals, lowering the risk of resistance [13].

Furthermore, essential oils are composed by low molecular weight molecules and are highly volatile. This property is of great interest, particularly when used on fresh products or during postharvest applications. However, this advantage, in terms of residue reduction, is also a major inconvenience for crop application, which has to be overcome by a formulation allowing to maintain the efficacy of the product [14].

In this study, the in vitro efficacy of 90 commercially available essential oils against 10 plant pathogens of agronomical importance has been assessed. This is, to our knowledge, the largest screening for antifungal and antibacterial activity of EOs made so far.

2. Materials and Methods

2.1. Essential Oils

The 90 essential oils (EOs) tested in our study were supplied by Pranarom International (Ghislenghien, Belgium) (Table 1).

Table 1. List of essential oils tested in this study.

Num. Code	Plant Species	Num. Code	Plant Species	Num. Code	Plant Species
1	*Allium sativum*	31	*Eucalyptus citriodora Ct citronnellal*	61	*Corydothymus capitatus*
2	*Trachyspermum amni*	32	*Eucalyptus globulus*	62	*Origanum heracleoticum*
3	*Anethum graveolens*	33	*Eucalyptus dives CT. Piperitone*	63	*Origanum compactum*
4	*Illicum verum*	34	*Eucalyptus smithii*	64	*Cymbopogon martini var. motia*
5	*Pimpinella anisum*	35	*Eucalyptus radiata ssp radiata*	65	*Citrus paradisi*
6	*Melaleuca alternifolia*	36	*Foeniculum vulgare*	66	*Citrus aurantium ssp amara*

Table 1. *Cont.*

Num. Code	Plant Species	Num. Code	Plant Species	Num. Code	Plant Species
7	*Ocimum basilicum ssp basilicum*	37	*Gaultheria fragrantissima*	67	*Pinus pinaster*
8	*Ocimum sanctum*	38	*Pelargonium x asperum*	68	*Pinus pinaster térébenthine*
9	*Copaifera officinalis*	39	*Zingiber officinale*	69	*Pinus sylvestris*
10	*Pimenta racemosa*	40	*Laurus nobilis*	70	*Piper nigrum*
11	*Styrax benzoe*	41	*Lavendula angustifolia ssp angustifolia*	71	*Cinnamomum camphora ct cinéole*
12	*Citrus bergamia*	42	*Lavendula x burnatii clone grosso*	72	*Rosmarinus officinalis ct camphre*
13	*Fokienia hodginsii*	43	*Cymbopogon citratus*	73	*Rosmarinus officinalis ct cinéole*
14	*Aniba rosaeodora var. amazonica*	44	*Leptospermum petersonii*	74	*Rosmarinus officinalis ct verbenone*
15	*Melaleuca cajputii*	45	*Citrus aurantifolia*	75	*Amyris balsamifera*
16	*Cinnamomum cassia*	46	*Litsea citrata*	76	*Abies alba*
17	*Cinnamomum zeylanicum*	47	*Citrus reticulata*	77	*Abies balsamea*
18	*Carum carvi*	48	*Cinnamosma fragrans*	78	*Abies sibirica*
19	*Cedrus atlantica*	49	*Origanum majorana ct thujanol*	79	*Salvia lanvandulifolia*
20	*Cedrus deodara*	50	*Thymus mastichina*	80	*Salvia officinalis*
21	*Juniperus virgiana*	51	*Mentha x citrata*	81	*Satureja hortensis*
22	*Apium graveolens var. dulce*	52	*Mentha arvensis*	82	*Satureja montana*
23	*Cymbopogon nardus*	53	*Mentha x piperita*	83	*Thymus satureioides*
24	*Cymbopogon winterianus*	54	*Mentha pulegium*	84	*Thymus vulgaris ct 1 à linalol*
25	*Cymbopogon giganteus*	55	*Monarda fistulosa*	85	*Thymus vulgaris ct thymol*
26	*Citrus limon*	56	*Myristica fragrans*	86	*Thuya occidentalis*
27	*Coriandrum sativum*	57	*Myrtus communis ct cinéole*	87	*Vanilla fragrans Auct*
28	*Cuminum cymincum*	58	*Myrtus communis ct acétate de myrtényle*	88	*Cymbopogon flexuosus*
29	*Cupressus sempervirens var. stricta*	59	*Melaleuca quinquenervia ct cinéole*	89	*Vetiveria zizanoides*
30	*Canarium luzonicum*	60	*Citrus sinensis*	90	*Eugenia caryophyllus*

2.2. Fungal and Bacterial Strains

The 10 host–pathogen combinations used in this study are listed in Table 2. All of the cultures were carried out at a 16D:8N photoperiod on the most appropriate solid media (see Table 2). The Potato dextrose agar (PDA) (Merck) medium was prepared according to the manufacturer's instructions (39 g of powder in 1 L of water). The Luria-Bertani-agar (LB-agar) medium was composed of 10 g/L of peptone 5g/L of yeast extract, 10g/L of NaCl, and 15 g/L of agar. The V8 medium was made with 100 mL/L of V8 juice, 200 mg/L of $CaCO_3$, and 20 g of agar. For in vitro screening procedures in liquid medium, pathogens have been cultured in the same media without the addition of agar. All of the media were autoclaved during 20 min at 120 °C.

Table 2. List of the pathogens tested in this study and their culture conditions.

Host Plant/Environment	Pathogen	Culture Conditions (Medium, Temperature (°C))
Wheat	*Fusarium graminearum*	PDA, 23 °C
	Fusarium culmorum	V8, 23 °C
Sugar beet	*Cercospora beticola*	V8, 23 °C
	Phytophthora infestans	V8, 16 °C
Potato (tuber)	*Pectobacterium carotovorum*	LB-Agar, 23 °C
	Pectobacterium atrosepticum	LB-Agar, 23 °C
Apple and pear (fruit)	*Botrytis cinerea*	PDA, 23 °C
	Penicillium expansum	PDA, 23 °C
Bean	*Colletotrichum lindemuthianum*	V8, 23 °C
Soils	*Pythium ultimum*	PDA, 23 °C

PDA (Potato dextrose agar); LB-Agar (Luria-Bertani-agar).

2.3. Making of a Stable EO Emulsion

EOs are not water soluble. In order to get homogenous and stable emulsions, a formulation was developed to get a final EO concentration of 1000 ppm (maximum dose tested in the in vitro screening). The EOs were first diluted in ethanol in a ratio of 16.7:83.3%. Half a milliliter of this solution was then mixed with 555 µL of Tween 20 and 26.71 mL of distilled water, in order to get an EO concentration of 0.3%. For the in vitro screening procedure, this emulsion was diluted to reach the desired final EO concentration (see Section 2.4.2).

2.4. In Vitro Screening Procedure

2.4.1. Determination of the Pathogens Kinetic Growth

The aim of this step was to determine the optimal growth conditions for each of the pathogens tested (exponential growth phase between 0 h and 48 h, followed by a growth plateau).

The kinetic growth of each pathogen in liquid media was determined using 96 wells ELISA microplates, following the method developed and validated by [15]. Three dilutions (3x, 30x, and 300x) of the medium and three concentrations (10^4, 10^5, and 10^6 spores/mL) of spores' suspensions were tested for each fungus (except for *P. infestans*, for which suspensions of 10^4, 10^5, and 0.3 10^6 spores/mL were tested). For bacteria, three dilutions of the medium (3x, 30x, 300x) and three bacterial suspensions (10^6, 10^7, and 10^8 bacteria/mL) were tested.

Each well was filled with one volume of culture medium, one volume of the pathogen suspension in culture medium, and one volume of water containing 2% of tween 20. The plates were then incubated in the dark at 23 °C. Pathogen growth was assessed by measuring the optic density at 630 nm with a spectrophotometer (Thermo, LabSystems Multiskan RC 351, Chantilly, VA, USA) every 24 h for 144 h. Sixteen replicates (wells) were made for each growing condition (medium and pathogen concentrations). Conditions giving the best pathogen growth are listed in Table 3, and will be the growth conditions selected to go further in the EO screening tests.

Table 3. Pathogen growth conditions selected for the screening tests.

Pathogen	Selected Growth Conditions
Fusarium graminearum	3 times diluted PDB/10^5 spores/mL
Fusarium culmorum	3 times diluted V8/10^5 spores/mL
Cercospora beticola	3 times diluted V8/10^5 spores/mL
Phytophthora infestans	300 times diluted V8/0.3 10^6 spores/mL
Pectobacterium carotovorum	3 times diluted LB/10^7 CFU/mL
Pectobacterium atrosepticum	3 times diluted LB/10^7 CFU/mL
Penicillium expansum	3 times diluted PDB/10^5 spores/mL
Botrytis cinerea	3 times diluted PDB/10^5 spores/mL
Colletotrichum lindemuthianum	3 times diluted V8/10^6 spores/mL
Pythium ultimum	3 times diluted PDB/10^5 spores/mL

2.4.2. Screening

The in vitro screening method in liquid medium is similar to the method used to determine pathogen kinetic growth (see Section 2.4.1). In 96-well ELISA plates—each well was filled with one volume of the selected medium at the optimal concentration (see Table 3), one volume of the pathogen at the optimal suspension (see Table 3), and one volume of the selected EO emulsion (see Section 2.3) at 500 and 1000 ppm (final concentration), except for *P. infestans*, for which EOs have only be tested at 1000 ppm. The plates were incubated in the dark at 23 °C. Growth was assessed by measuring the optic density at 630 nm with a spectrophotometer (Thermo, LabSystems Multiskan RC 351, Chantilly, VA, USA) after 24, 72, and 120 h (for fungi) or every 2 h during 12 h (for bacteria).

Figure 1 shows the wells repartition on the plate for an optimal screening procedure, minimizing the contaminations, following [16].

O1	O2	O3	O4	O5		water		T1	T5	
O1	O2	O3	O4	O5		water		T1	T5	
O1	O2	O3	O4	O5		water		T2	T'	
O1	O2	O3	O4	O5		water		T2	T'	
O1	O2	O3	O4	O5		water		T3		
O1	O2	O3	O4	O5		water		T3		
O1	O2	O3	O4	O5		water		T4		X'
O1	O2	O3	O4	O5		water		T4		X'

O6	O7	O8	O9	O10		water		T6	T10	
O6	O7	O8	O9	O10		water		T6	T10	
O6	O7	O8	O9	O10		water		T7	T'	
O6	O7	O8	O9	O10		water		T7	T'	
O6	O7	O8	O9	O10		water		T8		
O6	O7	O8	O9	O10		water		T8		
O6	O7	O8	O9	O10		water		T9		X'
O6	O7	O8	O9	O10		water		T9		X'

Figure 1. Objects repartition on the ELISA plate for an optimal screening procedure. O1 to O10 represent the tested objects (essential oils (Eos)), T1 to T10 represent negative controls (without pathogen), T' is the culture medium only and X' is the growth control (medium and pathogen). Eight replicates (wells) were made by object.

The efficacy of each EO against each pathogen was calculated using the following formula (1):

$$\text{Efficacy of treatment n (\%)} = \frac{(X' - X_0) - (X_n - X_{n0})}{(X' - X_0)} \times 100 \tag{1}$$

where X' is the optical density of the non-treated growth control at time "t", X_0 is the optical density of the non-treated growth control at time "0", X_n is the optical density of treatment "n" at time "t" and X_{n0} is the optical density of treatment "n" at time "0". The values of the negative control T_n (negative control for treatment n: EO and medium only) and T' (medium only) are also checked to be sure that no contaminations occurred. Heat maps were created using the "ggplot2" package of the R software using the mean of the eight replicates for each couple "EO x Pathogen x Time".

3. Results

3.1. Evaluation of the Effect of the 90 EOs on the 10 Pathogens

3.1.1. P. expansum

At 500 ppm, 20 compounds have shown an interesting effect on *P. expansum* growth (efficacy comprised between 67 and 100%) lasting at least 24 h (see Figure 2). In general, the efficacy of the EOs at 500 ppm do not last very long (around 24 h), with some exceptions, for which the activity lasts more than 120 h: *A. sativum*, *C. cassia*, *C. zeylanicum*, and *E. caryophyllus*.

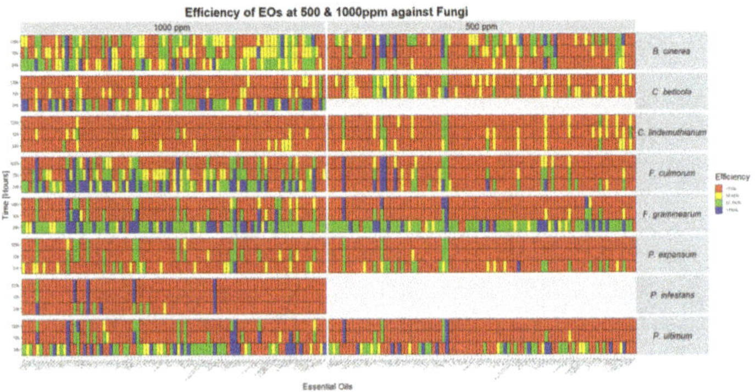

Figure 2. Heat map showing the efficiency of the 90 EOs at 500 and 1000 ppm on the growth of eight plant fungal pathogens after 24 to 120 h of contact in liquid medium in vitro. Red squares represent efficiencies below 50% of growth reduction. Yellow squares represent reduction of growth comprised between 50 and 66%. Efficiencies between 67 and 99% are represented by green squares, while blue squares show a complete inhibition of the organism.

At 1000 ppm, 20 compounds have shown an efficacy comprised between 67 and 100% lasting at least 24 h. In this case, there is also an increasing number of compounds keeping high efficiencies upon time: *A. sativum*, *C. cassia*, *C. zeylanicum*, *C. citratus*, *C. flexuosus*, *Leptospermum petersonii*, *L. citrata*, *C. capitatus*, *Origanum heracleoticum*, *Origanum compactum*, and *E. caryophyllus*.

In particular, EOs of *Monarda fistulosa* at 500 ppm and *O. heracleoticum* at 1000 ppm completely inhibited *P. expansum* during the first 24 h.

3.1.2. B. cinerea

At 500 ppm, 35 EOs have shown high activities (efficacies comprised between 67 and 100%) against *B. cinerea*, lasting at least 24 h. However, the growth inhibition was observed with a delay of at least 48 h for most of them (23/35). Moreover, EOs of *C. cassia*, *C. zeylanicum*, *C. citratus*, *C. flexuosus*, and *Pimpinella anisum* completely inhibited the pathogen growth from 72 h of contact, while EOs of *Myristica fragrans* and *Thymus vulgaris* ct. thymol showed 100% efficacies from 120 h of contact with the oil.

At 1000 ppm, the majority of the tested EOs (54) have shown efficacies between 67 and 100%. Among these, 34 showed efficiencies higher than 67%, lasting at least 72 h. EOs of *A. sativum*, *Cuminum cyminum*, *Eucalyptus dives*, *Lavendula angustifolia*, *Lavendula x burnetii*, and *Mentha pulegium* completely inhibited the growth of the pathogen the first 24 h and EO of *Copaifera officinalis* showed 100% of efficacy the first 72 h. In addition, oil from *Satureja hortensis* and *T. vulgaris* ct. thymol showed 100% efficacies from, respectively, 72 h and 120 h of contact with the EOs.

The pathogen was completely inhibited by EO of *C. capitatus* at 500 as well as 1000 ppm.

3.1.3. *C. beticola*

At 500 ppm, 14 EOs have shown efficacies between 67 and 100% against *C. beticola*, lasting at least 72 h. In particular, EOs of *A. sativum*, *C. cassia*, *C. zeylanicum*, *Canarium luzonicum*, *C. capitatus*, *C. flexuosus*, and *E. Caryophyllus* have shown activities lasting more than 120 h.

At 1000 ppm, 22 EOs have been highly efficient in reducing the pathogen growth (100% inhibition during the first 24 h). Moreover, 26 more have shown inhibition between 67 and 100%, lasting at least 24 h. However, only three EOs kept a high efficacy during the whole period of screening: *L. petersonii*, *Vetiveria zizanioides*, *E. caryophyllus*.

This is also the only pathogen for which EOs of *C. cassia* and *C. zeylanicum* have efficacies lower than 50% at 1000 ppm.

3.1.4. *F. culmorum*

At 500 ppm, 20 EOs showed maximal activities (67–100%) against the pathogen, lasting at least 24 h. In particular, EOs of *A. sativum*, *C. cassia*, *C. zeylanicum*, *C. citratus*, and *E. caryophyllus* completely inhibited the growth of *F. culmorum* up to 120 h of culture. EOs of *A. sativum*, *C. cassia*, and *C. citratus* completely inhibited the growth of *F. culmorum* for 120 h at this concentration, while EOs of *C. capitatus*, *C. zeylanicum*, *L. citrata*, and *O. heracleoticum* inhibited it completely during the first 24 h, and oil of *C. flexuosus* during the first 72 h.

At 1000 ppm, 61 EOs had efficacies comprised between 67 and 100% lasting at least 24 h. For 18 of these EOs the effect lasted for at least 120 h. Moreover, EOs of *A. sativum*, *C. cassia*, *C. flexuosus*, and *L. citrata* completely inhibited the growth of the pathogen during the 120 h of the test. In addition, 26 other EOs showed efficacies of 100% lasting at least 24 h.

3.1.5. *F. graminearum*

At 500 ppm, 75 of the 90 EOs tested had efficacies comprised between 67 and 100%, lasting at least 24 h. In addition, 21 EOs provided 100% of inhibition lasting at least 24 h, including *A. sativum*, *C. cassia*, and *C. zeylanicum*.

At 1000 ppm, almost all of the EOs (78) showed efficacies superior to 67%, lasting at least 24 h. Moreover, 29 EOs provided a complete inhibition of the pathogen, lasting at least 24 h, including EOs of *C. cassia* and *C. capitatus*.

Interestingly, EOs of *E. caryophyllus* at 500 ppm and of *C. capitatus*, at 1000 ppm, completely inhibited the growth of the pathogen during 120 h.

Some EOs (*C. sinensis* and *V. fragrans* auct., among others) have shown high activities (more than 67) during the first 24 h at 500 ppm, while their maximal efficacy at 1000 ppm never exceeded 50%.

3.1.6. *P. ultimum*

At 500 ppm, 37 EOs have shown efficacies between 67 and 100%, lasting at least 24 h. Interestingly, it can observed that EOs of *A. sativum* and *E. caryophyllus* completely inhibit the pathogen for at least 120 h.

At 1000 ppm, 61 EOs have efficacies greater than 67%, lasting at least 24 h, among which 12 have an activity lasting 120 h. EOs of *C. capitatus*, *C. citratus*, and *O. heracleoticum* completely inhibited the pathogen growth for at least 120 h.

3.1.7. *C. lindemuthianum*

At 500, only eight EOs have shown activities greater than 67%, lasting at least 24 h. EOs of *A. sativum*, *C. cassia*, *C. zeylanicum*, and *E. caryophyllus* showed the best results over time.

At 1000 ppm, three EOs have shown activities greater than 67%, lasting at least 24 h. EOs of *A. sativum*, *C. citratus*, and *L. citrata* are the most efficient EOs in this case.

None of the oils tested provided a total inhibition of the pathogen.

3.1.8. *P. infestans*

At 1000 ppm, 10 EOs showed efficacies higher than 67% lasting at least 24 h. Among these, only five EOs showed efficacies greater than 67% during 120 h. EOs of *C. cassia*, *C. flexuosus*, *C. zeylanicum*, and *M. pulegium* completely inhibited the pathogen for at least 120 h.

3.1.9. *P. carotovorum* (PCC)

At 500 ppm, four EOs are causing 100% inhibition, lasting at least 12 h: *A. sativum*, *C. capitatus*, *C. cassia*, and *C. citratus* (See Figure 3).

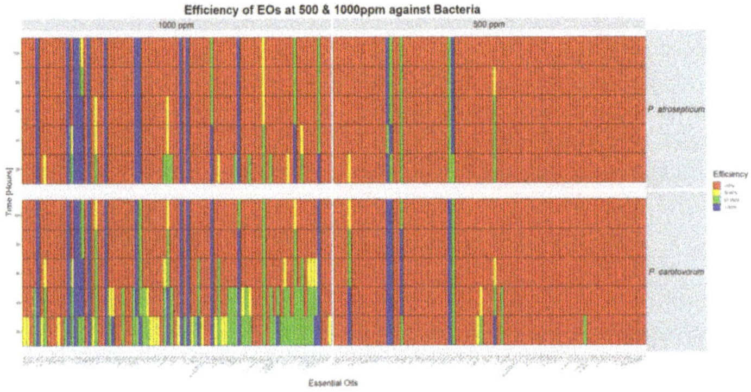

Figure 3. Heat map showing the efficiency of the 90 EOs at 500 and 1000 ppm on the growth of two plant bacterial pathogens after 2 to 12 h of contact in liquid medium in vitro. Red squares represent efficiencies below 50% of growth reduction. Yellow squares represent reduction of growth comprised between 50 and 66%. Efficiencies between 67 and 99% are represented by green squares, while blue squares show a complete inhibition of the organism.

At 1000 ppm, the same four EOs caused a complete inhibition of the pathogen, in addition to the one of *O. heracleoticum*.

3.1.10. *P. atrosepticum* (PCA)

Two EOs completely inhibited the bacterium at the two concentrations tested: *C. cassia* and *E. caryophyllus*.

At 1000 ppm, nine additional EOs caused a total inhibition: A. sativum, C. capitatus, C. citratus, C. flexuosus, Cymbopogon martini, C. zeylanicum, L. citrata, L. petersonii, and O. heracleoticum.

4. Discussion

In this study, the efficacy of 90 commercial essential oils against 10 plant pathogens of agronomical importance was studied.

Similar to the majority of the papers about antifungal and antibacterial effects of EOs [17,18], a dose dependent response was observed for almost all of the EOs tested in this study, the effects being stronger at 1000 ppm than at 500 ppm.

However, they were some exceptions. This is, for example, the case of *C. lindemuthianum* and *C. beticola*, for which most of the EOs showing activities were more effective at 500 ppm than at 1000 ppm. While this is not commonly found in the literature, there are some studies showing similar results [19,20]. Possible explanations are that diluted EOs could diffuse easier in aqueous environments, or that a higher rate of polymerization in concentrated EOs may reduce their antimicrobial activity [20,21].

In most of the cases, the comparison between screenings at 500 and 1000 ppm tend to show that the EO concentrations influence the time of their effectiveness on pathogens, with more concentrated formulations giving longer protection. This fact is certainly due to the high volatility of EOs.

Some EOs, such as the ones from *A. sativum*, *C. capitatus*, *C. cassia*, *C. zeylanicum*, *C. citratus*, *C. flexuosus*, *E. caryophyllus*, and *L. citrata*, have a generalist effect, and are active on several pathogens (between 7 and 10). These oils are rich in phenols, phenylpropanoids, organosulfur compounds, and/or aldehydes, known in the literature to have antifungal effects (thymol and carvacrol for *C. capitatus* [22]; neral and geranial for *C. citratus*, *C. flexuosus*, and *L. citrata* [23]; eugenol for *E. caryophyllus* and *C. zeylanicum* [24]; cinnamaldehyde for *C. cassia* and *C. zeylanicum* [25]; and diallyl di and tri-sulfide for *A. sativum* [26]).

Others, such as EOs from *C. sinensis*, *M. cajputii*, and *V. fragrans*, seem more specific, and are only active on one to three pathogens. These oils are rich in terpenes (limonene, myrcene, and pinenes for *C. sinensis* [27]; elemene, caryophyllene, terpinolene, humulene for *M. cajputii* [28]), and aldehydes (vanillin for *V. fragrans*) [29].

Some pathogens are more sensitive to the EOs tested, such as *B. cinerea* and the two *Fusarium* species. Some studies have already reported that fact [12,30].

Pathogens, such as *C. lindemuthianum* and *P. infestans*, seem less sensitive. Studies showing efficacies of EOs against *C. lindemuthianum* exist in the literature, but are indeed scarce: Khaledi and al [31] showed that EO of *Bunium persicum* was effective, while [32] showed effects for peppermint EO and winter green oil.

The moderate sensitivity of *P. infestans* to EOs was already reported in the literature [33] and could be explained by the fact that it is an oomycete, differing from fungi in cell wall composition and lifecycle, among others [34]. *P. ultimum*, another oomycete tested in our study, was affected by more EOs than *P. infestans*, but the observed effects were limited in time (lasting for only the first 24 h). All of the EOs having an effect on *P. infestans* also showed an activity on *P. ultimum*, except for *C. cyminum*.

For some pathogens, such as *F. graminearum*, *C. beticola*, and *P. ultimum*, the inhibition effect is very high the first 24 h, then it decreases or disappears. This result could indicate that these pathogens are more sensitive to EOs in the form of spores.

The opposite situation was observed with *B. cinerea*, where most of the efficient EOs only become active after at least 24 h of contact with the pathogen. This delayed efficacy could indicate that, in the case of this pathogen, the EOs are more efficient on the mycelium rather than on spores.

For bacteria, we observed that EOs are more efficient at 1000 ppm. PCC seem more sensitive. In general, after 10 h of contact, EOs showing an effect on PCA, which is less sensitive, are also acting on PCC. The most efficient EOs for bacteria are the same as the ones showing high activities for fungi (*C. cassia*, *E. caryophyllus*, *C. capitatus*, *A. sativum*, etc.). EOs rich in carvacrol, like the one of *C. capitatus*, were already found to be effective against PCC [35]. No oil showed specific activity against bacteria.

5. Conclusions

The number of studies available in the literature about fungicidal and fungistatic effects of essential oils, as well as their mechanism of action, is growing, and it is now commonly accepted that EOs have great potential in the development of new biopesticides [6,12].

In our study, 90 EOs were tested on eight fungal pathogens and two bacterial pathogens of agronomical importance. This is, to our knowledge, the largest in vitro screening of EOs made so far. This study allowed us to have a global vision of a large panel of EO efficacies, and to identify several interesting candidates, acting on a large range of pathogens: EOs of *A. sativum*, *C. capitatus*, *C. cassia*, *C. zeylanicum*, *C. citratus*, *C. flexuosus*, *E. caryophyllus*, and *L. citrata*. These oils could be promising candidates in the development of new biopesticides.

However, we have to be careful, as all of our tests have been made in vitro. The promising effects that we have observed need to be confirmed in vivo and, in particular, phytotoxic activities, which are often reported for Eos, will have to be studied [36]. We agree with [6], stating that more studies about the mode of action of EOs, the synergic effect among them or their components, and the identification of their more active components are required. More knowledge is also needed about the effect of these EO applications on the environment (beneficial organisms, soil microbiota, etc.), and on human health, even if the high volatility of EOs should minimize these effects.

Author Contributions: Conceptualization, O.P. and M.H.J.; methodology, O.P. and F.D.; formal analysis, O.P., C.D.C., and S.D.M.; investigation, C.D.C.; writing—original draft preparation, C.D.C. and S.D.M.; writing—review and editing, C.D.C., S.D.M., O.P., and M.H.J.; supervision, M.H.J.; funding acquisition, A.Z., M.H.J. All authors have read and agreed to the published version of the manuscript.

Funding: This research was funded by Pranarom International and Wallonia DGO6 (grant n°6282).

Conflicts of Interest: The authors declare no conflict of interest. The funders had no role in the design of the study; in the collection, analyses, or interpretation of data; in the writing of the manuscript, or in the decision to publish the results.

References

1. Antunes, M.D.C.; Cavaco, A.M. The use of essential oils for postharvest decay control. A review. *Flavour Fragr. J.* **2010**, *25*, 351–366. [CrossRef]
2. Andersen, E.J.; Ali, S.; Byamukama, E.; Yen, Y.; Nepal, M.P. Disease Resistance Mechanisms in Plants. *Genes (Basel)* **2018**, *9*, 339. [CrossRef]
3. World Health Organization; FAO. *Global Situation of Pesticide Management in Agriculture and Public Health: Report of A 2018 WHO-FAO Survey*; World Health Organization: Geneva, Switzerland, 2019; ISBN 9789241516884.
4. Bajpai, V.K.; Kang, S.; Xu, H.; Lee, S.G.; Baek, K.H.; Kang, S.C. Potential roles of essential oils on controlling plant pathogenic bacteria Xanthomonas species: A review. *Plant Pathol. J.* **2011**, *27*, 207–224. [CrossRef]
5. Bakkali, F.; Averbeck, S.; Averbeck, D.; Idaomar, M. Biological effects of essential oils—A review. *Food Chem. Toxicol.* **2008**, *46*, 446–475. [CrossRef]
6. Nazzaro, F.; Fratianni, F.; Coppola, R.; Feo, V. De Essential Oils and Antifungal Activity. *Pharmaceuticals (Basel)* **2017**, *10*, 86. [CrossRef]
7. Ben Hsouna, A.; Ben Halima, N.; Smaoui, S.; Hamdi, N. Citrus lemon essential oil: Chemical composition, antioxidant and antimicrobial activities with its preservative effect against Listeria monocytogenes inoculated in minced beef meat. *Lipids Health Dis.* **2017**, *16*, 146. [CrossRef] [PubMed]
8. Himed, L.; Merniz, S.; Monteagudo-Olivan, R.; Barkat, M.; Coronas, J. Antioxidant activity of the essential oil of citrus limon before and after its encapsulation in amorphous SiO2. *Sci. Afr.* **2019**, *6*, e00181. [CrossRef]
9. Morsy, N.F.S. Chapter 11-Chemical Structure, Quality Indices and Bioactivity of Essential Oil Constituents, Active Ingredients from Aromatic and Medicinal Plants. In *Active Ingredients from Aromatic and Medicinal Plant*; El-Shemy, H.A., Ed.; IntechOpen: London, UK, 2017; pp. 175–206. ISBN 978-953-51-2976-9.
10. Baser, K.H.C.; Buchbauer, G. *Handbook of Essential Oils: Science, Technology, and Applications*, 2nd ed.; CRC Press: New York, NY, USA, 2015; ISBN 9781466590472.
11. Ali, B.; Al-Wabel, N.A.; Shams, S.; Ahamad, A.; Khan, S.A.; Anwar, F. Essential oils used in aromatherapy: A systemic review. *Asian Pac. J. Trop. Biomed.* **2015**, *5*, 601–611. [CrossRef]
12. Raveau, R.; Fontaine, J.; Lounès-Hadj Sahraoui, A. Essential Oils as Potential Alternative Biocontrol Products against Plant Pathogens and Weeds: A Review. *Foods* **2020**, *9*, 365. [CrossRef]
13. Prasad, M.N.N.; Bhat, S.S.; Sreenivasa, M.Y. Antifungal activity of essential oils against Phomopsis azadirachtae—The causative agent of die-back disease of neem. *Int. J. Agric. Technol.* **2010**, *6*, 127–133.
14. Moretti, M.D.L.; Sanna-Passino, G.; Demontis, S.; Bazzoni, E. Essential oil formulations useful as a new tool for insect pest control. *AAPS PharmSciTech* **2002**, *3*, E13. [CrossRef] [PubMed]
15. Kouassi, K.H.S.; Bajji, M.; Brostaux, Y.; Zhiri, A.; Samb, A.; Lepoivre, P.; Jijakli, H. Development and application of a microplate method to evaluate the efficacy of essential oils against Penicillium italicum Wehmer, Penicillium digitatum Sacc. and Colletotrichum musea (Berk. & M.A. Curtis) Arx, three postharvest fungal pathogens of fruits. *Biotechnol. Agron. Société Environ.* **2012**, *16*, 325–336.
16. Kouassi, K.H.S.; Bajji, M.; Jijakli, H. The control of postharvest blue and green molds of citrus in relation with essential oil–wax formulations, adherence and viscosity. *Postharvest Biol. Technol.* **2012**, *73*, 122–128. [CrossRef]

17. Dianez, F.; Santos, M.; Parra, C.; Navarro, M.J.; Blanco, R.; Gea, F.J. Screening of antifungal activity of 12 essential oils against eight pathogenic fungi of vegetables and mushroom. *Lett. Appl. Microbiol.* **2018**, *67*, 400–410. [CrossRef]

18. Palfi, M.; Konjevoda, P.; Karolina Vrandečić, J.Ć. Antifungal activity of essential oils on mycelial growth of Fusarium oxysporum and Bortytis cinerea. *Emirates J. Food Agric.* **2019**, *31*, 544–554. [CrossRef]

19. Gakuubi, M.M.; Maina, A.W.; Wagacha, J.M. Antifungal Activity of Essential Oil of Eucalyptus camaldulensis Dehnh. against Selected Fusarium spp. *Int. J. Microbiol.* **2017**, *2017*, 8761610. [CrossRef]

20. Hood, J.R.; Wilkinson, J.M.; Cavanagh, H.M.A. Evaluation of Common Antibacterial Screening Methods Utilized in Essential Oil Research. *J. Essent. Oil Res.* **2003**, *15*, 428–433. [CrossRef]

21. Osée Muyima, N.Y.; Nziweni, S.; Mabinya, L.V. Antimicrobial and antioxidative activities of Tagetes minuta, Lippia javanica and Foeniculum vulgare essential oils from the Eastern Cape Province of South Africa. *J. Essent. Oil Bear. Plants* **2004**, *7*, 68–78. [CrossRef]

22. Chavan, P.S.; Tupe, S.G. Antifungal activity and mechanism of action of carvacrol and thymol against vineyard and wine spoilage yeasts. *Food Control* **2014**, *46*, 115–120. [CrossRef]

23. Leite, M.C.A.; Bezerra, A.P.d.B.; de Sousa, J.P.; Guerra, F.Q.S.; Lima, E.d.O. Evaluation of Antifungal Activity and Mechanism of Action of Citral against Candida albicans. *Evid. Based. Complement. Alternat. Med.* **2014**, *2014*, 378280. [CrossRef]

24. Schmidt, E.; Jirovetz, L.; Wlcek, K.; Buchbauer, G.; Gochev, V.; Girova, T.; Stoyanova, A.; Geissler, M. Antifungal Activity of Eugenol and Various Eugenol-Containing Essential Oils against 38 Clinical Isolates of Candida albicans. *J. Essent. Oil Bear. Plants* **2007**, *10*, 421–429. [CrossRef]

25. OuYang, Q.; Duan, X.; Li, L.; Tao, N. Cinnamaldehyde Exerts Its Antifungal Activity by Disrupting the Cell Wall Integrity of Geotrichum citri-aurantii. *Front. Microbiol.* **2019**, *10*, 55. [CrossRef] [PubMed]

26. Li, W.-R.; Shi, Q.-S.; Dai, H.-Q.; Liang, Q.; Xie, X.-B.; Huang, X.-M.; Zhao, G.-Z.; Zhang, L.-X. Antifungal activity, kinetics and molecular mechanism of action of garlic oil against Candida albicans. *Sci. Rep.* **2016**, *6*, 22805. [CrossRef] [PubMed]

27. González-Mas, M.C.; Rambla, J.L.; López-Gresa, M.P.; Blázquez, M.A.; Granell, A. Volatile Compounds in Citrus Essential Oils: A Comprehensive Review. *Front. Plant Sci.* **2019**, *10*, 12. [CrossRef]

28. Sharif, Z.M.; Kamal, A.F.; Jalil, N.J. Chemical composition of melaleuca cajuputi powell. *Int. J. Eng. Adv. Technol.* **2019**, *9*, 3479–3483.

29. Sun, R.; Sacalis, J.N.; Chin, C.K.; Still, C.C. Bioactive aromatic compounds from leaves and stems of Vanilla fragrans. *J. Agric. Food Chem.* **2001**, *49*, 5161–5164. [CrossRef]

30. Simas, D.L.R.; de Amorim, S.H.B.M.; Goulart, F.R.V.; Alviano, C.S.; Alviano, D.S.; da Silva, A.J.R. Citrus species essential oils and their components can inhibit or stimulate fungal growth in fruit. *Ind. Crops Prod.* **2017**, *98*, 108–115. [CrossRef]

31. Khaledi, N.; Hassani, F. Antifungal activity of Bunium persicum essential oil and its constituents on growth and pathogenesis of Colletotrichum lindemuthianum. *J. Plant Prot. Res.* **2018**, *58*, 431–441.

32. Kumar, A.; Kudachikar, V.B. Antifungal properties of essential oils against anthracnose disease: A critical appraisal. *J. Plant Dis. Prot.* **2018**, *125*, 133–144. [CrossRef]

33. Quintanilla, P.; Rohloff, J.; Iversen, T.-H. Influence of essential oils onPhytophthora infestans. *Potato Res.* **2002**, *45*, 225–235. [CrossRef]

34. Clavaud, C.; Aimanianda, V.; Latge, J.-P. Chapter 4.3—Organization of Fungal, Oomycete and Lichen (1,3)-β-Glucans. In *Chemistry, Biochemistry, and Biology of 1-3 Beta Glucans and Related Polysaccharides*; Bacic, A., Fincher, G.B., Stone, B.A., Eds.; Academic Press: San Diego, CA, USA, 2009; pp. 387–424. ISBN 978-0-12-373971-1.

35. Hajian-Maleki, H.; Baghaee-Ravari, S.; Moghaddam, M. Efficiency of essential oils against Pectobacterium carotovorum subsp. carotovorum causing potato soft rot and their possible application as coatings in storage. *Postharvest Biol. Technol.* **2019**, *156*, 110928. [CrossRef]

36. Werrie, P.-Y.; Durenne, B.; Delaplace, P.; Fauconnier, M.-L. Phytotoxicity of Essential Oils: Opportunities and Constraints for the Development of Biopesticides. A Review. *Foods* **2020**, *9*, 1291. [CrossRef] [PubMed]

Article

Use of Essential Oils to Increase the Safety and the Quality of Marinated Pork Loin

Lorenzo Siroli [1,†], Giulia Baldi [1,†], Francesca Soglia [1], Danka Bukvicki [2], Francesca Patrignani [1,3], Massimiliano Petracci [1,3] and Rosalba Lanciotti [1,3,*]

[1] Department of Agricultural and Food Sciences, University of Bologna, Piazza Goidanich 60, 47521 Cesena, Italy; lorenzo.siroli2@unibo.it (L.S.); giulia.baldi4@unibo.it (G.B.); francesca.soglia2@unibo.it (F.S.); francesca.patrignani@unibo.it (F.P.); m.petracci@unibo.it (M.P.)

[2] Institute of Botany and Botanical Garden "Jevremovac", Faculty of Biology, University of Belgrade, Takovska 43, 11000 Belgrade, Serbia; dankabukvicki@gmail.com

[3] Interdepartmental Center for Industrial Agri-food Research, University of Bologna, Via Quinto Bucci 336, 47521 Cesena (FC), Italy

* Correspondence: rosalba.lanciotti@unibo.it; Tel.: +39-0547-338132

† Co-first author, these authors contributed equally to this work.

Received: 8 July 2020; Accepted: 21 July 2020; Published: 24 July 2020

Abstract: This study aimed at evaluating the effects of the addition of an oil/beer/lemon marinade solution with or without the inclusion of oregano, rosemary and juniper essential oils on the quality, the technological properties as well as the shelf-life and safety of vacuum-packed pork loin meat. The results obtained suggested that, aside from the addition of essential oils, the marination process allowed to reduce meat pH, thus improving its water holding capacity. Instrumental and sensorial tests showed that the marination also enhanced the tenderness of meat samples, with those marinated with essential oils being the most positively perceived by the panelists. In addition, microbiological data indicated that the marinated samples showed a lower microbial load of the main spoiling microorganisms compared to the control samples, from the 6th to the 13th day of storage, regardless of the addition of essential oils. Marination also allowed to inhibit the pathogens *Salmonella enteritidis*, *Listeria monocytogenes* and *Staphylococcus aureus*, thus increasing the microbiological safety of the product. Overall outcomes suggest that the oil/beer/lemon marinade solution added with essential oils might represent a promising strategy to improve both qualitative and sensory characteristics as well as the safety of meat products.

Keywords: essential oil; marinating solution; pork loin; quality; safety

---

1. Introduction

In the past decade, global consumer demand for marinated meat products has significantly increased [1,2]. The reasons behind this scenario are mainly related to the nutritional characteristics, the extended shelf-life as well as the improvement of sensorial and textural traits of this kind of commodity [2,3]. In addition, marination technology allows to diversify meat products and, conferring them peculiar sensorial traits, to offer a broader choice to the consumers [4]. Marination is a widely used process in the meat industry consisting in the injection or immersion of meat cuts into aqueous solutions containing a wide range of ingredients such as water, salt, vinegar, lemon juice, wine, soy sauce, brine, essential oils, tenderizers, herbs, spices and organic acids [5,6]. Depending on the selected ingredients, a huge variety of marinade solutions, either alkaline or acid, exists. The firsts contain phosphates, while the seconds are usually prepared with the addition of organic acids or their salts [7,8]. Another type of marinade solution are the water/oil emulsions. Overall, the addition of marinade solutions to a meat cut is usually performed to improve the production yields (i.e., by increasing the moisture content of the product), improve the organoleptic characteristics of the final product and, eventually, limit (or at least retard) the occurrence of oxidative reactions [9–11]. In addition, recent studies have reported that marinade solutions including "natural" ingredients (e.g., spices, herbs, essential oils, etc.) can exert an antimicrobial effect against pathogenic and spoilage microorganisms in poultry, beef and pork meat [5,12,13]. Aside from their ability to improve the safety and the shelf-life of marinated meat [14], the utilization of ingredients such as essential oils may also enhance consumers' willingness to buy, in light of the recent increasing attitude towards the consumption of clean-label products [15].

The use of essential oils or of their components (extracted from flowers, fruits, roots, buds and leaves through distillation processes) is widespread in the food industry, precisely in light their organoleptic, antimicrobial and antioxidant properties [16–18]. Within this context, a remarkable antimicrobial effect of several essential oils (included during processing) has been recently highlighted. To cite some examples, the use of rosemary essential oil (0.05%) on beef and chicken meat was found to be able to inhibit the growth of *Listeria monocytogenes*, *Escherichia coli* and *Staphylococcus aureus* [19,20]. On the other hand, the inclusion of thyme essential oil (0.08%) allowed to inhibit the growth of both spoiling microorganisms such as *Pseudomonas* spp. and pathogens such as *Staphylococcus aureus* [16]. Oregano essential oil has been found to exert antimicrobial effects against various pathogenic microorganisms such as *Escherichia coli*, *Listeria monocytogenes* and *Salmonella enteritidis* on both beef and pork meat [16]. However, it is noteworthy to mention that, as essential oils have low sensory thresholds [17], their sensory compatibility as well as their impact on the sensory profile of the final product should be carefully considered [21,22].

In addition, the flavor innovation represents a marketing strategy aimed at keeping up with the continually changing food trends [23]. Within this context, creating appealing alternatives for the consumers represents an important challenge for the meat industry. As a matter of fact, the possibility to set-up a marinade solution with typical ingredients of the Mediterranean area could certainly offer an added value to the final product and differentiate it from the alternatives currently existing on the market. In this framework, the purpose of this research was to evaluate the effect of the addition of a marinade solution composed by olive oil, beer and lemon (i.e., typical ingredients from Mediterranean area) with or without the inclusion of a mixture of essential oils on the shelf-life, the safety as well as the sensory and quality traits of vacuum-packed pork loin slices.

2. Materials and Methods

2.1. Preliminary Tests: Selection of the Marinade Solution's Composition and Essential Oils Mixture

Preliminary tests were performed on pork loin slices (weighing about 60 g) in order to set the best combination and concentration of ingredients in the marinade solution in terms of either organoleptic traits (taste, smell, tenderness) and technological properties (absorption of the marinade solution, tenderness, color, etc.). In detail, 8 ingredients (i.e., water, lemon juice, olive oil, balsamic vinegar, red wine, white wine, beer and mustard) have been tested through different combinations and ratios as well as percentage of marinade solution (*w/w*) added to the meat slices, as reported in Table 1.

Table 1. Marinade solutions tested in the preliminary trials.

Marinade Solution	Ingredients Ratio	% of Marinade Solution (*w/w*)
Water/lemon juice	1:1	10
Water/lemon juice	1:1	5
Olive oil/lemon juice	1:2	5
Olive oil/lemon juice	1:2	10
Olive oil/balsamic vinegar	1:1	5

Table 1. *Cont.*

Marinade Solution	Ingredients Ratio	% of Marinade Solution (*w/w*)
Olive oil/balsamic vinegar	1:1	10
Olive oil/red wine	1:2	10
Olive oil/red wine	1:3	10
Olive oil/white wine	1:2	10
Olive oil/white wine	1:3	10
Olive oil/beer	1:2	10
Olive oil/beer	1:3	10
Olive oil/beer/lemon juice	1:2:1	10
Olive oil/mustard/lemon juice	1:1:1	10
Olive oil/mustard/lemon juice	1:1:1	5

With the aim to obtain homogenous solutions, the ingredients of each combination were mixed with an Ultraturrax (IKA–WERKE, Labortechnik, Staufen, Germany) (13,000 rpm, 30 s, in ice). To each slice of pork loin, 1% NaCl (*w/w*) was added in the marinated product. The samples were placed in heat-resistant plastic bags, in which the marinating solution was directly added. The slices were then vacuum packaged (99.9%) and placed in a small-scale tumbler (model MHG-20, VakonaQualitat, Lienen, Germany) under vacuum conditions (−0.95 bar) and at a temperature of 2–4 °C. Tumbling was performed in 60 min at a speed of 20 rpm including two working cycles (25 min per cycle) and a 10 min pause cycle.

Subsequently, in order to select the combination allowing to obtain the best organoleptic properties of the product without altering its flavor, the addition of essential oils to the marinade solutions was tested. The essential oils considered during the preliminary tests were thyme, rosemary, oregano, and juniper, in different combinations and concentrations (0.02, 0.04, 0.06 and 0.08% on the final product). The evaluation was done by an untrained panel of 20 panellist taking into consideration the sensory parameters such as color, odour, overall accettability before and after cooking.

On the basis of preliminary results (data not shown), the marinade solution selected for the main experiment was composed by olive oil/beer/lemon juice (1:2:1, 10% *w/w*) with a mixture of oregano (0.02%), rosemary (0.03%) and juniper (0.03%) essential oils.

2.2. Ingredients and Microorganisms Used

The pork loin slices used in this work were obtained from a local retailer the same day of the trial and kept at refrigerated temperatures (4 ± 1 °C) until the analyses. The marinade solution was composed as follows: the bock style beer Moretti la rossa (7.2% ABV) (Heineken Italia S.p.A., Pollein, AO, Italy), extra virgin olive oil (Monini, Spoleto, PG, Italy) and concentrated lemon juice (LIMMI, Perugia, PG, Italy). The essential oils used in this experimentation were oregano, rosemary, and juniper (Flora, Pisa, PI, Italy).

The strains used in the challenge test trial, *Listeria monocytogenes* Scott A, *Salmonella enteritidis* E5 and *Staphylococcus aureus* SR41 belonged to the Department of Agricultural and Food Sciences of Bologna University. The strains were maintained at −80 °C before experiments and before inoculation they were cultured twice in Brain Heart Infusion broth (BHI, Oxoid Ltd. Basingstoke, UK) at 37 °C for 24 h.

2.3. Preparation of the Samples and Shelf-Life Trials

The experiment was carried out on a total of 81 slices of pork loin (having an average weight of 60 g), divided into 3 groups (27 slices/group) as follows:

1. Control group (non-marinated) added with 1% NaCl (C);
2. Marinade solution beer/olive oil/concentrated lemon juice (2/1/1; 10% *w/w*) (M);
3. Marinade solution beer/olive oil/concentrated lemon juice (2/1/1; 10% *w/w*) added with a mixture of essential oils (oregano 0.02%, rosemary 0.03% and juniper 0.03% essential oils) (M + E).

As previously mentioned, the marinade solution was realized by mixing bock style beer, concentrated lemon juice and extra virgin olive oil at a 2:1:1 ratio using an Ultraturrax (IKA–WERKE, Labortechnik, Staufen, Germany) (13,000 rpm, 30 s, in ice). Part of this solution was used for samples belonging to the experimental group M, while the remaining was added of a mixture of essential oils (0.08% of the final weight) consisting of oregano (0.02%), rosemary (0.03%) and juniper (0.03%) and included in the samples M + E. Each pork loin slice (about 60 g), was added of 1% NaCl, calculated on the final weight of the marinated product. Subsequently, the samples were placed in heat-resistant plastic bags, in which the marinating solution was directly added, with the only exception of the samples belonging to the control group to which only 1% NaCl was included. The amount of marinade solution added to the samples corresponded to 10% (*w/w*) of the final product. The slices were then vacuum packaged (99.9%) and placed in a small-scale tumbler (model MHG-20, VakonaQualitat, Lienen, Germany) under vacuum conditions (−0.95 bar) and at a temperature of 2–4 °C. Tumbling was performed in 60 min at a speed of 20 rpm including two working cycles (25 min per cycle) and a 10 min pause cycle. The vacuum-tumbled loin slices were then stored at 4 °C and used for analytical determinations after 3, 9 and 15 days of storage.

2.3.1. pH

The pH of the samples was determined by taking an aliquot of meat (avoiding fat and connective tissue) according to Jeacocke [24]. About 2.5 g of finely chopped meat were homogenized for 30 s by Ultraturrax in 25 mL of a solution 5 mM of sodium iodoacetate and 150 mM of KCl at pH 7.0. The pH was determined by pH meter (mod. Jenway 3510; Electrode 924001, Cole-Parmer, Stone, UK) previously calibrated. The pH determination was performed after 3, 6 and 15 days of refrigerated storage on raw meat samples.

2.3.2. Color

Color was assessed by a Minolta® CR-400 colorimeter (Milan, Italy), previously calibrated using a standard white ceramic tile, in standardized illuminant (C) and observation angle (0° with respect to an area of 8 mm in diameter) conditions. The CIELAB system [25] was utilized and the parameter of lightness (L*), redness (a*) and yellowness (b*) were used to objectively define color. The color determination was performed, for each group, after 3, 9 and 15 days of refrigerated storage on raw meat samples.

2.3.3. Marinade Uptake

Marinade uptake (i.e., the ability of meat to bind the saline solution added) was calculated by the difference in weight of the samples before and after the marination process. The amount of marinade solution absorbed was calculated as a percentage of the initial weight of the meat sample, according to the formula:

Marinade uptake (%) = [(Weight after marination − Initial weight)/Initial weight] × 100

2.3.4. Cooking Loss

After 3, 9 and 15 days of storage, samples were cooked in a in a stone grill (model GL-33, Fimar, Rimini, Italy) in standardized conditions (200 °C, 190 s) and the cooking loss (amount of liquid lost after cooking) was calculated as a percentage of the initial weight of the sample according to the formula:

$$\text{Cooking loss (\%)} = [(\text{Raw weight} - \text{Cooked weight})/\text{Raw weight}] \times 100$$

2.3.5. Shear Force

Shear force was assessed by a texture analyzer TA-HDi 500 (Stable Micro System, Godalming, Surrey, UK) equipped with a 5-kg load cell and a Warner-Bratzler shear probe. From each cooked sample, sub-samples (having the dimension of $4 \times 1 \times 0.5$ cm) were excised and placed inside the load cell. The resulting shear force was expressed as kg/cm^2.

2.3.6. Sensory Analysis

Panel tests were performed after 3, 9 and 15 days of refrigerated storage on cooked samples in order to test their visual appearance, olfactory acceptability and taste. The analysis was carried out by 20 untrained panelists who evaluated on a 1 to 5 scale the following parameters: meat odor intensity, spicy odor intensity, color intensity, flavor intensity, tenderness, overall assessment and finally favorite sample.

2.3.7. Microbiological Analysis

During storage at 4 °C, the cell count over time of lactic acid bacteria, yeasts, total aerobic mesophilic bacteria, total aerobic psychrotrophic bacteria, *Pseudomonas* spp. and *Brochotrix thermosphacta* was evaluated by plate counting in specific agar media. Aerobic mesophilic and psychotrophic bacteria were detected on Plate Count Agar (PCA, Oxoid Ltd., Basingstoke, UK), lactic acid bacteria on de Man Rogosa and Sharpe Agar (MRS, Oxoid Ltd. Basingstoke, UK) with added 0.05% cycloheximide (Sigma-Aldrich, St. Louis, US), yeasts on Sabouraud Dextrose Agar (SAB, Oxoid Ltd. Basingstoke, UK), added to 0.02% chloramphenicol (Sigma-Aldrich, St. Louis, US), *Pseudomonas* spp. on Pseudomonas Agar Base (PAB, Oxoid Ltd. Basingstoke, UK) supplemented with Pseudomonas CFC selective agar supplement (Oxoid Ltd. Basingstoke, UK) and *Brochotrix thermosphacta* on STAA Agar base (Oxoid Ltd. Basingstoke, UK) supplemented with STAA selective supplement (Oxoid Ltd. Basingstoke, UK). To perform microbiological analyses, 10 g of meat sample were diluted into 90 mL of physiological solution (0.9% (*w/v*) NaCl), homogenized by a BagMixer 400 P (Interscience, St Nom la Bretèche, France), followed by serial dilution in physiological solution. The MRS agar plates were incubated 24 h at 37 °C, the PCA plates for the detection of psychrotrophic bacteria were incubated at 10 °C for 7 days, all the other agar media were incubated at 30 °C for 24–48 h.

2.4. Challenge-Test Trials

The preparation of marinated pork loin was done similarly to what reported in paragraph 2.3. The experiment was carried out on a total of 60 slices of pork loin (having an average weight of 60 g), divided into 3 groups (20 slices/group). Three groups of samples were obtained:

1. Control group (non-marinated) + pathogens (*L. monocytogenes*, *S. enteritidis* and *S. aureus*), (C+P) inoculated at a level of 4.0 log CFU/g;
2. Marinade solution beer/olive oil/concentrated lemon juice (2/1/1) used at 10% + pathogens (*L. monocytogenes*, *S. enteritidis* and *S. aureus*), (M + P) inoculated at a level of 4.0 log CFU/g;
3. Marinade solution beer/olive oil/concentrated lemon juice (2/1/1) used at 10%; added with essential oils (oregano 0.02%, rosemary 0.03% and juniper 0.03% essential oils) + pathogens (*L. monocytogenes*, *S. enteritidis* and *S. aureus*), (M + E + P) inoculated at a level of 4.0 log CFU/g.

Listeria monocytogenes Scott A, *Salmonella enteritidis* E5 and *Staphylococcus aureus* SR231, used in the challenge test belongs to the Department of Agricultural and Food Sciences (DISTAL, University of Bologna) collection. The bacterial strains were cultured overnight two times in Brain Heart Infusion (Oxoid Ltd., Basigstone, UK) at 37 °C. The pathogens were directly inoculated on the loin slices through 0.5 mL of physiological solution for the control group, while for the groups M + P and M + E + P were added to the marinating solution before the addition to the product. The inoculum was done in order to have an initial cell load of the pathogens, on the product, of approximately 4.0 log CFU/g. After the addition of the marinating solution the product was packaged and churned as reported in paragraph 2.3. The samples were stored at 4 °C and used for microbiological analyses immediately after the marinating and after 3, 6, 9, 13 and 15 days.

2.5. Microbiological Analysis

During the storage microbiological analyses were performed in order to detect the cell loads of the inoculated *L. monocytogenes*, *S. enteritidis* and *S. aureus*. Specifically, the entire slice of loin (about 60 g) was placed in sterile bags and added with sterile physiological solution in a 1:2 (*w/w*) ratio and then homogenized for 2 min by a BagMixer 400 P (Interscience, St Nom la Bretèche, France) followed by serial dilution in physiological solution. *L. monocytogenes*, *S. enteritidis* and *S. aureus* were detected in specific selective agar media. Listeria Selective Agar (LSA, Oxoid Ltd., Basingstoke, UK) supplemented with Listeria selective supplement (SR0140, Oxoid Ltd., Basingstoke, UK) for the enumeration of *L. monocytogenes*; Bismuth Sulphite Agar (BSA, Oxoid Ltd., Basingstoke, UK) for the detection of *S. enteritidis*, while Baird-Parker Agar base (BPA, Oxoid Ltd., Basingstoke, UK) added with Egg Yolk Tellurite Emulsion (SR0054, Oxoid Ltd., Basingstoke, UK) for the enumeration of *S. aureus*. The agar plates were then incubated at 37 °C for 24 h.

2.6. Statistical Analysis

Data were analyzed using the one-way ANOVA option of Statistica software (version 8.0; StatSoft., Tulsa, Oklahoma, USA) in order to test the effect of the addition of a marinade solution (with or without the inclusion of essential oils) at each sampling time (3, 9 and 15 days). Following, mean values were separated through Tukey honest significant difference (HSD) test, by considering a significance level of $p < 0.05$.

3. Results

3.1. Shelf-Life Trials

3.1.1. pH and Color

As reported in Figure 1, at each sampling time, control samples showed significantly higher pH than the marinated ones ($p < 0.05$) which, in their turn, exhibited similar values. A slight decrease in pH following refrigerated storage was observed for all the experimental groups, with C samples showing the greatest pH decline. In more detail, control samples exhibited an average pH decrease of 0.32 units, while M and M + E decreased of 0.19 and 0.17, respectively.

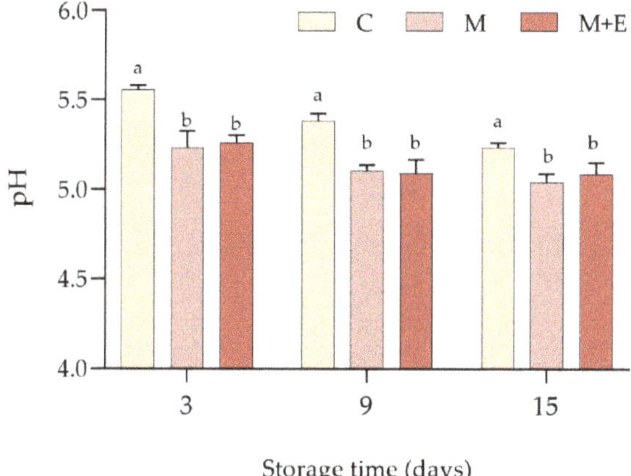

Figure 1. Average pH values of non-marinated (C), marinated (M) and marinated with essential oils (M + E) pork loin slices at 3, 9 and 15 days of refrigerated storage. Data represent means ± SD. a, b = average values lacking a common letter significantly differ among the same sampling time.

Results concerning the evolution of color parameters (lightness—L*, redness—a*, yellowness—b*) during the refrigerated storage are reported in Figure 2. Overall, regardless the storage time, no significant differences were found either in L* or a* values among the experimental groups. Although these differences were not statistically significant, non-marinated samples showed noticeably higher a* values at both 9 and 15 days of storage. On the contrary, marination treatment exploited a remarkable effect on yellowness (b*): at each storage time, both M and M + E exhibited significantly higher b* values if compared to the control ($p < 0.05$).

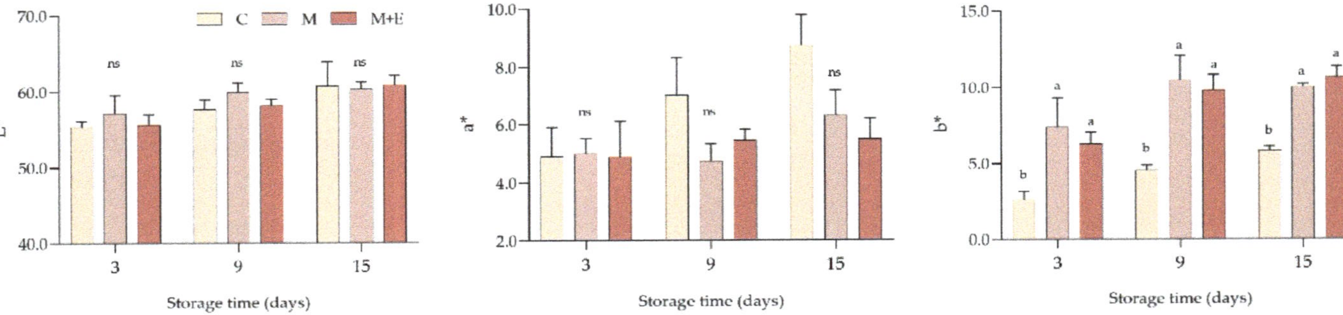

Figure 2. Average lightness (L*), redness (a*) and yellowness (b*) values of non-marinated (C), marinated (M) and marinated with essential oils (M + E) pork loin slices at 3, 9 and 15 days of refrigerated storage. Data represent means ± SD. a, b = average values lacking a common letter significantly differ among the same sampling time. At the same storage time, ns indicates no significant differences among the samples.

3.1.2. Marinade Uptake and Cooking Loss

Data concerning the marinade uptake during the refrigerated storage are shown in Figure 3. Albeit any difference has been detected among the experimental groups at 9 and 15 days of storage, at day 3, a significantly ($p < 0.05$) higher marinade uptake has been observed in M + E samples in comparison to M (7.8 vs. 7.3%, respectively).

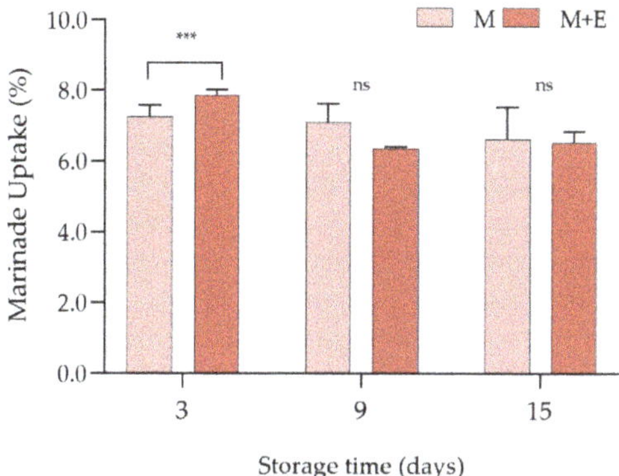

Figure 3. Average marinade uptake (%) values of marinated (M) and marinated with essential oils (M + E) pork loin slices at 3, 9 and 15 days of refrigerated storage. Data represent means ± SD. *** = $p < 0.001$. At the same storage time, ns indicates no significant differences among the samples.

Results concerning the cooking losses at different storage times are shown in Figure 4. After 3 days of refrigerated storage, C (non-marinated samples) showed significantly higher liquid losses if compared to M + E samples ($p < 0.001$), while M group exhibited intermediate values. However, different results were observed at day 9 with the C group showing significantly lower values if compared to M, whereas no significant differences were found at day 15.

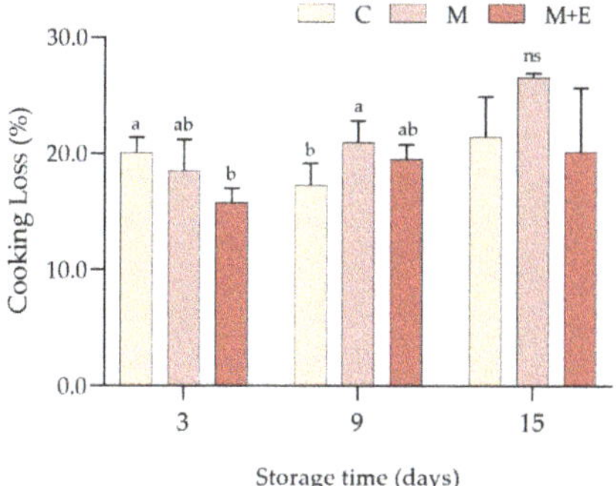

Figure 4. Average cooking loss values (%) of non-marinated (C), marinated (M) and marinated with essential oils (M + E) pork loin slices at 3, 9 and 15 days of refrigerated storage. Data represent means ± SD. a, b = average values lacking a common letter significantly differ among the same sampling time. At the same storage time, ns indicates no significant differences among the samples.

3.1.3. Shear Force

Results concerning the shear force of cooked pork loin samples after 3, 9 and 15 days of refrigerated storage are displayed in Figure 5. After 3 days of refrigerated storage, non-marinated samples showed significantly higher shear forces than the marinated ones (M and M + E) ($p < 0.05$), with M + E group exhibiting the lowest values. Albeit no statistical difference has been detected at 9 and 15 days likely due to the high variability of data, M + E samples showed the lowest shear force values, thus suggesting that the effect of essential oils on improving meat tenderness is considerable in particular in the first days of storage.

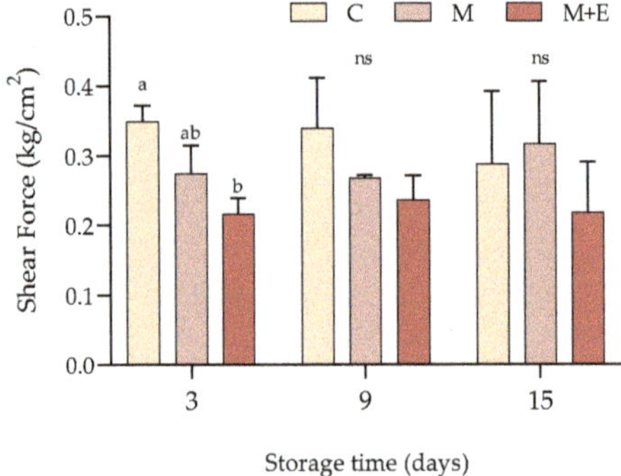

Storage time (days)

Figure 5. Average shear force values (kg/cm^2) of non-marinated (C), marinated (M) and marinated with essential oils (M + E) pork loin slices at 3, 9 and 15 days of refrigerated storage. Data represent means ± SD. a, b = average values lacking a common letter significantly differ among the same sampling time. At the same storage time, ns indicates no significant differences among the samples.

3.1.4. Sensory Analysis

Panel tests were performed on pork loin samples after 3, 9 and 15 days of storage with the aim of determining the acceptability of the product by the consumers. The results of the panel tests are shown in Figure 6a–c.

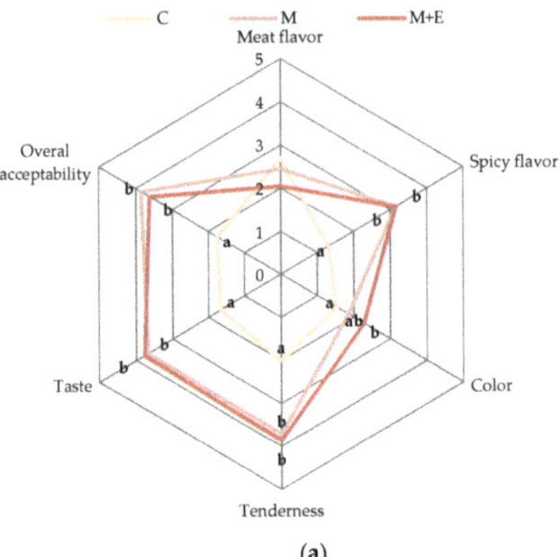

(a)

Figure 6. *Cont.*

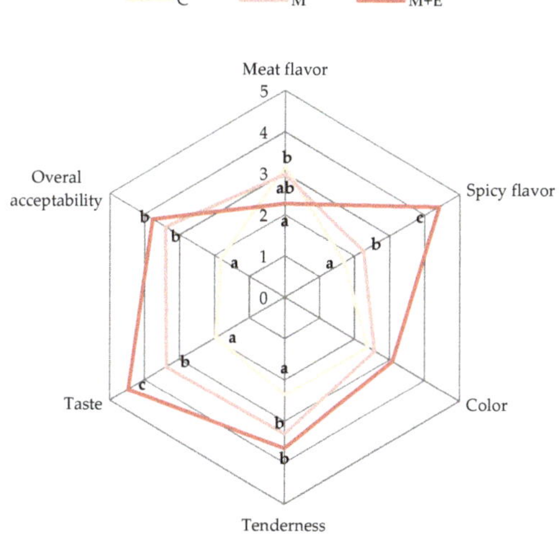

(b)

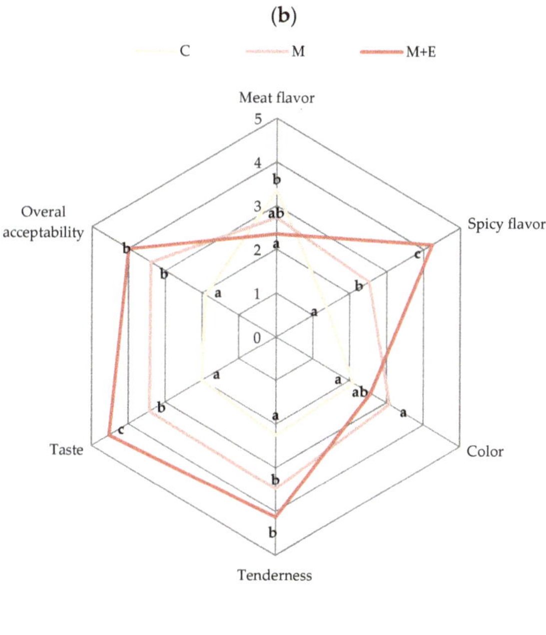

(c)

Figure 6. Sensory data of pork loin slices after 3 days (**a**), 9 days (**b**) and 15 days (**c**) of storage in relation to the sample (Control (C), Marinated (M), marinated with essential oils (M + E)). Data represent means ± SD. a, b, c = average values of each sensorial parameter lacking a common letter significantly differ among the same sensory parameter.

The results showed that, regardless the sampling time, the marinated meat, and especially that with essential oils (M + E), exhibited better scores compared to the non-marinated one, with the only exception of meat flavor intensity parameter. In addition, the marinated samples showed a greater intensity of flavor and taste, positively perceived by the panelists. In particular, marinated meat slices showed higher tenderness, color, flavor and taste intensities for the whole storage period, resulting in an overall improved acceptability compared to the controls. Considering the effect of essential oils, no differences between M and M + E samples were observed after 3 days of storage. However, starting from the second panel test (day 9), M + E samples showed higher scores for spicy flavor and taste intensity compared to M samples. The differences among M and M + E samples intensified at the end of storage (day 15), when the M + E group showed the highest scores for overall acceptability, thus being the preferred sample for over 60% of panelists.

3.1.5. Microbiological Analysis

The microbiological analyses were aimed to detect various microbiological groups frequently associated with the spoilage of processed meat products. In particular, during the refrigerated storage of the samples, the cell loads of total aerobic mesophilic and psychotropic bacteria, lactic acid bacteria, yeasts, *Pseudomonas* spp., total coliforms and *Brochotrix thermosphacta* were detected.

In Figure 7, the cell loads of mesophilic aerobic bacteria, lactic acid bacteria, yeasts, *Pseudomonas* spp., total coliforms and *Brochotrix thermosphacta* are reported.

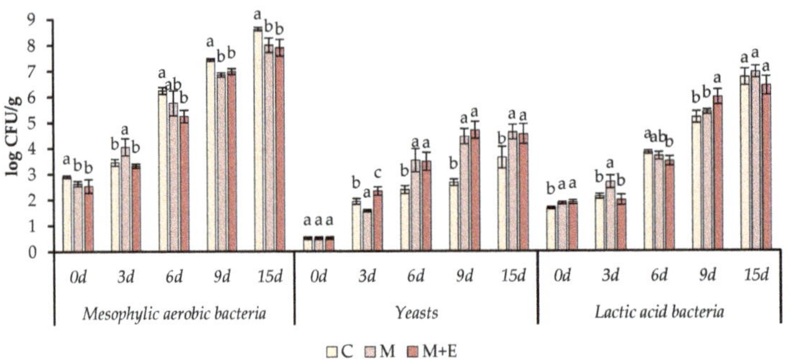

(a)

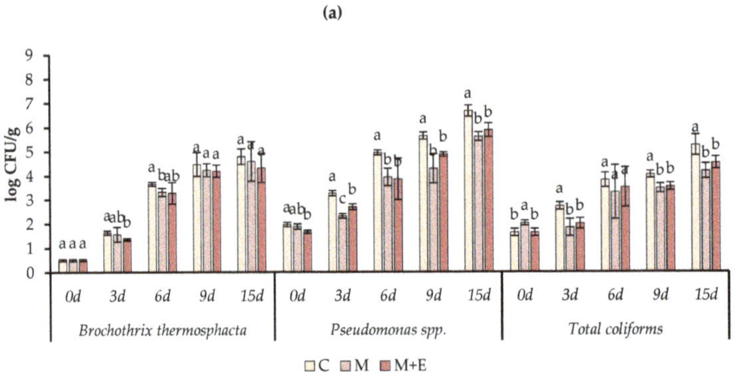

(b)

Figure 7. Cell load (log CFU/g ± SD), during the refrigerated storage, of total aerobic mesophilic bacteria, yeast and lactic acid bacteria (**a**) *Brochothrix thermosphacta*, *Pseudomonas* spp. and total coliforms (**b**) in different pork loin slices: Control (C), Marinated (M), marinated with essential oils (M + O). Data represent means ± SD. a-b-c = for each microorganism, at the same time of storage, average values lacking a common letter significantly differ among the same sampling time ($p < 0.05$).

The data obtained indicated a satisfactory microbiological quality of the raw meat. In fact, the initial cell load of the main spoiling microorganisms was below 3.0 log CFU/g, independently on the use of marinade solution or the addition of essential oils. As expected, the mesophilic bacteria represented the main microbial spoiling group. In fact, a fast increase of the cell load of this group was observed in all the samples starting from the sixth day of refrigerated storage. However, from day 6 of storage, samples M and M + E showed significant lower cell loads compared to C, while no differences were observed between M and M + E samples. The C samples were the only ones found to exceed 8.0 log CFU/g after 15 d of storage. The same trend was observed for psychotropic aerobic bacteria.

A similar tendency was observed for *Pseudomonas* spp. Starting from day 3 of storage C samples showed significant higher cell loads compared to M and M + E samples. No significant differences regarding the cell load of *Pseudomonas* spp. were observed between M and M + E samples, with the only exception of day 3. At the end of the storage *Pseudomonas* spp. resulted 6.67, 5.61 and 5.88 log CFU/g respectively in C, M and M + E samples. Total coliforms resulted significantly lower in M and M + E samples compared to C ones, excepted at day 3 of storage. The greatest differences were observed at day 15 when coliforms were 5.25, 4.18 and 4.22 log CFU/g respectively in samples C, M and M + E. In general, the highest inhibition due to marination and the addition of essential oils was observed against the Gram-negative bacteria *Pseudomonas* spp. and total coliforms. Otherwise, minor differences were observed considering *B. thermosphacta* since no significant differences were observed starting from day 9 of storage. However, depending on the sample, this microorganism reached a cell load ranging between 4.4 and 4.8 log CFU/g.

A different trend was observed for yeasts and lactic acid bacteria. In fact, starting from day 6 of storage, yeasts resulted significantly higher in samples M and M + E compared to the control. However, yeasts never exceed 5.0 log CFU/g for the whole period of storage. In case of lactic acid bacteria, no significant differences were detected at the end of the storage among the samples.

3.2. Challenge Test

In order to evaluate the effects of the marinade solution with or without essential oils on the safety of vacuum packed pork loin slices, a challenge test inoculating *Listeria monocytogenes* Scott A, *Salmonella enteritidis* E5 and *Staphylococcus aures* SR31 was performed. Figure 8a–c shows the cell loads of the pathogen microorganisms during the refrigerated storage.

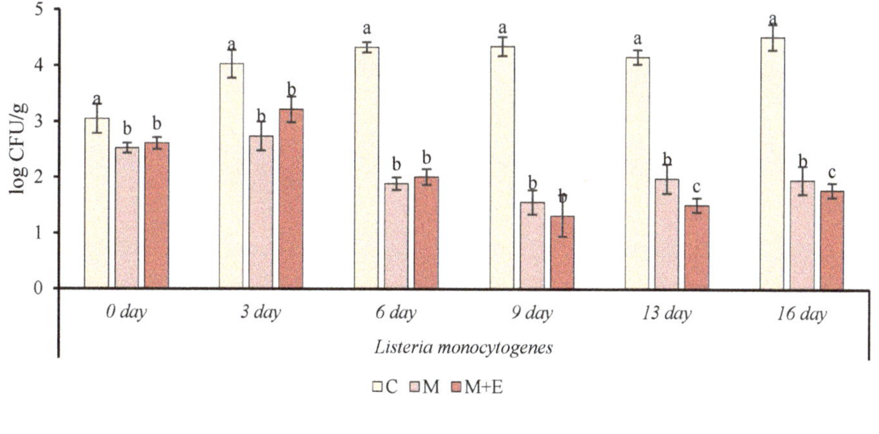

(**a**)

Figure 8. *Cont.*

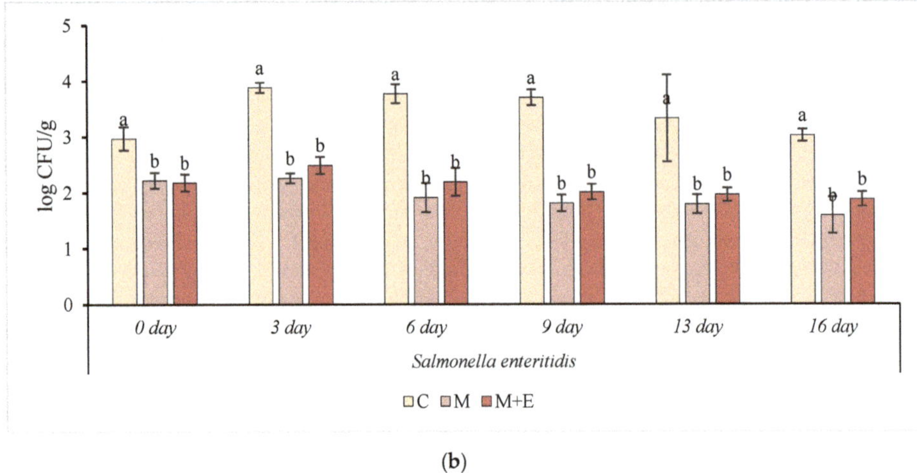

(b)

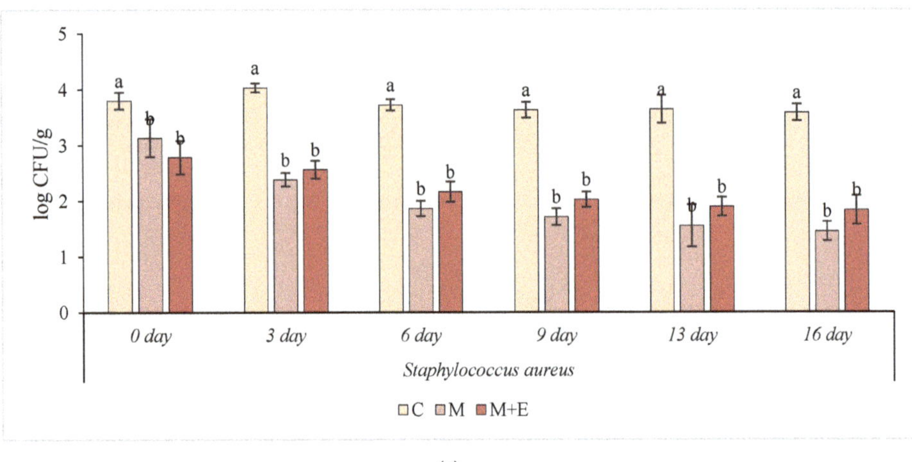

(c)

Figure 8. Cell load (log CFU/g ± SD), during refrigerated storage, of *Listeria monocytogenes* (**a**), *Salmonella enteritidis* (**b**), and *Staphylococcus aureus* (**c**). in different pork loin samples: Control (C), Marinated (M), marinated with essential oils (M + E). Data represent means ± SD. a, b, c = for each microorganism, at the same time of storage, average values lacking a common letter significantly differ among the same sampling time ($p < 0.05$).

It is noteworthy to mention that marination allowed a significant ($p < 0.05$) reduction of the initial microbial cell load of all the pathogens, regardless of the presence or absence of essential oils. The highest initial cell load reduction, compared to control samples, was observed for *S. aureus*, and ranged between 0.7 and 1.0 log CFU/g, followed by *S. enteritidis* (0.7–0.8 log CFU/g) and *L. monocytogenes* (0.5–0.6 log CFU/g). In all cases, the differences in the pathogen levels between marinated and not marinated samples increased during the storage period. At the end of the storage, M and M + E samples showed cell loads lower than 2.0 logarithmic cycles for *L. monocytogenes* and *S. aureus* and lower than 1.5 logarithmic cycles in the case of *S. enteritidis*. On the contrary, an increase of the level of all the pathogens in C samples, greater in the case of *L. monocytogenes*, was observed during storage. Contrarily, a decrease of the pathogen loads in the marinated products was observed during the storage but without allowing their complete inactivation. Considering the effect of the addition of essential oils, no significant differences were found between the samples M and M + E for *S. enteritis* and *S. aureus* while in the case of *L. monocytogenes* the samples M + E showed a significantly lower cell load with respect to samples M starting from day 13 of storage. The greatest antimicrobial effect from marinating was observed against *S. aureus*. In fact, a reduction of more than 3.5 log CFU/g at the end of storage compared to the initial load of C samples was observed.

4. Discussion

The marinade solution prepared with extra virgin olive oil, beer, concentrated lemon juice and a mixture of essential oils used within this study was selected based on the findings of preliminary trials. Considering that offering a marinated product including typical ingredients and flavors belonging to the Mediterranean diet may represent an added value to product itself, all the marinade ingredients and essential oils chosen in this work derive from plants commonly used in the traditional recipes of this area. The selected marinade solution was then tested with the aim of exploring its effect on the shelf-life, safety and quality traits of pork loin slices during refrigerated storage.

Aside from the inclusion of essential oils, the addition of the marinade solution significantly reduced the pH of vacuum-packed pork loin. These outcomes might be ascribed to the addition of an acid marinade solution in which the inclusion of beer (pH = 3.96) and concentrated lemon juice (pH = 2.26) results in a remarkable reduction in pH. This might be desirable for several reasons. First, meat pH exerts a direct effect on its water holding capacity (WHC), since it is generally held that the ability of meat to retain water progressively improves above and below pH values corresponding to the isoelectric point of meat proteins (i.e., 5.5 in the case of pork meat) [26]. Furthermore, processed meat products with a low pH are less likely to develop pathogen microbial growth and off-odors, thus having an improved safety and shelf-life [27,28]. Lastly, reduced pH values might also be advantageous to facilitate the action of collagenases and other proteolytic enzymes responsible for meat tenderization during the refrigerated storage [29].

The addition of the marinade solution, regardless of the use of essential oils, also exerted a significant effect on the yellowness (b*) of meat samples, while lightness (L*) and redness (a*) were not affected. The higher b* values detected for marinated samples might be likely due to the presence of coloring compounds in the solution itself (i.e., extra virgin olive oil, beer and concentrated lemon juice) which might have increased the yellowness of samples. However, the increase in b* values did not negatively affect the sensory evaluation by panelists who associated to the marinated samples in general, and to those including essential oils in particular, a better color retention if compared to the control.

Beside all, the marinating process is a widely used procedure at industrial level implemented with the aim to improve not only the sensory and eating qualities of meat products but also their technological properties, with a special reference to WHC [30,31]. Accordingly, satisfactory marinade uptakes (of more than 7%) were observed for both marinated pork loin groups after 3 days of storage. Albeit little literature is available concerning the effects of essential oils to improve the technological properties of meat, the remarkable improvement in marinade uptakes might be ascribed to the acid pH of the marinade solution. Indeed, as lemon juice contains citric acid, this ingredient is often included within the marinade solution to improve meat WHC by lowering its pH [32]. These outcomes are in agreement with those reported by other authors that observed a marinade uptake ranging between 4.6 and 9.7% in acidic marinated *Longissimus dorsi* muscles [33]. However, it is noteworthy to remember that the marinade uptake is strongly related to the meat type, marination technique as well as the duration of the process [34].

The marination process allowed to remarkably reduce the cooking losses compared to control samples after 3 days of refrigerated storage. This trend is in agreement with what reported by Gao et al. [35] who assessed the effect of marination on the main quality aspects of vacuum-packed pork loin meat. However, after both 9 and 15 days of storage, marinated meat (either M or M + E) exhibited slightly higher cooking losses if compared to the control group. This trend might be likely due to the greater marinade uptake measured during the storage period, which might have resulted in a higher loss of fluids during cooking. Therefore, it is reasonable that raw meat, added with salt without the inclusion of marinade solution, presented reduced cooking losses after a week of refrigerated storage.

Several authors have reported an increase in tenderness of marinated poultry, pork and beef [11,32,35]. Accordingly, the addition of marinade solution with or without essential oils allowed to reduce the shear forces of pork loin meat of about 40% and 22.8%, respectively, just after 3 days of refrigerated storage. These outcomes suggest the effectiveness of an acidic marinade solution to improve the tenderness of meat samples, as previously reported by Miller [36]. Accordingly, several studies have reported that acidic substances in the marinating solution (including lemon juice) can play a crucial role in the tenderization of marinated meat, leading to meat fibers swelling and enhancing proteolysis [37,38].

The sensory analysis data, according to the available literature, suggested that the marinated samples, and in particular those in which essential oils were added to the marinade, were tender and characterized by better color, flavor and taste intensity compared to the control samples. On the other hand, the positive effect of acidic marinade solutions on tenderness and other quality characteristics of different types of meat is widely reported in the literature [2,39]. The addition of essential oils strongly increased the overall acceptability of the samples, especially at the end of the storage, resulting in the preference of the consumers. Recently, many studies have reported an improvement of the sensory qualities and an extended shelf life of meat and meat products supplemented with different essential oils including, rosemary, thyme, oregano, basil, coriander, ginger, garlic, clove, juniper and fennel, used alone or in combination [40,41]. In addition, essential oils are widely reported as characterized by a strong antioxidant activity [42,43]. A wide literature reports a reduction of the lipid oxidation of meat and meat products added with essential oils during storage [40,44,45]. A better sensory quality and a longer shelf-life is normally associated to the reduction of lipid oxidation [45,46].

The predominant spoiling bacteria associated to refrigerated pork and beef, are *Pseudomonas* spp. during storage in aerobic conditions and lactic acid bacteria belonging to the genus *Lactobacillus* spp., *Leuconostoc* spp. and *Carnobacterium* spp. but also *Brochothrix thermosphacta*, *Enterobacteriaceae* and psychrophilic *Clostridium* spp. in case of anaerobic conditions [47,48]. Meat defects due to off-odors and off-flavors normally linked to a discoloration, gas production and acidification are generally associated to the growth of these microorganisms [49–51]. Our results indicate a satisfactory initial microbiological quality of the pork loin used in the present study. In fact, for all the main microbiological spoilage agents considered, the cell load was lower than 3.0 log CFU/g. During storage, an increase of the total viable mesophilic and psychotropic bacteria, *Pseudomonas* spp. lactic acid bacteria and *B. thermosphacta* was observed. The enumeration of total viable mesophilic and psychotropic microorganisms represents one of the most widely used and recognized criteria for evaluating the microbiological quality of meat [52]. Generally, the product is considered acceptable when the cell load of these microorganisms is lower than 7.0 log CFU/g [53] and this level is generally taken at industrial level as the upper threshold to determine the product expiry date. Our results showed that marinated samples overcome this limit only after 15 days of storage while control samples exceeded the limit after 9 days of refrigerated storage. The marination, regardless the addition of essential oils, showed the highest inhibition against the Gram-negative bacteria *Pseudomonas* spp. and total coliforms. Several literature data showed that species belonging to *Pseudomonas* and other psychotropic microorganisms are the predominant cause of alteration of fresh packaged meat [54]. Several *Pseudomonas* spp. are responsible for the formation of superficial patinas and off-flavor when their concentration reaches levels between 7–8 log CFU/g in chilled meat products [55].

Currently, foodborne outbreaks caused by foodborne pathogens transmitted from meat product still represent a significant public health challenge [56]. Considering the last 10–15 years the most important foodborne bacterial pathogens associated to meat belong to *Salmonella* spp., *Escherichia coli*, *Campylobacter jejuni* and *Staphylococcus aureus* [57–59]. Our results showed a clear inhibitory effect of the tested marinades on the growth kinetic of *Listeria monocytogenes*, *Salmonella enteritidis* and *Staphylococcus aureus* resulting in an increased safety of the product. In particular, the tested marinating solution proved an immediate inhibitory effect against all the pathogens. In addition, an increase of pathogens cell load during storage was observed in control samples, while the marinated products induced a more or less marked decrease of the pathogens load without allowing their complete deactivation. Regarding the addition of essential oils, a significant additional antimicrobial effect, compared to marinated samples, was observed only against *Listeria monocytogenes*. The antimicrobial activities of essential oils and their bioactive components are well known and reviewed in a wide literature even if strongly affected by microbial species, strains, and physico-chemical and process variables [17,18,21,60,61]. Although strain dependent and affected by application conditions, the greatest resistance of Gram-negative bacteria, due to the presence of the outer membrane acting as a barrier to hydrophobic molecules, to many essential oils is well known [62]. Among the Gram-positive bacteria, the very high resistance of *Staphylococcus aureus* to many stress factors and antimicrobials including essential oils and their components is well documented [62,63]. Also the action mechanisms of several essential oil components against many microorganisms, including the target microorganisms taken into consideration in the present research, have been clarified by molecular tools [64–67]. The limited antimicrobial effects of the essential oils in the present work is probably due to the masking effect of ethanol and its synergistic effects with low pH values and NaCl of marinade. In fact, as shown by Lanciotti et al. [68] studying the boundary between the growth and no growth of *Salmonella enteritidis*, *Bacillus cereus* and *Staphylococcus aureus* in the presence of different growth controlling factors through probabilistic models, the effects of ethanol on the limitation of growth of the considered species was significant also at concentration of about 1% and not merely additive with temperature and NaCl concentration. Also, the presence of organic acids and the pH reduction by marinade contribute to mask the effects of the essential oils on the target microorganisms considered [69].

Several authors have reported the antimicrobial effect of marinating solution components [10,12]. In particular the antimicrobial effect of some acidic marinade solutions containing alcoholic drinks is associated to the presence of ethanol but also to phenolic derivatives and organic acids, contributing the last to the reduction of the pH of the product [10,70,71]. In addition, the combination of organic acids, ethanol and sodium chloride can strongly inhibit several microorganisms including pathogens like *Salmonella*, *Listeria monocytogenes*, *Escherichia coli* and *Staphylococcus aureus* [72,73].

5. Conclusions

The results of the present study highlighted that the marination of pork loin slices using a solution (formulated with typical ingredients from Mediterranean area) with a mix of extra virgin olive oil, beer and lemon juice (in the presence/absence of essential oils) allows to obtain an overall improvement of the technological and sensory properties of meat. In particular, panel test results suggest a clear preference for marinated products with the addition of essential oils. Furthermore, the tested marinade solution exerted a remarkable meat pH reduction and significant antimicrobial activity both towards the common spoiling microflora normally present on the product and on pathogenic microorganisms deliberately inoculated, improving product safety and shelf-life. The use of marinade allowed the extension of the shelf-life of six days. In addition, offering a marinated product formulated with typical ingredients and flavors belonging to the Mediterranean diet may represent an added value to product itself. However, the addition of essential oils did not lead to a further increase of the antimicrobial activity exerted by the marinade solution. Though, the results obtained in this study suggest that an optimization of the concentration and type of essential oils used for the marination of pork loin could further increase its antimicrobial activity.

Author Contributions: Conceptualization, M.P., R.L. and F.P.; methodology, L.S., M.P., R.L., F.P. and F.S.; software, G.B., L.S. and F.S.; validation, G.B., L.S. and F.S.; formal analysis, G.B. and L.S.; investigation, G.B., D.B., L.S. and F.S.; resources, M.P. and R.L.; data curation, G.B., L.S., D.B. and F.S.; writing—original draft preparation, G.B., L.S. and F.S.; writing—review and editing, L.S., F.P., M.P., D.B. and R.L.; visualization, G.B. and D.B.; supervision, F.P., M.P. and R.L.; project administration, M.P. and R.L. All authors have read and agreed to the published version of the manuscript.

Funding: This research received no external funding.

Conflicts of Interest: The authors declare no conflict of interest.

References

1. Barbut, S.; Sosnicki, A.A.; Lonergan, S.M.; Knapp, T.; Ciobanu, D.C.; Gatcliffe, L.J.; Huff-Lonergan, E.; Wilson, E.W. Progress in reducing the pale, soft and exudative (PSE) problem in pork and poultry meat. *Meat Sci.* **2008**, *79*, 46–63. [CrossRef] [PubMed]

2. Yusop, S.M.; O'Sullivan, M.G.; Kerry, J.P. Marinating and enhancement of the nutritional content of processed meat products. In *Processed Meats: Improving Safety, Nutrition and Quality*; Elsevier Ltd.: Amsterdam, The Netherlands, 2011; pp. 421–449, ISBN 9781845694661.

3. King, D.A.; Wheeler, T.L.; Shackelford, S.D.; Koohmaraie, M. 3 Fresh meat texture and tenderness BT—Improving the Sensory and Nutritional Quality of Fresh Meat. In *Woodhead Publishing Series in Food Science, Technology and Nutrition*; Woodhead Publishing: Cambridge, UK; Sawston, UK, 2009; pp. 61–88, ISBN 978-1-84569-343-5.

4. García-Márquez, I.; Cambero, M.I.; Ordóñez, J.A.; Cabeza, M.C. Shelf-life extension and sanitation of fresh pork loin by e-beam treatment. *J. Food Prot.* **2012**, *75*, 2179–2189. [CrossRef]

5. Kargiotou, C.; Katsanidis, E.; Rhoades, J.; Kontominas, M.; Koutsoumanis, K. Efficacies of soy sauce and wine base marinades for controlling spoilage of raw beef. *Food Microbiol.* **2011**, *28*, 158–163. [CrossRef] [PubMed]

6. Barbut, S. *The Science of Poultry and Meat Processing*; University of Guelph: Guelph, ON, Canada, 2015; Self-published online book; ISBN 9780889556256.

7. Warriss, P.D. *Meat Science: An Introductory Text*; CABI Publishing: New York, NY, USA, 2000; ISBN 978-0-85199-424-6.

8. Balestra, F.; Petracci, M. Technofunctional ingredients for meat products. In *Sustainable Meat Production and Processing*; Elsevier: Amsterdam, The Netherlands, 2019; pp. 45–68.

9. Alvarado, C.; McKee, S. Marination to improve functional properties and safety of poultry meat. *J. Appl. Poult. Res.* **2007**, *16*, 113–120. [CrossRef]

10. Vlahova-Vangelova, D.; Dragoev, S. Marination: Effect on meat safety and human health. A review. *Bulg. J. Agric. Sci.* **2014**, *20*, 503–509.

11. Yusop, S.M.; O'Sullivan, M.G.; Kerry, J.F.; Kerry, J.P. Effect of marinating time and low pH on marinade performance and sensory acceptability of poultry meat. *Meat Sci.* **2010**, *85*, 657–663. [CrossRef]

12. Björkroth, J. Microbiological ecology of marinated meat products. In *Meat Science*; Elsevier Ltd.: Amsterdam, The Netherlands, 2005; Volume 70, pp. 477–480.

13. Pathania, A.; McKee, S.R.; Bilgili, S.F.; Singh, M. Antimicrobial activity of commercial marinades against multiple strains of Salmonella spp. *Int. J. Food Microbiol.* **2010**, *139*, 214–217. [CrossRef]

14. Karam, L.; Roustom, R.; Abiad, M.G.; El-Obeid, T.; Savvaidis, I.N. Combined effects of thymol, carvacrol and packaging on the shelf-life of marinated chicken. *Int. J. Food Microbiol.* **2019**, *291*, 42–47. [CrossRef]

15. Asioli, D.; Aschemann-Witzel, J.; Caputo, V.; Vecchio, R.; Annunziata, A.; Næs, T.; Varela, P. Making sense of the "clean label" trends: A review of consumer food choice behavior and discussion of industry implications. *Food Res. Int.* **2017**, *99*, 58–71. [CrossRef]

16. Grant, A.; Parveen, S. All natural and clean-label preservatives and antimicrobial agents used during poultry processing and packaging. *J. Food Prot.* **2017**, *80*, 540–544. [CrossRef]

17. Patrignani, F.; Siroli, L.; Serrazanetti, D.I.; Gardini, F.; Lanciotti, R. Innovative strategies based on the use of essential oils and their components to improve safety, shelf-life and quality of minimally processed fruits and vegetables. *Trends Food Sci. Technol.* **2015**, *46*, 311–319. [CrossRef]

18. Siroli, L.; Patrignani, F.; Gardini, F.; Lanciotti, R. Effects of sub-lethal concentrations of thyme and oregano essential oils, carvacrol, thymol, citral and trans-2-hexenal on membrane fatty acid composition and volatile molecule profile of Listeria monocytogenes, Escherichia coli and Salmonella enteritidis. *Food Chem.* **2015**, *182*, 185–192. [CrossRef] [PubMed]

19. Lucera, A.; Costa, C.; Conte, A.; Del Nobile, M.A. Food applications of natural antimicrobial compounds. *Front. Microbiol.* **2012**, *3*, 287. [CrossRef] [PubMed]

20. Ruiz, A.; Williams, S.K.; Djeri, N.; Hinton, A.; Rodrick, G.E. Nisin, rosemary, and ethylenediaminetetraacetic acid affect the growth of Listeria monocytogenes on ready-to-eat turkey ham stored at four degrees Celsius for sixty-three days. *Poult. Sci.* **2009**, *88*, 1765–1772. [CrossRef]

21. Lanciotti, R.; Gianotti, A.; Patrignani, F.; Belletti, N.; Guerzoni, M.; Gardini, F. Use of natural aroma compounds to improve shelf-life and safety of minimally processed fruits. *Trends Food Sci. Technol.* **2004**, *15*, 201–208. [CrossRef]

22. Ghaderi-Ghahfarokhi, M.; Barzegar, M.; Sahari, M.A.; Ahmadi Gavlighi, H.; Gardini, F. Chitosan-cinnamon essential oil nano-formulation: Application as a novel additive for controlled release and shelf life extension of beef patties. *Int. J. Biol. Macromol.* **2017**, *102*, 19–28. [CrossRef]

23. Albertsen, L.; Wiedmann, K.P.; Schmidt, S. The impact of innovation-related perception on consumer acceptance of food innovations—Development of an integrated framework of the consumer acceptance process. *Food Qual. Prefer.* **2020**, *84*, 103958. [CrossRef]

24. Jeacocke, R.E. Continuous measurements of the pH of beef muscle in intact beef carcases. *Int. J. Food Sci. Technol.* **2007**, *12*, 375–386. [CrossRef]

25. Pointer, M.R. A comparison of the CIE 1976 colour spaces. *Color Res. Appl.* **2009**, *6*, 108–118. [CrossRef]

26. Aberle, E.; Forrest, J.; Gerrard, D.; Mills, E. *Principles of Meat Science*; Kendall Hunt Publishing Company: Dubuque, IA, USA, 2012.

27. Allen, C.D.; Russell, S.M.; Fletcher, D.L. The relationship of raw broiler breast meat color and pH to cooked meat color and pH. *Poult. Sci.* **1997**, *76*, 1042–1046. [CrossRef]

28. Odeyemi, O.A.; Alegbeleye, O.O.; Strateva, M.; Stratev, D. Understanding spoilage microbial community and spoilage mechanisms in foods of animal origin. *Compr. Rev. Food Sci. Food Saf.* **2020**, *19*, 311–331. [CrossRef]

29. Jones, J. Meat science: An introductory text. *Br. Poult. Sci.* **2000**, *41*, 521.

30. Lunde, K.; Egelandsdal, B.; Choinski, J.; Mielnik, M.; Flåtten, A.; Kubberød, E. Marinating as a technology to shift sensory thresholds in ready-to-eat entire male pork meat. *Meat Sci.* **2008**, *80*, 1264–1272. [CrossRef] [PubMed]

31. Gao, T.; Li, J.; Zhang, L.; Jiang, Y.; Yin, M.; Liu, Y.; Gao, F.; Zhou, G. Effect of different tumbling marination methods and time on the water status and protein properties of prepared pork chops. *Asian-Australas. J. Anim. Sci.* **2015**, *28*, 1020–1027. [CrossRef]

32. Mozuriene, E.; Bartkiene, E.; Krungleviciute, V.; Zadeike, D.; Juodeikiene, G.; Damasius, J.; Baltusnikiene, A. Effect of natural marinade based on lactic acid bacteria on pork meat quality parameters and biogenic amine contents. *LWT Food Sci. Technol.* **2016**, *69*, 319–326. [CrossRef]

33. Fadiloğlu, E.E.; Serdaroğlu, M. Effects of pre and post-rigor marinade injection on some quality parameters of Longissimus dorsi muscles. *Korean J. Food Sci. Anim. Resour.* **2018**, *38*, 325–337.

34. Gamage, H.G.C.L.; Mutucumarana, R.K.; Andrew, M.S. Effect of marination method and holding time on physicochemical and sensory characteristics of broiler meat. *J. Agric. Sci.* **2017**, *12*, 172–184. [CrossRef]

35. Gao, T.; Li, J.; Zhang, L.; Jiang, Y.; Ma, R.; Song, L.; Gao, F.; Zhou, G. Effect of different tumbling marination treatments on the quality characteristics of prepared pork chops. *Asian-Australas. J. Anim. Sci.* **2015**, *28*, 260–267. [CrossRef]

36. Miller, R. Functionality Of Non-Meat Ingredients Used In Enhanced Pork—Hogs, Pigs, and Pork. Available online: https://swine.extension.org/functionality-of-non-meat-ingredients-used-in-enhanced-pork/ (accessed on 30 April 2020).

37. Berge, P.; Ertbjerg, P.; Larsen, L.M.; Astruc, T.; Vignon, X.; Møller, A.J. Tenderization of beef by lactic acid injected at different times post mortem. *Meat Sci.* **2001**, *57*, 347–357. [CrossRef]

38. Komoltri, P.; Pakdeechanuan, P. Effects of marinating ingredients on physicochemical, microstructural and sensory properties of golek chicken. *Int. Food Res. J.* **2012**, *19*, 1449–1455.

39. Brooks, C. *Marinating of Beef for Enhancement*; National Cattlemen's Beef Association: Centennial, CO, USA, 2007.

40. Jayasena, D.D.; Jo, C. Essential oils as potential antimicrobial agents in meat and meat products: A review. *Trends Food Sci. Technol.* **2013**, *34*, 96–108. [CrossRef]

41. Van Haute, S.; Raes, K.; Van der Meeren, P.; Sampers, I. The effect of cinnamon, oregano and thyme essential oils in marinade on the microbial shelf life of fish and meat products. *Food Control* **2016**, *68*, 30–39. [CrossRef]

42. Hyldgaard, M.; Mygind, T.; Meyer, R.L. Essential oils in food preservation: Mode of action, synergies, and interactions with food matrix components. *Front. Microbiol.* **2012**, *3*, 12. [CrossRef] [PubMed]

43. Prakash, B.; Kedia, A.; Mishra, P.K.; Dubey, N.K. Plant essential oils as food preservatives to control moulds, mycotoxin contamination and oxidative deterioration of agri-food commodities—Potentials and challenges. *Food Control* **2015**, *47*, 381–391. [CrossRef]

44. Aminzare, M.; Hashemi, M.; Hassanzad Azar, H.; Hejazi, J. The use of herbal extracts and essential oils as a potential antimicrobial in meat and meat products: A review. *J. Hum. Environ. Health Promot.* **2016**, *1*, 63–74. [CrossRef]

45. Pateiro, M.; Barba, F.J.; Domínguez, R.; Sant'Ana, A.S.; Mousavi Khaneghah, A.; Gavahian, M.; Gómez, B.; Lorenzo, J.M. Essential oils as natural additives to prevent oxidation reactions in meat and meat products: A review. *Food Res. Int.* **2018**, *113*, 156–166. [CrossRef]

46. Domínguez, R.; Pateiro, M.; Gagaoua, M.; Barba, F.J.; Zhang, W.; Lorenzo, J.M. A comprehensive review on lipid oxidation in meat and meat products. *Antioxidants* **2019**, *8*, 429. [CrossRef]

47. Borch, E.; Kant-Muermans, M.L.; Blixt, Y. Bacterial spoilage of meat and cured meat products. *Int. J. Food Microbiol.* **1996**, *33*, 103–120. [CrossRef]

48. Doulgeraki, A.I.; Ercolini, D.; Villani, F.; Nychas, G.-J.E. Spoilage microbiota associated to the storage of raw meat in different conditions. *Int. J. Food Microbiol.* **2012**, *157*, 130–141. [CrossRef]

49. Casaburi, A.; Piombino, P.; Nychas, G.J.; Villani, F.; Ercolini, D. Bacterial populations and the volatilome associated to meat spoilage. *Food Microbiol.* **2015**, *45*, 83–102. [CrossRef]

50. Cauchie, E.; Delhalle, L.; Taminiau, B.; Tahiri, A.; Korsak, N.; Burteau, S.; Fall, P.A.; Farnir, F.; Baré, G.; Daube, G. Assessment of spoilage bacterial communities in food wrap and modified atmospheres-packed minced pork meat samples by 16S rDNA metagenetic analysis. *Front. Microbiol.* **2020**, *10*, 3074. [CrossRef] [PubMed]

51. Zhao, F.; Zhou, G.; Ye, K.; Wang, S.; Xu, X.; Li, C. Microbial changes in vacuum-packed chilled pork during storage. *Meat Sci.* **2015**, *100*, 145–149. [CrossRef] [PubMed]

52. Ercolini, D.; Russo, F.; Nasi, A.; Ferranti, P.; Villani, F. Mesophilic and psychrotrophic bacteria from meat and their spoilage potential in vitro and in beef. *Appl. Environ. Microbiol.* **2009**, *75*, 1990–2001. [CrossRef] [PubMed]

53. FCD. Available online: http://www.fcd.fr/media/filer_public/7e/51/7e51261b-e18d-4c9f-be28-5dad72e4cfd0/1200_fcd_criteres_microbiologiques_2016_ateliers_rayons_coupe_28012016.pdf (accessed on 5 April 2020).

54. Wickramasinghe, N.N.; Ravensdale, J.; Coorey, R.; Chandry, S.P.; Dykes, G.A. The predominance of psychrotrophic Pseudomonas on aerobically stored chilled red meat. *Compr. Rev. Food Sci. Food Saf.* **2019**, *18*, 1622–1635. [CrossRef]

55. Nychas, G.J.E.; Skandamis, P.N.; Tassou, C.C.; Koutsoumanis, K.P. Meat spoilage during distribution. *Meat Sci.* **2008**, *78*, 77–89. [CrossRef]

56. Omer, M.K.; Álvarez-Ordoñez, A.; Prieto, M.; Skjerve, E.; Asehun, T.; Alvseike, O.A. A systematic review of bacterial foodborne outbreaks related to red meat and meat products. *Foodborne Pathog. Dis.* **2018**, *15*, 598–611. [CrossRef]

57. Fegan, N.; Jenson, I. The role of meat in foodborne disease: Is there a coming revolution in risk assessment and management? *Meat Sci.* **2018**, *144*, 22–29. [CrossRef]

58. Mor-Mur, M.; Yuste, J. Emerging bacterial pathogens in meat and poultry: An overview. *Food Bioproc. Technol.* **2010**, *3*, 24–35. [CrossRef]

59. Kamdem, S.S.; Patrignani, F.; Guerzoni, M.E. Shelf-life and safety characteristics of Italian Toscana traditional fresh sausage (Salsiccia) combining two commercial ready-to-use additives and spices. *Food Control* **2007**, *18*, 421–429. [CrossRef]

60. Burt, S. Essential oils: Their antibacterial properties and potential applications in foods—A review. *Int. J. Food Microbiol.* **2004**, *94*, 223–253. [CrossRef]

61. Alirezalu, K.; Pateiro, M.; Yaghoubi, M.; Alirezalu, A.; Peighambardoust, S.H.; Lorenzo, J.M. Phytochemical constituents, advanced extraction technologies and techno-functional properties of selected Mediterranean plants for use in meat products. A comprehensive review. *Trends Food Sci. Technol.* **2020**, *100*, 292–306. [CrossRef]

62. Lanciotti, R.; Patrignani, F.; Bagnolini, F.; Guerzoni, M.E.; Gardini, F. Evaluation of diacetyl antimicrobial activity against Escherichia coli, Listeria monocytogenes and Staphylococcus aureus. *Food Microbiol.* **2003**, *20*, 537–543. [CrossRef]

63. Montanari, C.; Serrazanetti, D.I.; Felis, G.; Torriani, S.; Tabanelli, G.; Lanciotti, R.; Gardini, F. New insights in thermal resistance of staphylococcal strains belonging to the species Staphylococcus epidermidis, Staphylococcus lugdunensis and Staphylococcus aureus. *Food Control* **2015**, *50*, 605–612. [CrossRef]

64. Lanciotti, R.; Braschi, G.; Patrignani, F.; Gobbetti, M.; De Angelis, M. How Listeria monocytogenes Shapes Its Proteome in Response to Natural Antimicrobial Compounds. *Front. Microbiol.* **2019**, *10*, 437. [CrossRef] [PubMed]

65. Siroli, L.; Braschi, G.; de Jong, A.; Kok, J.; Patrignani, F.; Lanciotti, R. Transcriptomic approach and membrane fatty acid analysis to study the response mechanisms of *Escherichia coli* to thyme essential oil, carvacrol, 2-(E)-hexanal and citral exposure. *J. Appl. Microbiol.* **2018**, *125*, 1308–1320. [CrossRef] [PubMed]

66. Braschi, G.; Serrazanetti, D.I.; Siroli, L.; Patrignani, F.; De Angelis, M.; Lanciotti, R. Gene expression responses of Listeria monocytogenes Scott A exposed to sub-lethal concentrations of natural antimicrobials. *Int. J. Food Microbiol.* **2018**, *286*, 170–178. [CrossRef] [PubMed]

67. Braschi, G.; Patrignani, F.; Siroli, L.; Lanciotti, R.; Schlueter, O.; Froehling, A. Flow cytometric assessment of the morphological and physiological changes of Listeria monocytogenes and Escherichia coli in response to natural antimicrobial exposure. *Front. Microbiol.* **2018**, *9*, 2783. [CrossRef]

68. Lanciotti, R.; Sinigaglia, M.; Gardini, F.; Vannini, L.; Guerzoni, M.E. Growth/no growth interfaces of Bacillus cereus, Staphylococcus aureus and Salmonella enteritidis in model systems based on water activity, pH, temperature and ethanol concentration. *Food Microbiol.* **2001**, *18*, 659–668. [CrossRef]

69. Castellano, P.; Pérez Ibarreche, M.; Blanco Massani, M.; Fontana, C.; Vignolo, G. Strategies for pathogen biocontrol using lactic acid bacteria and their metabolites: A focus on meat ecosystems and industrial environments. *Microorganisms* **2017**, *5*, 38. [CrossRef]

70. Nisiotou, A.; Chorianopoulos, N.G.; Gounadaki, A.; Panagou, E.Z.; Nychas, G.J.E. Effect of wine-based marinades on the behavior of salmonella typhimurium and background flora in beef fillets. *Int. J. Food Microbiol.* **2013**, *164*, 119–127. [CrossRef]

71. Vaquero, M.J.R.; Alberto, M.R.; de Nadra, M.C.M. Antibacterial effect of phenolic compounds from different wines. *Food Control* **2007**, *18*, 93–101. [CrossRef]

72. Theron, M.M.; Lues, J.F.R. Organic acids and meat preservation: A review. *Food Rev. Int.* **2007**, *23*, 141–158. [CrossRef]

73. Friedman, M.; Henika, P.R.; Levin, C.E.; Mandrell, R.E. Recipes for antimicrobial wine marinades against Bacillus cereus, Escherichia coli O157:H7, Listeria monocytogenes, and Salmonella enterica. *J. Food Sci.* **2007**, *72*, M207–M213. [CrossRef] [PubMed]

 *foods*

Communication

Composition of the Essential Oil and Insecticidal Activity of *Launaea taraxacifolia* (Willd.) Amin ex C. Jeffrey Growing in Nigeria

Moses S. Owolabi [1,*]**, Akintayo L. Ogundajo** [1]**, Azeezat O. Alafia** [2]**, Kafayat O. Ajelara** [2] **and William N. Setzer** [3,4,*]

[1] Department of Chemistry, Lagos State University, P.M.B. 001, LASU, Lagos 102001, Nigeria; ogundajotayo@yahoo.com

[2] Department of Zoology and Environmental Biology, Lagos State University, P.M.B. 001, LASU, Lagos 102001, Nigeria; honeylafia@yahoo.com (A.O.A.); kajelara@yahoo.com (K.O.A.)

[3] Department of Chemistry, University of Alabama in Huntsville, Huntsville, AL 35899, USA

[4] Aromatic Plant Research Center, 230 N 1200 E, Suite 100, Lehi, UT 84043, USA

* Correspondence: sunnyconcept2007@yahoo.com (M.S.O.); wsetzer@chemistry.uah.edu (W.N.S.)

Received: 12 June 2020; Accepted: 9 July 2020; Published: 11 July 2020

Abstract: The rice weevil (*Sitophilus oryzae*) is a pest of stored grain products such as rice, wheat, and corn. Essential oils represent a green environmentally-friendly alternative to synthetic pesticides for controlling stored-product insect pests. *Launaea taraxacifolia* is a leafy vegetable plant found in several parts of Nigeria. The leaves are eaten either fresh as a salad or cooked as a sauce. The essential oil obtained from fresh leaves of *L. taraxacifolia* was obtained by hydrodistillation and analyzed by gas chromatography/mass spectrometry (GC-MS). Twenty-nine compounds were identified, accounting for 100% of the oil composition. The major component classes were monoterpene hydrocarbons (78.1%), followed by oxygenated monoterpenoids (16.2%), sesquiterpene hydrocarbons (2.1%), oxygenated sesquiterpenoids (0.3%), and non-terpenoid derivatives (3.3%). The leaf essential oil was dominated by monoterpene hydrocarbons including limonene (48.8%), sabinene (18.8%), and (*E*)-β-ocimene (4.6%), along with the monoterpenoid aldehyde citronellal (11.0%). The contact insecticidal activity of *L. taraxacifolia* essential oil against *Sitophilus oryzae* was carried out; median lethal concentration (LC_{50}) values of topical exposure of *L. taraxacifolia* essential oil were assessed over a 120-h period. The LC_{50} values ranged from 54.38 µL/mL (24 h) to 10.10 µL/mL (120 h). The insecticidal activity of the *L. taraxacifolia* essential oil can be attributed to major components limonene (48.8%), sabinene (18.8%), and citronellal (11.0%), as well as potential synergistic action of the essential oil components. This result showed *L. taraxacifolia* essential oil may be considered as a useful alternative to synthetic insecticides.

Keywords: essential oil composition; limonene; sabinene; citronellal; *Sitophilus oryzae*

---

1. Introduction

Insects such as *Callosobruchus maculatus* (Fabr.) (bruchid beetle), *Sitophilus granarius* (L.) (wheat weevil), *S. oryzae* (L.) (rice weevil), *S. zeamais* (Motsch.) (maize weevil), and *Tribolium castaneum* (Herbst) (red flour beetle), are important pests that attack stored grains, causing widespread economic losses [1–3]. The long-term use of synthetic insecticides to control these pests has become problematic, however. Compounds such as chlorinated hydrocarbons, organophosphates, carbamates, etc., tend to be toxic to non-target organisms such as mammals, birds, and fish [4–6], they are persistent in the environment [7–10], and many stored-grain insect pests have developed insecticide resistance [11–13]. Essential oils have emerged as viable alternatives to synthetic pesticides for control of stored-grain insect pests; they are generally non-toxic to mammals, birds, fish, or humans, have limited persistence, are readily biodegradable, and are renewable resources [14–17].

Launaea taraxacifolia (Willd.) Amin ex. C. Jeffrey (syn. *Lactuca taraxacifolia* (Willd.) Schumach, wild lettuce) is a leafy vegetable plant belonging to the Asteraceae (Compositae). The family consists of roughly 1100 genera, and 20,000 species distributed across several countries including Mexico, West Indies, Central and South America, Europe, North Africa, and tropical West African countries like Ghana, Senegal, Benin, and Nigeria [18]. *L. taraxacifolia* is a wild erect perennial herb that grows up to 1–3 m in height with 3–5 pinnately lobed leaves at the base of the stem in a rosette form. The plant is found singly or in clusters of rocky soil, but it is also cultivated in small open gardens near homes for family consumption. The leaves are eaten fresh as a salad or cooked as sauces [18–24]. The plant is known as 'efo yanrin' among the Yorubas of the southwestern part of Nigeria, 'ugu' among the Ibos of the eastern part of Nigeria, and 'nonon barya' among the Hausas of the northern part of Nigeria. Minerals, proteins, flavonoids, fatty acids, and vitamins have been reported to be found in the leaves of *L. taraxacifolia* [25,26]. The nutritional aspects of *L. taraxacifolia* have been reviewed [27,28]. The antioxidant and antiviral activities as well as the use of *L. taraxacifolia* leaves in treatment and control of blood cholesterol levels, blood pressure, and diabetes have been reported [29,30]. Phytochemical studies of *L. taraxacifolia* revealed that the plant possesses chemical classes such as phenolic glycosides, flavonoids, saponins and triterpenoids, which are known to have phytotherapeutic value for humans [25,31–34]. To the best of our knowledge, there is little or no information on the composition of the essential oil or the insecticidal activity of *L. taraxacifolia*. Therefore, the present research was undertaken with the aim of investigating the essential oil composition and evaluating the insecticidal potential of *L. taraxacifolia* leaves from southwestern Nigeria.

2. Materials and Methods

2.1. Plant Materials

The leaves of *L. taraxacifolia* were collected from Ipara, Badagry (6°4′54.07″ N and 2°52′52.75″ E) Lagos state, Nigeria. Botanical identification was done at the Herbarium, University of Lagos, Nigeria, where a voucher specimen (LUH: 7959) was deposited. Fresh leaves of *L. taraxacifolia* were cut into pieces, air dried, and pulverized in a blender to increase the surface area. A 450-g sample of blended *L. taraxacifolia* was hydrodistilled for 4 h in an all-glass modified Clevenger-type apparatus according to British Pharmacopoeia [35]. The obtained essential oil was stored in a sealed glass bottle with a screw lid cover under refrigeration at 4 °C until ready for use. Oil yield was calculated on a dry weight basis.

2.2. Gas Chromatographic–Mass Spectral Analysis

The chemical composition of *L. taraxacifolia* essential oil was determined by gas chromatography–mass spectrometry (GC-MS) using a Shimadzu GCMS-QP2010 Ultra operated in the electron impact (EI) mode (electron energy = 70 eV), scan range = 40–400 atomic mass units, with a scan rate of 3.0 scans per s, with GC-MS solution software. The GC column was a ZB-5 fused silica capillary column (30 m length × 0.25 mm inner diameter) with a 5% phenyl-polymethylsiloxane stationary phase and a film thickness of 0.25 μm. Helium gas was used as a carrier gas with column head pressure of 552 kPa at a flow rate of 1.37 mL/min. The injector temperature was 250 °C and the ion source temperature was 200 °C. The oven temperature of 50 °C was initially programmed for the GC and gradually increased at 2 °C/min to 260 °C. The sample (5% w/v) was dissolved in dichloromethane and 0.1 μL of the solution was injected using a split injection technique (30:1). Identification of the essential oil components was achieved by comparing the retention indices determined with respect to a homologous series of *n*-alkanes, and by comparison of the mass spectral fragmentation patterns with those stored in the MS databases [36–39].

2.3. Insecticidal Activity Screening

The essential oil was screened for insecticidal activity based on the method of Ilboudo and co-workers [40] with modifications. *Sitophilus oryzae* (L.) (rice weevil) were reared on whole rice (10:1 w/w). Adult insects, 1–7 days old, were used for contact toxicity tests. The insects were cultured in a dark growth chamber at a temperature of $27 \pm 1\ ^\circ C$ with relative humidity of $65 \pm 5\%$. The insecticidal activity of *L. taraxacifolia* oil against *S. oryzae* (rice weevil) was evaluated by treatment of Whatman No. 1 filter paper discs with the essential oil diluted in ethanol. The required quantities of oil (0.10, 0.20, 0.30, and 0.40 µL) were diluted to 1 mL with ethanol and applied to filter paper discs, respectively. Permethrin (0.6% w/w) and ethanol were used as positive and negative controls, respectively. The solvent was allowed to evaporate from the filter paper, which was then placed into polyethylene cups (80 mm diameter). Ten well-fed mixed sex adult *S. oryzae* were introduced into the polyethylene cups, containing 20 g uninfected rice grains, and covered with a muslin cloth, held in place with rubber bands. Each treatment was replicated four times. Control experiments were set up as described as above without the essential oil. The experiment was arranged in a complete randomized design on a laboratory bench. The insect was considered dead when the legs or antennae were observed to be immobile. Insect mortalities were investigated by observing the recovery of immobilized insects after 24 h intervals for 120 h and the percentage of insect mortality was corrected using the Abbott formula [41]. Probit analysis [42] using XLSTAT version 2018.1.1.60987 (Addinsoft™, Paris, France) was used to estimate median lethal concentration (LC_{50}) values and insect toxicity data were analyzed using one-way ANOVA Tukey's honestly significant difference test.

3. Results and Discussion

3.1. Essential Oil Composition

The essential oil from *L. taraxacifolia* was obtained by hydrodistillation with a yield of 1.68% as a pale-yellow essential oil, which was analyzed by GC-MS. The chemical composition of the leaf volatile oil of *L. taraxacifolia* is listed in Table 1. A total of 29 compounds were identified, accounting for 100% of the essential oil composition. The major chemical classes were monoterpene hydrocarbons (78%) and oxygenated monoterpenoids (16.2%), followed by sesquiterpene hydrocarbons (2.1%), oxygenated sesquiterpenoids (0.3%), and non-terpenoid derivatives (3.3%). The leaf essential oil was dominated by monoterpene hydrocarbons including limonene (48.8%), sabinene (18.8%), and (*E*)-β-ocimene (4.6%), along with the monoterpenoid aldehyde citronellal (11.0%). The chemical constituents of *L. taraxaciflora* essential oil have not been previously reported to the best of our knowledge. However, a phytochemical study and antioxidant and bacterial screening of the leaf extract of *L. taraxacifolia* have been reported [43].

Table 1. The chemical constituents of *Launaea taraxacifolia* leaf essential oil.

Constituents	RI_{calc} [1]	RI_{db} [2]	Relative Abundance (%)
α-Pinene	941	933 [37]	0.9
Sabinene	976	971 [37]	18.8
Myrcene	993	991 [37]	2.2
α-Terpinene	1018	1018 [37]	0.6
Limonene	1032	1030 [37]	48.8
(*Z*)-β-ocimene	1042	1034 [37]	0.9
(*E*)-β-ocimene	1052	1045 [37]	4.6
γ-Terpinene	1062	1058 [37]	1.0
Terpinolene	1088	1086 [36]	0.4
Linalool	1101	1099 [38]	3.1
Citronellal	1155	1151 [38]	11.0
Terpinen-4-ol	1178	1180 [37]	1.4
1-Dodecene	1192	1192 [39]	0.5

Table 1. *Cont.*

Constituents	RI$_{calc}$ [1]	RI$_{db}$ [2]	Relative Abundance (%)
n-Dodecane	1200	1200 [36]	0.5
Neryl acetate	1366	1366 [39]	0.7
1-Tetradecene	1392	1388 [36]	0.5
n-Tetradecane	1400	1400 [36]	0.2
β-Caryophyllene	1420	1417 [36]	1.5
α-Humulene	1456	1452 [36]	0.1
Bicyclogermacrene	1495	1497 [38]	0.3
Germacrene B	1556	1559 [36]	0.2
Caryophyllene oxide	1581	1582 [36]	0.3
1-Hexadecene	1592	1588 [36]	0.7
Pentadecanal	1712	1715 [38]	1.0
Monoterpene hydrocarbons			78.1
Oxygenated monoterpenoids			16.2
Sesquiterpene hydrocarbons			2.1
Oxygenated sesquiterpenoids			0.3
Non-terpene derivatives			3.3
Total identified (%)			100

[1] RI$_{calc}$ = Kovats retention index determined with respect to a homologous series of *n*-alkanes on a ZB-5 column.
[2] RI$_{db}$ = Retention index from the databases [36–39].

3.2. Insecticidal Activity

The contact toxicity of *L. taraxacifolia* against *S. oryzae* revealed considerable differences in insect mortality rate to the essential oil with different concentrations and different exposure times. Table 2 shows that at a dose of 10.00 µL/mL, the volatile oil produced 25.00% mortality after 48 h (not significantly different than the negative EtOH control) and 52.50% after 120 h (significantly higher toxicity than the EtOH control). The essential oil produced 30.00%, 47.50%, 60.00%, and 75.00% mortality after 48, 72, 96, and 120 h at a dose of 20.00 µL/mL, respectively, while a dose of 30.00 µL/mL yielded a mortality rate of 42.50%, 57.50%, 75.00%, and 75.00%, respectively, over the same period of time. With longer contact times (≥48 h), 20 µL/mL and 30 µL/mL concentrations of *L. taraxacifolia* essential oil was significantly more toxic than the EtOH control, but less toxic than the permethrin positive control. The highest concentration of 40.00 µL/mL produced a mortality of 97.50%, and 100.00% after 96 and 120 h, respectively, which is significantly comparable to the permethrin positive control. Permethrin (0.6% w/w) against *S. oryzae* caused 40.0% mortality with 24 h of exposure and 100.0% mortality after 48 h. The negative control showed no appreciable activity against *S. oryzae* until after 120 h.

Table 2. Contact insecticidal effects of *Launaea taraxacifolia* essential oil on adult mortality of *Sitophilus oryzae* reared on rice grains 120 h after treatment.

Concentration (µL/mL)	Mean % Mortality (±SE) [1]				
	24 h	48 h	72 h	96 h	120 h
10.00	7.50 ± 5.00 [c,d]	25.00 ± 12.91 [c,d]	25.00 ± 12.91 [d,e]	25.00 ± 12.91 [c]	52.50 ± 17.08 [c]
20.00	15.00 ± 5.77 [c,d]	30.00 ± 14.14 [c]	47.50 ± 17.08 [c,d]	60.00 ± 14.14 [b]	75.00 ± 5.77 [b]
30.00	22.50 ± 9.57 [b,c]	42.50 ± 12.58 [b,c]	57.50 ± 9.57 [b,c]	75.00 ± 5.77 [b]	75.00 ± 5.77 [b]
40.00	45.00 ± 17.32 [a]	65.00 ± 12.91 [b]	75.00 ± 5.77 [b]	97.50 ± 5.00 [a]	100.00 ± 0.00 [a]
EtOH control	2.50 ± 5.00 [d]	5.00 ± 5.77 [d]	10.00 ± 8.16 [e]	12.50 ± 9.57 [c]	25.00 ± 5.77 [d]
Permethrin	40.00 ± 0.00 [a,b]	100.00 ± 0.00 [a]	100.00 ± 0.00 [a]	100.00 ± 0.00 [a]	100.00 ± 0.00 [a]
F-value, DF [2]	15.08, 5	37.44, 5	39.69, 5	62.21, 5	51.13, 5

[1] Mean followed by different letters in a column is significantly different at ($p < 0.05$). Insect toxicity data were analyzed using one-way ANOVA followed by Tukey's test. [2] Degrees of freedom.

Median lethal concentration (LC_{50}) values at 95% confidence limits over exposure of *L. taraxacifolia* essential oil were assessed and are shown in Table 3. After 120 h of exposure with an increase in concentration at regular intervals of 24 h, the LC_{50} values were 54.38, 31.64, 21.48, 16.38, and 10.10 µL/mL, respectively. In this study, the essential oil of *L. taraxacifolia* demonstrated contact toxicity to *S. oryzae*, since it had higher insecticidal activity with increasing essential oil concentration and exposure time. This result showed *L. taraxacifolia* essential oil to have promising insecticidal activity against *S. oryzae* and therefore may be considered as a useful, environmentally benign alternative to synthetic insecticides.

Table 3. Median lethal concentrations (LC_{50}, µL/mL, and 95% confidence limits) of *Launaea taraxacifolia* essential oil against *Sitophilus oryzae*.

	Contact Time				
	24 h	**48 h**	**72 h**	**96 h**	**120 h**
LC_{50}	54.38	31.64	21.48	16.38	10.10
(95% confidence limits)	(39.26–133.8)	(23.86–55.67)	(16.62–27.21)	(13.56–18.78)	(5.67–13.31)

To best of our knowledge, there have been no previous literature reports on the insecticidal activity of *L. taraxacifolia* essential oil against *S. oryzae* insect pest. However, contact toxicity of both limonene and sabinene, the major chemical components in this present study, have shown insecticidal activity against *S. oryzae* [44]. Limonene has been previously reported to have a moderate contact effect against *S. zeamais* (LD_{50} values of 198.66 µg/cm^2) and *S. oryzae* (with LD_{50} of 260.18 µg/cm^2) [45] as well as fumigant toxicity against *S. oryzae* (24-h LC_{50} 61.5 µL/L) [46]. Garcia et al. reported that limonene showed contact toxicity against *T. castaneum* [47]. Sabinene, on the other hand, demonstrated weaker insecticidal activity against *S. oryzae* (24-h LC_{50} 463 µL/L) [44]. Interestingly, the *S. oryzae* fumigant insecticidal activities of limonene and sabinene parallel the acetylcholinesterase (AChE) inhibitory activities; AChE IC_{50} = 9.57 µL/mL and 85.03 µL/mL, respectively, for limonene and sabinene [48]. Furthermore, the binary combination of limonene + sabinene showed synergistic AChE inhibition [48]. The insecticidal activity of the *L. taraxacifolia* essential oil could be attributed to those known major components and the resulting synergistic action of the monoterpene hydrocarbons limonene (48.8%) and sabinene (18.8%).

The major aldehyde essential oil component, citronellal (11.0%), has also shown contact insecticidal activity against *Musca domestica* [49] and *S. oryzae* [50] and fumigant insecticidal activity against *T. castaneum* [51] and *S. zeamais* [52]. (−)-Citronellal has also shown AChE inhibitory activity with IC_{50} of 18.4 mM [50]. The contact toxicities of bornyl acetate, (+)-limonene, myrcene, α-phellandrene, α-pinene, sabinene, and terpinolene, essential oil constituents obtained from leaves of *Chamaecyparis obtusa*, against *Callosobruchus chinensis* (L.) and *Sitophilus oryzae* (L.) have been reported [44]. The insecticidal activity of the essential oil components 1,8-cineole, *p*-cymene, α-pinene, and limonene has been previously reported with the order of activity 1,8-cineole > *p*-cymene > α-pinene > limonene [46]. Abdelgaleil et al. reported a comparative study of eleven monoterpenes contact and fumigant toxicity: camphene, (+)-camphor, (−)-carvone, 1-8-cineole, cuminaldehyde, (L)-fenchone, geraniol, (−)-limonene, (−)-linalool, (−)-menthol, and myrcene, against two important stored products insects, *S. oryzae*, and *T. castaneum*, and discovered that the toxicity varied according to insect pest with *S. oryzae* more susceptible to most of the components than *T. castaneum* [53].

4. Conclusions

This study investigated the essential oil composition and evaluated the insecticidal potential of *L. taraxacifolia* leaves for the first time as a potential substitute to synthetic insecticides. *L. taraxacifolia* offers an advantage in Nigeria due to its accessibility and renewability. Despite many advantages of medicinal plants, especially the essential oils, further studies need to be conducted to ascertain the safety of this essential oil before its practical use as an insecticide for controlling stored product insect pests. In addition, while the insecticidal properties of *L. taraxacifolia* essential oil are promising, this work is preliminary and future investigations extrapolating the use of the essential oil under grain-storage conditions should be pursued. In addition, studies on the controlled-release formulations of the essential oil could be examined to curb some of the challenges of essential oil treatments such as rapid degradation, volatility, and low bioavailability of the essential oils.

Author Contributions: Conceptualization, M.S.O.; methodology, M.S.O, K.O.A., and W.N.S.; validation, W.N.S., formal analysis, K.O.A. and W.N.S.; investigation, M.S.O., A.L.O., A.O.A., K.O.A., and W.N.S.; data curation, M.S.O.; writing—original draft preparation, M.S.O.; writing—review and editing, M.S.O. and W.N.S.; supervision, M.S.O.; project administration, M.S.O. All authors have read and agreed to the published version of the manuscript.

Funding: This research received no external funding.

Acknowledgments: W.N.S. participated in this work as part of the activities of the Aromatic Plant Research Center (APRC, https://aromaticplant.org/).

Conflicts of Interest: The authors declare no conflict of interest.

References

1. Ahmed, H. Losses incurred in stored food grains by insect pests—A review. *Pak. J. Agric. Res.* **1983**, *4*, 198–207.
2. Agoda, S.; Atanda, S.; Usanga, O.E.; Ikotun, I.; Isong, I.U. Post-harvest food losses reduction in maize production in Nigeria. *Afr. J. Agric. Res.* **2011**, *6*, 4833–4839.
3. Jones, M.; Alexander, C.; Widmar, N.O.; Ricker-Gilbert, J.; Lowenberg-DeBoer, J.M. Do insect and mold damage affect maize prices in Africa? Evidence from Malawi. *Mod. Econ.* **2016**, *7*, 1168–1185. [CrossRef]
4. Vandekar, M.; Plestina, R.; Wilhelm, K. Toxicity of carbamates for mammals. *Bull. World Health Organ.* **1971**, *44*, 241–249. [PubMed]
5. Hermanutz, R.O. Endrin and malathion toxicity to flagfish (*Jordanella floridae*). *Arch. Environ. Contam. Toxicol.* **1978**, *7*, 159–168. [CrossRef]
6. Toś-Luty, S.; Obuchowska-Przebirowska, D.; Latuszyńska, J.; Tokarska-Rodak, M.; Haratym-Maj, A. Dermal and oral toxicity of malathion in rats. *Ann. Agric. Environ. Med.* **2003**, *10*, 101–106. [PubMed]
7. Nash, R.G.; Woolson, E.A. Persistence of chlorinated hydrocarbon insecticides in soils. *Science* **1967**, *157*, 924–927. [CrossRef] [PubMed]
8. Talekar, N.S.; Sun, L.-T.; Lee, E.-M.; Chen, J.-S. Persistence of some insecticides in subtropical soil. *J. Agric. Food Chem.* **1977**, *25*, 348–352. [CrossRef]
9. Al-Makkawy, H.K.; Madbouly, M.D. Persistence and accumulation of some organic insecticides in Nile water and fish. *Resour. Conserv. Recycl.* **1999**, *27*, 105–115. [CrossRef]
10. Bondarenko, S.; Gan, J.; Haver, D.L.; Kabashima, J.N. Persistence of selected organophosphate and carbamate insecticides in waters from a coastal watershed. *Environ. Toxicol. Chem.* **2004**, *23*, 2649–2654. [CrossRef]
11. Haubruge, E.; Arnaud, L. Fitness consequences of malathion-specific resistance in red flour beetle (Coleoptera: Tenebrionidae) and selection for resistance in the absence of malathion. *J. Econ. Entomol.* **2009**, *94*, 552–557. [CrossRef] [PubMed]
12. Ribeiro, B.M.; Guedes, R.N.C.; Oliveira, E.E.; Santos, J.P. Insecticide resistance and synergism in Brazilian populations of *Sitophilus zeamais* (Coleoptera: Curculionidae). *J. Stored Prod. Res.* **2002**, *39*, 21–31. [CrossRef]
13. Attia, M.A.; Wahba, T.F.; Shaarawy, N.; Moustafa, F.I.; Guedes, R.N.C.; Dewer, Y. Stored grain pest prevalence and insecticide resistance in Egyptian populations of the red flour beetle *Tribolium castaneum* (Herbst) and the rice weevil *Sitophilus oryzae* (L.). *J. Stored Prod. Res.* **2020**, *87*, 101611. [CrossRef]
14. Isman, M.B. Plant essential oils for pest and disease management. *Crop Prot.* **2000**, *19*, 603–608. [CrossRef]
15. Koul, O.; Walia, S.; Dhaliwal, G. Essential oils as green pesticides: Potential and constraints. *Biopestic. Int.* **2008**, *4*, 63–84.
16. Reis, S.L.; Mantello, A.G.; Macedo, J.M.; Gelfuso, E.A.; Da Silva, C.P.; Fachin, A.L.; Cardoso, A.M.; Beleboni, R.O. Typical monoterpenes as insecticides and repellents against stored grain pests. *Molecules* **2016**, *21*, 258. [CrossRef]
17. Campolo, O.; Giunti, G.; Russo, A.; Palmeri, V.; Zappalà, L. Essential oils in stored product insect pest control. *J. Food Qual.* **2018**, *2018*. [CrossRef]

18. Burkill, H.M. *The Useful Plants of West. Tropical Africa*; Volume 1: Families A-D; Royal Botanic Gardens: Kew, UK, 1985.

19. Cronquist, A. *An Integrated System of Classification of Flowering Plants*; Columbia University Press: New York, NY, USA, 1981.

20. Adebisi, A.A. *Launaea taraxacifolia* (Willd.) Amin ex C. Jeffrey. In *Plant Resources of Tropical Africa*; Grubben, G.J.H., Denton, O.A., Eds.; Backhuys Publishers: Leiden, The Netherlands, 2004; Volume 2, pp. 103–264.

21. Adetutu, A.; Olorunnisola, O.S.; Owoade, A.O.; Adegbola, P. Inhibition of In Vivo growth of *Plasmodium berghei* by *Launaea taraxacifolia* and *Amaranthus viridis* in mice. *Malar. Res. Treat.* **2016**, *2016*. [CrossRef]

22. Koukoui, O.; Senou, M.; Agbangnan, P.; Seton, S.; Koumayo, F.; Azonbakin, S.; Adjagba, M.; Laleye, A.; Sezan, A. Effective In Vivo cholesterol and triglycerides lowering activities of hydroethanolic extract of *Launaea taraxacifolia* leaves. *Int. J. Pharm. Sci. Res.* **2017**, *8*, 2040.

23. Borokini, F.B.; Labunmi, L. In Vitro investigation of antioxidant activities of *Launea taraxacifolia* and *Crassocephalum rubens*. *Int. J. Food Stud.* **2017**, *6*, 82–94. [CrossRef]

24. Owoeye, O.; Arinola, G.O. A vegetable, *Launaea taraxacifolia*, mitigated mercuric chloride alteration of the microanatomy of rat brain. *J. Diet. Suppl.* **2017**, *14*, 613–625. [CrossRef]

25. Adinortey, M.B.; Sarfo, J.K.; Quayson, E.T.; Weremfo, A.; Adinortey, C.A.; Ekloh, W.; Ocran, J. Phytochemical screening, proximate and mineral composition of *Launaea taraxacifolia* leaves. *Res. J. Med. Plants* **2012**, *6*, 171–191. [CrossRef]

26. Dickson, R.A.; Annan, K.; Fleischer, T.C.; Amponsah, I.K.; Nsiah, K.; Oteng, J.A. Phytochemical investigations and nutritive potential of eight selected plants from Ghana. *J. Pharm. Nutr. Sci.* **2012**, *2*, 172–177. [CrossRef]

27. Adinortey, M.B.; Sarfo, J.K.; Kwarteng, J.; Adinortey, C.A.; Ekloh, W.; Kuatsienu, L.E.; Nyarko, A.K. The ethnopharmacological and nutraceutical relevance of *Launaea taraxacifolia* (Willd.) Amin ex C. Jeffrey. *Evidence-Based Complement. Altern. Med.* **2018**, *2018*. [CrossRef] [PubMed]

28. Bello, O.M.; Abiodun, O.B.; Oguntoye, S.O. Insight into the ethnopharmacology, phytochemistry, pharmacology of *Launaea taraxacifolia* (Willd.) Amin ex C. Jeffrey as an underutilized vegetable from Nigeria: A review. *Ann. Univ. Dunarea Jos Galati* **2018**, *42*, 137–152.

29. Arawande, J.O.; Amoo, I.A.; Lajide, L. Chemical and phytochemical composition of wild lettuce (*Launaea taraxacifolia*). *J. Appl. Phytotechnol. Environ. Sanit.* **2013**, *2*, 25–30.

30. Dansi, A.; Vodouhè, R.; Azokpota, P.; Yedomonhan, H.; Assogba, P.; Adjatin, A.; Loko, Y.L.; Dossou-Aminon, I.; Akpagana, K. Diversity of the neglected and underutilized crop species of importance in Benin. *Sci. World J.* **2012**, *2012*. [CrossRef]

31. Gbadamosi, I.T.; Okolosi, O. Botanical galactogogues: Nutritional values and therapeutic potentials. *J. Appl. Biosci.* **2013**, *61*, 4460–4469. [CrossRef]

32. Olugbenga, D.J.; Ukpanukpong, R.U.; Ngozi, U.R. Phytochemical screening, proximate analysis and acute toxicity study of *Launaea taraxacifolia* ethanolic extract on albino rats. *Int. J. Sci. Technoledge* **2015**, *3*, 199–202.

33. Koukoui, O.; Agbangnan, P.; Boucherie, S.; Yovo, M.; Nusse, O.; Combettes, L.; Sohounhloué, D. Phytochemical study and evaluation of cytotoxicity, antioxidant and hypolipidemic properties of *Launaea taraxacifolia* leaves extracts on cell lines HepG2 and PLB985. *Am. J. Plant Sci.* **2015**, *6*, 1768–1779. [CrossRef]

34. Ruffina, A.N.; Maureen, C.O.; Esther, A.E.; Chisom, I.F. Phytochemical analysis and antibacterial activity of *Launaea taraxacifolia* ethanolic leave extract. *Sch. Acad. J. Biosci.* **2016**, *4*, 193–196.

35. *British Pharmacopoeia*; H. M. Starionery Office: London, UK, 1993; Volume I.

36. Adams, R.P. *Identification of Essential Oil Components by Gas Chromatography/Mass Spectrometry*, 4th ed.; Allured Publishing: Carol Stream, IL, USA, 2007.

37. Mondello, L. *FFNSC 3*; Shimadzu Scientific Instruments: Columbia, MD, USA, 2016.

38. Satyal, P. Development of GC-MS Database of Essential Oil Components by the Analysis of Natural Essential Oils and Synthetic Compounds and Discovery of Biologically Active Novel Chemotypes in Essential Oils. Ph.D. Thesis, University of Alabama in Huntsville, Huntsville, AL, USA, 2015.

39. *NIST17*; National Institute of Standards and Technology: Gaithersburg, MD, USA, 2017.

40. Ilboudo, Z.; Dabiré, L.C.B.; Nébié, R.C.H.; Dicko, I.O.; Dugravot, S.; Cortesero, A.M.; Sanon, A. Biological activity and persistence of four essential oils towards the main pest of stored cowpeas, Callosobruchus maculatus (F.) (Coleoptera: Bruchidae). *J. Stored Prod. Res.* **2010**, *46*, 124–128. [CrossRef]

41. Abbott, W.S. A method of computing the effectiveness of an insecticide. *J. Econ. Entomol.* **1925**, *18*, 265–267. [CrossRef]

42. Finney, D. *Probit Analysis*, reissue ed.; Cambridge University Press: Cambridge, UK, 2009; ISBN 978-0521135900.

43. Ololade, Z.S.; Kuyooro, S.E.; Ogunmola, O.O.; Abiona, O.O. Phytochemical, antioxidant, anti-arthritic, anti-inflammatory and bactericidal potentials of the leaf extract of *Lactuca teraxacifolia*. *Glob. J. Med. Res. B Pharma, Drug Discov. Toxicol. Med.* **2017**, *17*, 18–28.

44. Park, I.K.; Lee, S.G.; Choi, D.H.; Park, J.D.; Ahn, Y.J. Insecticidal activities of constituents identified in the essential oil from leaves of *Chamaecyparis obtusa* against *Callosobruchus chinensis* (L.) and *Sitophilus oryzae* (L.). *J. Stored Prod. Res.* **2003**, *39*, 375–384. [CrossRef]

45. Wang, D.C.; Qiu, D.R.; Shi, L.N.; Pan, H.Y.; Li, Y.W.; Sun, J.Z.; Xue, Y.J.; Wei, D.S.; Li, X.; Zhang, Y.M.; et al. Identification of insecticidal constituents of the essential oils of *Dahlia pinnata* Cav. against *Sitophilus zeamais* and *Sitophilus oryzae*. *Nat. Prod. Res.* **2015**, *29*, 1748–1751. [CrossRef] [PubMed]

46. Lee, B.-H.; Choi, W.-S.; Lee, S.-E.; Park, B.-S. Fumigant toxicity of essential oils and their constituent compounds towards the rice weevil, *Sitophilus oryzae* (L.). *Crop Prot.* **2001**, *20*, 317–320. [CrossRef]

47. García, M.; Donadel, O.J.; Ardanaz, C.E.; Tonn, C.E.; Sosa, M.E. Toxic and repellent effects of *Baccharis salicifolia* essential oil on *Tribolium castaneum*. *Pest Manag. Sci.* **2005**, *61*, 612–618. [CrossRef]

48. Liu, T.-T.; Chao, L.K.-P.; Hong, K.-S.; Huang, Y.-J.; Yang, T.-S. Composition and insecticidal activity of essential oil of *Bacopa caroliniana* and interactive effects of individual compounds on the activity. *Insects* **2020**, *11*, 23. [CrossRef]

49. Samarasekera, R.; Kalkari, K.S.; Weerasinghe, I.S. Insecticidal activity of essential oils of Ceylon *Cinnamomum* and *Cymbopogon* species against *Musca domestica*. *J. Essent. Oil Res.* **2006**, *18*, 352–354. [CrossRef]

50. Saad, M.M.G.; Abou-Taleb, H.K.; Abdelgaleil, S.A.M. Insecticidal activities of monoterpenes and phenylpropenes against *Sitophilus oryzae* and their inhibitory effects on acetylcholinesterase and adenosine triphosphatases. *Appl. Entomol. Zool.* **2018**, *53*, 173–181. [CrossRef]

51. Bossou, A.D.; Ahoussi, E.; Ruysbergh, E.; Adams, A.; Smagghe, G.; De Kimpe, N.; Avlessi, F.; Sohounhloue, D.C.K.; Mangelinckx, S. Characterization of volatile compounds from three *Cymbopogon* species and *Eucalyptus citriodora* from Benin and their insecticidal activities against *Tribolium castaneum*. *Ind. Crops Prod.* **2015**, *76*, 306–317. [CrossRef]

52. Yildirim, E.; Emsen, B.; Kordali, S. Insecticidal effects of monoterpenes on *Sitophilus zeamais* Motschulsky (Coleoptera: Curculionidae). *J. Appl. Bot. Food Qual.* **2013**, *86*, 198–204.

53. Abdelgaleil, S.A.M.; Mohamed, M.I.E.; Badawy, M.E.I.; El-Arami, S.A.A. Fumigant and contact toxicities of monoterpenes to *Sitophilus oryzae* (L.) and *Tribolium castaneum* (Herbst) and their inhibitory effects on acetylcholinesterase activity. *J. Chem. Ecol.* **2009**, *35*, 518–525. [CrossRef] [PubMed]

Article

Thyme Antimicrobial Effect in Edible Films with High Pressure Thermally Treated Whey Protein Concentrate

Iulia Bleoancă, Elena Enachi and Daniela Borda *

Faculty of Food Science and Engineering, Dunarea de Jos University of Galati, 800201 Galati, Romania; Iulia.Bleoanca@ugal.ro (I.B.); Elena.Ionita@ugal.ro (E.E.)
* Correspondence: Daniela.Borda@ugal.ro; Tel.: +40-336-130-177

Received: 5 June 2020; Accepted: 26 June 2020; Published: 30 June 2020

Abstract: Application of high pressure-thermal treatment (600 MPa and 70 °C, 20 min) for obtaining edible films functionalized with thyme extracts have been studied in order to evaluate the antimicrobial capacity of films structure to retain and release the bioactive compounds. The high pressure-thermally treated films (HPT) were compared with the thermally treated (TT) ones (80 ± 0.5 °C, 35 min). The film structures were analyzed and the sorption isotherms, water vapor permeability, antimicrobial activity and the volatile fingerprints by GC/MS were performed. The HPT film presented more binding sites for water chemi-sorption than TT films and displayed significantly lower WVP than TT films ($p < 0.05$). TT films displayed slightly, but significant higher, antimicrobial activity ($p < 0.05$) against *Geotrichum candidum* in the first day and against *Bacillus subtilis* in the 10th day of storage. The HPT film structure had ~1.5-fold higher capacity to retain volatiles after drying compared to TT films. From the HPT films higher amount of p-cymene and α-terpinene was volatilized during 10 days of storage at 25 °C, 50% RH while from the TT films higher amount of caryophyllene and carvacrol were released. During storage HPT films had a 2-fold lower capacity to retain monoterpenes compared to TT films.

Keywords: thyme; essential oil; edible films; high pressure thermal treatment; ultrasonication; antimicrobial; thymol; carvacrol; food safety

---

1. Introduction

Consumers' increasing demand for minimally processed food products led to increased researchers' attention towards new ways to valorize the potential of plant-based extracts as preservatives for extending food shelf-life and insuring food safety. Essential oils (EOs), used conventionally as flavorings by the food industry are considered for new applications as antimicrobials and antioxidants and are generally recognized as safe (GRAS) by the United States Food and Drug Administration [1]. In the EU, Regulation 1334/2008 sets the maximum levels of certain substances present as flavorings in or on foods, EOs included.

The biological properties of EOs are determined by its components, which are typically low molecular weight terpenes and terpenoids, nonetheless other aromatic and aliphatic molecules could be present. From the aromatic plants volatile profile, terpenes (C10) are representing 90% of the EOs but sesquiterpenes (C15) are also frequently present [2]. Even though EOs have antimicrobial effect against a wide range of food related spoilage and pathogenic microorganisms, the required concentration is often too high and their intense odor may negatively interfere with food quality and consumers' acceptance. One solution to reduce the negative effect of EOs on food flavor is the inclusion of EOs into edible packaging, such as films and coatings.

Edible films (EF) are thin layers of edible materials (polysaccharides, proteins and lipids, and the combination of two or more of the above), which once formed can be placed on or between food components [3].

EF functionalized with EOs act as antimicrobial and antioxidant carriers, enabling their release at the interface between packaging and food product while maintaining the antimicrobial effect [4] and preserving food quality for longer periods of time.

Recently it has been demonstrated that lemongrass EOs have succeeded to limit the extent of depolymerization in chia mucilage emulsion and prevented autooxidation [5]. To overcome the EOs inherent photo-, thermal-sensitivity coupled with their high volatility micro and nanoencapsulation methods have been employed [6–9].

In the same time, edible packagings are an environment- friendly solution as their constituents are fully biodegradable and in some cases they valorize industrial waste, as is the case of whey protein recovered from the cheese- making process. Nonetheless, EOs incorporation into EF change their most relevant properties, such as the continuity of polymer matrix, weakening the film structure, reducing its transparency, while improving water barrier properties [3]. In this regard it is necessary to investigate the specific interactions between the polymer matrix and the EOs composition in order to determine the effectiveness of EOs as active ingredients. Application of whey proteins with EOs in EF has been investigated by several researchers showing the EOs' antimicrobial effect, the excellent oxygen barrier properties, transparency but the relatively low water vapor permeability of the films [10–12]. To favor film formations, the whey proteins should undergo thermal denaturation, above 70 °C. Further, the unfolded globular whey proteins, expose the buried SH groups and hydrophobic groups that can react forming inter- and intra-molecular bonding during film drying [13]. Besides thermal treatment, ultrasound and addition of transglutaminase have been tested for protein denaturation prior EF drying [13]. These alternative methods could also dictate the capacity of film structure to retain and gradually release the volatiles compounds, but also influence the mechanical properties and water vapor permeability of films.

High pressure processing (HPP) is an alternative to thermal treatment that can induce structural changes in macromolecules which are distinct from those of conventional thermal treatment [14]. However, to favor the protein film formation, a combination of high pressure with thermal treatment is required. Due to the different mechanism involved in protein denaturation, high pressure thermal processing (HPT) could result in the formation of a protein-based network with different properties compared to thermal treatment (TT).

In this study, combined high pressure at 600 MPa with thermal treatment (70 °C) was employed as original alternative to thermal treatment alone for whey protein aggregation, promoting intermolecular interactions between film forming substances, which are crucial for film forming step [15]. For obtaining a homogenous film forming emulsion, ultrasound treatment was used here [16].

The objective of this research was to obtain a homogenous, flexible, resistant film formulae made of whey proteins and functionalized with thyme EO (TEO) as antimicrobial agent. The films obtained by casting were further characterized to assess their potential for food packaging applications, in terms of mechanical, physico-chemical and antimicrobial properties. The capacity of HPT and TT films to retain and release the EOs trapped in the films structure was assessed over time in relation with their antimicrobial activity.

2. Materials and Methods

2.1. Materials

Whey protein concentrate, ProMilk 852FB1 was kindly offered by KUK-Romania (composition on dry-weight basis: 86% protein, 1% total fat, 11% lactose, 2.9% total ash, 5% moisture). Anhydrous glycerol (98% purity) purchased from Redox SRL (Bucharest, Romania). Tween 20, was purchased from Sigma- Aldrich (Bucharest, Romania). Thyme (*Thymus vulgaris*) EO, kindly provided by SC Hofigal SRL (Bucharest, Romania).

2.2. Film Preparation

The film was prepared by dispersing 7.6% (*w/w*) WPC powder, into distilled water under continuous magnetic stirring (180 rpm, 15 min) following a method previously optimized by Bleoanca et al. [17]. The pH was adjusted to 7.0 using 2 N NaOH [18]. In order to transform the protein solution in a flexible film either thermal crosslinking (80 ± 0.5 °C, for 35 min) or combined HPT denaturation (600 MPa, 70 °C, for 20 min) were applied.

2.2.1. Thermal Treatment

The thermal inactivation was applied in a thermostatic water bath at 80 ± 0.5 °C for 35 min. Timing was started after the temperature measured inside the sample has reached 80 °C, as measured by type K thermocouple in one of the glass vials. Immediately after finishing the thermal treatment the samples were cooled in iced-water to stop the thermal effect.

2.2.2. Combined Mild-Thermal High Pressure Treatment

Combined pressure- temperature treatments were conducted in a multivessel (4 vessels of 100 mL) high-pressure equipment (Resato, Roden, The Netherlands). As a pressure transmitting fluid, a mixture of water and propylene glycol (TR15, Resato) was used. The sample, approximately 30 mL, was first heated at 65 °C, and then filled without air into Teflon cylinders and placed into the HPP vessels to avoid temperature gradients. Compression started when the temperature was equal to target temperature, 70 °C, up to 600 MPa, and 20 min holding times. The compression rate was of approximately 10 MPa/s, until the preset pressure was reached, whereupon the valves of the individual vessels were closed and the central circuit was decompressed. An additional one-minute equilibration period was taken into account to ensure constant temperatures. Temperature inside the samples was monitored during the treatment with a thermocouple placed in the upper part of the Teflon cylinders. Decompression of the vessels was almost instantaneously (~5 s). After the pressure-temperature treatment, the samples were immediately transferred into iced water.

After forced cooling on ice, into the resulting film forming mixture obtained either by thermal or combined high pressure- thermal denaturation, anhydrous glycerol was added at a concentration of 8.0% (*w/w*) as plasticizer to reduce the brittleness of the WPC films, thus improving its mechanical properties. As surfactant for reducing the surface tension, tween 20 was used in a concentration of 0.9% (*w/w*). Then, thyme (*Thymus vulgaris*) EO, was added in the mixture as antimicrobial compound in a concentration of 2.5% (*w/w*). This plant EO was chosen due to its high content of carvacrol, thymol and p-cymene, all known to be efficient antimicrobials [19] and considering the results of previous tests performed by our research group [20].

Further the mix was homogenized by ultrasonication with equipment Sonoplus HD3100 Bandelin, Germany equipped with a sonication probe of 8 mm diameter, at 35% amplitudes, for 3 min. The sonication probe was immersed 1 cm below the liquid surface and the temperature of the film forming emulsion was kept at 23 ± 2 °C during sonication by placing the tube in an iced water bath [16].

The film forming emulsion was then poured onto silicone trays (diameter 5 cm). To control film thickness, the same amount (11 mL) of film forming mixture was poured. The spread solutions were allowed to dry at room temperature, approximately 22 °C, for 48 h at 50% RH [21,22], then easily peeled off.

Considering the hydrophilic nature of the protein film, therefore its susceptibility to absorb humidity from the environment, a standardization of the films was necessary to ensure that the mechanical properties of the film are not impaired. For this reason, prior to all investigations, the films were preconditioned by storing them in a controlled temperature- humidity environment, at 50 ± 3% RH and 25 ± 1 °C, for at least 72 h [23].All the experiments were performed in triplicate.

2.3. Film Characterization

2.3.1. Film Thickness

A digital micrometer (Digimatic Micrometer, Mitutoyo, Japan) was used to measure film thickness to the nearest 0.0001 mm. The mean thickness was calculated from five measurements taken randomly at different locations on each film.

2.3.2. Moisture Content

The moisture content (MC) of the whey protein films was determined after oven drying at 105 ± 1 °C for 24 h until a constant weight was attained. After adequate conditioning, 3.4 cm diameter discs were cut from the edible film and weighed in order to be compared to the ones after drying. The moisture content values were determined as percentage of initial film weight loss during drying [24].

$$\text{MC} = \frac{w_1 - w_2}{w_1 - w_0} \times 100 \ [\%] \tag{1}$$

w_0 is the weight of empty and dry weighing glass bottle, (g); w_1 is the weight of weighing glass bottle with film, before drying, (g); w_2 is the weight of weighing glass bottle with film, after drying, (g).

2.3.3. Water Activity

The water activity (a_w) of preconditioned edible films was measured with a (Fast lab water activity meter; GBX, Loire, France), using discs of films (4 ± 0.1 cm diameter).

2.3.4. Moisture Sorption Isotherms

Moisture sorption isotherms were determined by static gravimetric method [25]. Dried film samples were first conditioned for 5–10 days into a controlled humidity environment at a constant temperature until equilibrium has been reached. Samples discs of 49.58 ± 0.31 mm were placed into desiccators, each containing one saturated salt solution giving various RH at 25 °C: LiCl for an a_w of 0.114, $MgCl_2$ giving a 0.331 a_w, KI giving an a_w of 0.700, NaCl for an a_w of 0.755, KCl giving an a_w of 0.851 and KNO_3 for an a_w of 0.935. Film samples were equilibrated at each environment for 5–10 days at 25 ± 0.5 °C; following removal from desiccators they were immediately weighed, the a_w was determined and moisture content was measured gravimetrically as described above. The Guggenheim-Anderson-de-Boer and Halsey models [26] as indicated by Tudose et al. [27] were applied by nonlinear regression analysis (SAS, 2009):

$$M = \frac{M_o \times C \times K \times a_w}{(1 - K \times a_w) \times (1 - K \times a_w + C \times K \times a_w)} \tag{2}$$

where M is the equilibrium moisture content (% dry basis); M_0 is the monolayer moisture content (% dry basis); C—Guggenheim constant; K—corrective constant; a_w is the water activity (dimensionless);
The Halsey equation is:

$$a_w = \exp\left(-\frac{k}{M^n}\right) \tag{3}$$

where k and n are model constants.

2.3.5. Water Vapor Permeability

Water vapor permeability (WVP) was estimated gravimetrically according to ASTM E96 [28], adapted for edible films. Film discs of 49.58 ± 0.31 mm diameter equilibrated at 25 °C, 50% RH for 48 h with saturate salt solution ($Mg(NO_3)_2$) were cut and mounted on glass cups filled with distilled water to 10 mm below the film underside. The glass cups had 46 mm diameter and 150 mm depth. The steady-state films water- vapor flow was measured at certain intervals for 48 h by digital-balance nearest to 0.0001 g. Films permeability was calculated according to the method described by Zinoviadou et al. [11]. The weight loss was monitored and expressed by the slopes calculated using linear regressions equations where $R^2 > 0.99$. At least five replicates were tested for WVP estimation.

$$WVP = \frac{Slope \times x}{A \times \Delta p} \ (\text{g·mm}/\text{m}^2 \text{·s·Pa}) \tag{4}$$

where *slope* is the weight loss of the cup per second, (g/s); x is the average film thickness, (mm); A is the area of exposed film, (m^2); Δp is the difference in vapor pressure across the test film (Pa).

2.3.6. Microstructural Analysis of The Film Forming Mixtures

A scanning electron microscope (Quanta 250, Thermo Fisher Scientific) (Waltham, MA 02451, USA) was used to determine the microstructure of thermal treated (TT) and combination of high pressure with temperature treatment (HPT) whey protein film samples with an accelerating voltage of 12.5 kV in a low vacuum environment. A magnification of 400×–1400× was used to scan each film sample.

2.4. Antimicrobial Assay

The antimicrobial effect of edible films was tested against three target microorganisms, *Bacillus subtilis*, *Geotrichum candidum* and *Torulopsis stellata*, all part of MIUG collection from Dunarea de Jos University of Galati- Romania. The antimicrobial effectiveness of the edible films was tested 10 days after the films were obtained, by vapor phase test [29]. This specific indirect contact assay for testing the antimicrobial activities was chosen to assess the protection provided by the thyme antimicrobial volatiles under no direct contact between the food product and the packaging. To perform vapor-phase diffusion tests, edible films of approx. 50 mm diameter discs were placed on the lids of Petri dishes, with previously spread 10^6 cfu/mL microbial inoculum. The inoculated agar plate was inverted with discs on the top of each lid containing antimicrobial film. Parafilm was used to tightly seal the edge of each Petri dish. Sealed and inverted Petri dishes were incubated at 27 °C for evaluation of anti-*Torulopsis* and anti- *Geotrichum* activity and at 37 °C for anti-*Bacillus* activity. Growth of each test microorganism was evaluated after two days of incubation. The inhibition radius (absence of growth) on each Petri dish was measured with a digital caliper and the inhibition area was calculated and expressed as mm^2. The negative control, represented by whey protein EF without TEO, were also tested under the same conditions. The vapor phase inhibition test was performed in duplicate, in two separate experimental runs.

2.5. Solid-Phase Micro-Extraction (SPME)

Before analysis the HPT and TT films were placed in desiccators of 6 L capacity with Mg(NO$_3$)$_2$ salt at 25 °C and 50% RH and stored for maximum 10 days. Each film had a 19.65 cm^2 surface exposed to the environment and 3 discs were present in each desiccator for all the duration of the experiments. From each film discs with 34 mm diameter were cut, weight, introduced in sealed vials and maintained at 40 °C for 10 min for equilibration before concentration by SPME on a CAR/PDMS fiber. The extraction of the volatiles under isothermal conditions at 40 °C was made over 30 min followed by 5 min of desorption into the GC injection port.

2.6. Gas Chromatography-Mass Spectrometric (GC-MS) Analysis

The volatiles fingerprints of the edible film samples were analyzed using a Trace GC-MS Ultra equipment with ionic trap- ITQ 900 from Thermo Scientific (USA). The GC column was a TG-WAX capillary column (60 m × 0.25 mm, i.d. 0.25 μm). The carrier gas was helium (99.996% purity, Messer S.A., Bucharest, Romania) that ran at a flow rate of 1 mL/min. The temperature ramp selected for the analysis was: 40 °C isothermal treatment for 4 min followed by an increase to 50 °C at 5 °C/min and to 100 °C with 7 °C/min, to 150 °C at 10 °C/min and finally to 230 °C at 12 °C/min, when temperature was kept constant for 2 min. The temperature of the transfer line in MS was set to 270 °C. Mass spectra were obtained from the full scan of the positive ions resulted with a scanning in the 35 to 450 m/z range and operated with an electron impact (EI)-mode of 200 eV. The compounds were identified in comparison with the mass spectra from Wiley and Nist 08 library database available with Xcalibur 2.1 software. The retention indices (RI) of each compound were calculated by using n-alkane series from C8-C40 (Sigma Aldrich Chemie GmbH, Steinheim, Germany) under the same conditions. Each analysis was performed in triplicate, in the first and the 10th day of storage.

The volatile organic compounds (VOCs) were estimated semi-quantitatively using n-octanol as internal standard (IS) and Equation (5) [20,30]:

$$VOC_{conc} = IS_{conc} \times \left(VOC_{peak\ area} \big/ IS_{peak\ area} \right) \qquad (5)$$

where $VOC_{peak\ area}$ is the area of the integrated individual peak, $IS_{peak\ area}$ is the area of 2-octanol in the spiked samples and IS_{conc} is the concentration of internal standard (2-octanol).

2.7. Statistical Analysis

Data were expressed as mean ± standard deviation (SD). The statistical analysis was carried out using analysis of variance (ANOVA) and Tuckey' s post-hoc test was applied to evaluate significant differences among groups ($p < 0.05$).

The quality of the sorption isotherms models' fit applied was evaluated by the regression coefficient ($R^2{}_{adj}$) and the mean relative percentage deviation (%E):

$$E = \frac{100}{N} \sum_{i=1}^{N} \frac{|m_i - m_{pi}|}{m_i} \qquad (6)$$

where m_i and m_{pi} are the experimental and predicted values, respectively, and N is the population of the experimental data.

$$R^2_{adj} = 1 - \left(\frac{n_t - 1}{n_t - n_p} \right) \cdot \frac{SSE}{SSTO} \qquad (7)$$

Principal component analysis (PCA) was performed using the Unscrambler software (Version 9.7; CAMO, Norway). PCA was performed with the peak list resulting from SPME GC/MS analysis for all the volatile compounds. The data matrix was formed by $n = 6$ cases and 25 variables defined as the VOCs peak areas obtained for each individual component. Data were transformed by unit vector normalization prior to statistical analysis.

3. Results and Discussion

3.1. Film Appearance

Appearance of the two sides of the WPC film was similar for HPT and TT films. The film side facing the casting plate was shiny, while the other was dull; this is likely an indication of some phase separation occurring in the mixture during drying. HPT and TT types of film were easily separated from the casting plates. During the TT the three dimensional structure of proteins was unfolded and the internal sulfhydrilic groups were exposed, later forming intermolecular disulfide bonds while hydrophobic groups interactions also might have occurred during film drying [18,31]. Combination of HPP and TT resulted in both denaturation via above referred mechanism and by forcing the water molecules inside the protein matrix, that exposed the hydrophobic core, followed by protein unfolding [32,33].

Films manufactured from WPC with 7.6%(w/w) protein showed a thickness of 0.133–0.193 mm, close to those reported by other researchers [34–36]. Neither one of the HPT and TT WPC-based films functionalized with thyme essential oils (TEO) did not exhibit any statistically significant differences either ($p < 0.05$) (Table 1).

Table 1. Thickness and WVP of TT and HPT films [#].

Films	Thickness (mm)	ΔRH (%)	WPV·10^{-11} (g/s·m·Pa)
Control_TT	0.171 ± 0.163 [a] [*]	46	24.867 ± 2.855 [a]
TT	0.193 ± 0.052 [a]	46	19.557 ± 2.109 [b]
Control_HPT	0.156 ± 0.043 [a]	46	13.852 ± 1.137 [b,c]
HPT	0.133 ± 0.071 [a]	46	10.178 ± 1.690 [c]

[#] mean results ± stdev; [*] different letters indicate significant differences ($p < 0.05$) among columns by post-hoc Tuckey test. WVP: Water vapor permeability; TT: thermal treated; HPT: high pressure-thermally treated; RH: relative humidity.

3.2. Sorption Isotherms

Sorption isotherms were studied at 25 °C and equilibrium moisture content. The data obtained confirmed the distribution on a sigmoidal shaped curve, characteristic of type II isotherms observed for most of the biopolymer materials and foods.

Table 2 presents the GAB and Halsey model parameters estimated for the two films formulation TT and HPT. The values indicating the goodness of the model fit to the experimental data are showing that both GAB and Halsey models are adequate for the data with %E 0.197–1.21 and a good agreement between experimental and predicted data ($R^2_{adj} = 0.89 \div 0.99$).

Table 2. Estimated parameters of the GAB and Halsey model fit to experimental data of sorption isotherms for TT and HPT films at 25 °C.

Film	GAB Model			Halsey Model	
	K	**C**	$\mathbf{M_0}$	k	n
		(g water/100 g dw [b])			
TT	0.822 ± 0.061 [a]	10.871 ± 0.036	12.581 ± 3.617	1582.915 ± 327.473	2.422 ± 0.195
	$R^2_{adj} = 0.992$	E(%) = 1.212		$R^2_{adj} = 0.807$	E(%) = 0.856
HPT	0.692 ± 0.096	3.331 ± 0.292	23.045 ± 2.681	260.816 ± 29.281	1.925 ± 0.176
	$R^2_{adj} = 0.999$	E(%) = 0.831		$R^2_{adj} = 0.891$	E(%) = 0.197

[a] mean results ± stdev; [b] dw = dry weight.

However, the R^2_{adj} and E-values for GAB have better values than for Halsey. GAB has the advantage of providing information on the monolayer water content (M_0) that indicates the number of the sorbing sites and the maximum amount of water that can be absorbed [37,38]. The values indicated by the current study are close to the ones reported by Wang et al. [39] who demonstrated that WPC films are able to adsorb more moisture than casein films. Similar values were also registered by Silva et al. [40] and Huntrakul and Harnkarnsujarit [37] and lower values were recorded by Zinoviadou et al. [11] for the whey protein isolate films with oregano compared to this study.

An almost twice higher value was obtained for the HPT film compared to TT thus it can be presumed that HPT films had more binding sites for water chemi-sorption than TT films and this could make more susceptible to swelling.

3.3. Water Vapor Permeability (WPV)

The water vapor permeability (WVP) and the film thickness are presented in Table 2. WVP of food packaging is an important parameter that gives information on sorption, diffusion and adsorption. Low values of WVP are desired for the edible films since one of the required characteristics of the edible film is to retard moisture transfer between the food product and the environment [41]. The WVP of the films with TEO treated by HPP have a significant lower permeability compared to the TT films. The values reported in this study for the 46–100% RH are in the same range, however slightly lower than the ones with those reported by Kokoszka et al. [34] for whey protein isolates and by [42] for WPC. However, compared to Kokoszka et al. [34] in our case WPC, tween and TEO was added in film formulation. The WPV was lower than the values reported by Bahram et al. [42], but the amount of essential oil used in this case was higher (2.5%) than the maximum amount used in the films with cinnamon oil (1.5%). Compared to control (control TT and control HPT) the films with TEO added (HPT and TT) had a significantly ($p < 0.05$) lower WVP (Table 2), explained by the increase in hydrophobicity and observed also by other researchers when EOs were added to the film structure [42,43].

3.4. Scanning Electron Microscopy

Figure 1 illustrates the SEM micrographs of TT (1) and HPT (2) TEO WPC films surface. The microstructure of the films reveals the structural arrangement of its components that influence both physical and mechanical properties of the films [44].Microscopy images of edible films surface show continuous, compact and homogenous structures, without any irregularities such as air bubbles or cracks. Nonetheless, the TT films are more homogenous and exhibit a smoother film surface compared to the HPT ones, that could be due to different intermolecular interactions mechanisms. At a higher magnification, the TEO droplets can be easily observed in the HPT films compared to the TT films. Moreover, the TEO droplets are scarcely observable in the TT samples, which could be related to their better integration in the thermal denatured whey protein matrix compared to the case of HPT protein denaturation.

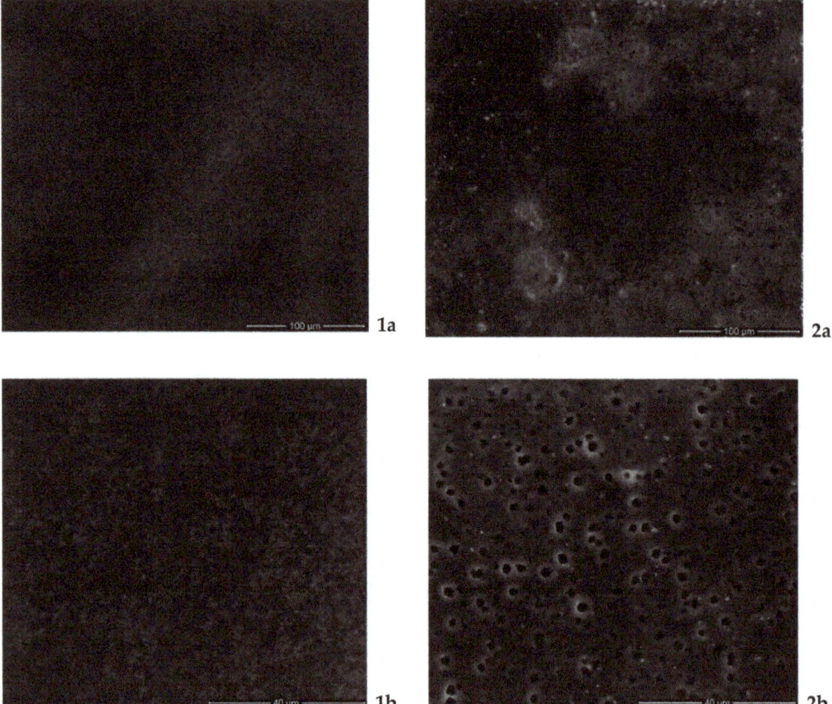

Figure 1. Surface morphology of TEO WPC EF. The film forming mixture was denatured either by TT (**1**) or by HPT (**2**). Surfaces viewed at magnification of 400× (**a**) and 1400× (**b**). TEO: thyme EO; EF: Edible films ; TT: thermal treated; HPT: high pressure-thermally treated.

Previous researches have shown that film microstructure is also correlated with mechanical and optical properties of the EFs [44]; however this properties were not investigated by the current study.

3.5. Antimicrobial Effect of PFunctionalizing the WPC-EF

The antimicrobial activity of EOs has been intensively studied and is well recognized. The growing published evidence towards a more effective antimicrobial activity of EOs in vapor phase compared to EOs in liquid form applied by direct contact [45–47] led to identification of new applications for EOs vapors, including those in the food industry [46,47]. One plausible explanation for the different antimicrobial effectiveness is the mechanism presented by the group of researchers Nadjib et al. [48] indicating formation of micelles from association of lipophilic molecules in the aqueous phase which negatively interfere with the EOs attachment to the microorganisms, while the EO vapors allow free attachment to microorganism's cells.

The current study evaluated the antimicrobial effect by vapor phase diffusion method of TEO functionalizing the WPC-EF against three test microorganisms. The current TEO WPC-EF is intended to function as an active antimicrobial food packaging providing microbial surface protection of the fresh food product by effectively controlling the growth of aerobic microorganisms through the volatile antimicrobials released into the food package headspace.

Due to the absence of direct contact between the test microorganisms and TEO WPC-EF, this method allowed the detection of the antimicrobial potency of volatile components exclusively. Results of the antimicrobial activity of 2.5% (*w/w*) TEO WPC-EF through vapor phase test are presented in Figure 2 and Table 3.

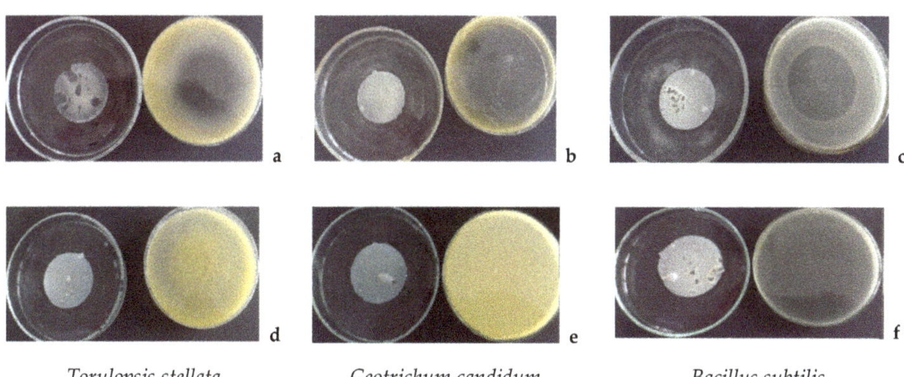

Figure 2. Sample pictures of vapor phase test (**a–c**) of TEO WPC- EF on test microorganisms *Torulopsis stellata*, *Geotrichum candidum* and *Bacillus subtilis*. **d–f** are control WPC-EF without TEO.

Table 3. Inhibition and growth reduction zones provided by thyme volatiles functionalizing WPC-EF. Results are expressed in mm, as mean ± standard deviation.

	TT		HPT	
	Day 1	Day 10	Day 1	Day 10
Torulopsis stellata	10.50 ± 0.50 [b,B,*]	17.50 ± 0.71 [a,B]	9.00 ± 1.41 [b,B]	15.00 ± 1.41 [a,C]
Geotrichum candidum	16.00 ± 1.41 [b,A]	19.50 ± 0.71 [a,B]	10.50 ± 0.71 [c,B]	20.00 ± 0.00 [a,B]
Bacillus subtilis	16.50 ± 0.71 [c,A]	39.00 ± 1.41 [a,A]	15.50 ± 0.71 [c,A]	35.00 ± 0.00 [b,A]

* Superscripts with different letters indicate significant differences ($p < 0.05$) between the rows values (small caps) and between the column values (capital letters). by post-hoc Tuckey test; TT—Thermal treatment of film forming mixture; HPT—High pressure & thermal treatment of film forming mixture.

In vitro assessment of sensitivity to thyme volatiles of three spoilage test microorganisms of environmental origin was evaluated by vapor phase assay. TEO functionalizing both types of EFs, TT and HPT, showed effective antimicrobial activity based on the inhibition zones against all three fresh products spoilage microorganisms. For *Torulopsis stellata* inhibition zones ranged between 9.00 to 17.50 mm, with no significant statistical differences between TT and HPT EFs during the 10 days tested. *Geotrichum candidum* produced inhibition halos higher than *Torulopsis stellata*, up to 20.00 mm after 10 days for HPT-EF. Significant differences in terms of thyme antimicrobial efficacy against *Geotrichum candidum* were observed only for the first day of test, higher for TT-EFs. *Bacillus subtilis* proved to be the most sensitive of all three tested microorganisms, with inhibition halos ranging from 15.50 to 39.00 mm.

When comparing protein denaturation treatments, TT with HPT, the antimicrobial activity of the HPT- EF against *Torulopsis stellata* after 10 days of storage, no significantly differences ($p > 0.05$) compared to the other samples, with higher inhibition radius for TT-EF. Thyme antimicrobial effect against *Geotrichum candidum* is significantly higher in TT-EF in the beginning, on day 1 compared to day 10, however no significant differences was registered after 10 days of storage between the TT and HPT films. For *Bacillus subtilis* the antimicrobial efficacy has no significant differences ($p > 0.05$) in the first day between TT and HPT films, however throughout the 10 days evaluation the TT films displayed a slightly higher antimicrobial effectiveness ($p < 0.05$) compare to the HPT films.

Two main characteristics greatly influence the volatility of EOs components in general, here thyme in particular: one is the molecular weight of their constituents; each chemical compounds from the mixture forming EOs has a different volatility according to its molecular weight, which influences their diffusion rate when EO is introduced in a non-saturated environment, as is the case with the sealed Petri dishes used for the this diffusion assay. The other TEO characteristic is related to the denaturation treatment of proteins from film forming mixture which influences the entrapment of the EOs in the WPC matrix, as well as promoting the release of TEO out of the proteic matrix.

It is fully understood that the antimicrobial activity of the essential oils in vapor phase is closely related to its composition in the headspace [49]. However, it should be mentioned that in the case of antimicrobial activity, an additive day-by-day effect of the VOCs was evaluated on the tested microorganisms, produced by the gradual release of the VOCs from the film matrix during storage in a contained environment created by the Petri dishes.

3.6. Gas-Chromatography Fingerprint

The individual chromatograms of the tested sample are shown in the Supplementary Materials (Figures S1–S4) and the VOCs entrapped in the 2-types of matrices tested (TT, HPT) are presented in Table 4. A total number of 25 volatiles were tentatively identified using NIST library and the compounds were present in different concentrations in all the film structures analyzed where thyme has been added (Figure 3). The most abundant VOCs were the ones regularly present in TEOs [20,50], namely thymol, p-cymene, α-terpinene, and carvacrol (Table 4). Often, p-cymene and γ- terpinene are reported as precursors of thymol and carvacrol that occur in variable proportions in plants [20,51]. In this case, in the film's matrices, only α-terpinene was identified. In all the edible films formulae p-cymene was present in high concentrations, however thymol had the highest concentrations in all films, while there were no significant differences ($p < 0.05$) in the concentrations of this compound between the two formulations (TT and HPT) (Table 4).

While in all the initially prepared emulsions the concentration of TEO added was the same, the capacity of the dried films structure to retain the VOCs can be judged as a function of the pretreatment applied. Immediately after drying, the film structure able to retain the highest concentration of the main VOCs was the HPT film that displayed in general ~1.5-fold better capacity to retain the VOCs compared to the TT film. The better capacity of HPT film to trap the VOCs compared to TT could be related to the different mechanisms involved in whey protein denaturation [14] and consequently related to the different film structure capacity to retain volatiles. High pressure treatment can be used as a tool to tailor unique properties of food structures, which may not be forthcoming through other ways of processing [14,52,53]. High-pressure predispose the whey proteins to changes in their tertiary and quaternary structures towards formation of small aggregates dominated by side-by-side interactions, enabling a narrower size distribution than thermal treatment. Usually, the changes are also associated with an increase in the apparent viscosity of the pressurized systems [54]. During HPT treatment no gelation occurred, however the samples displayed higher viscosity than the TT ones.

Table 4. The GC/MS SPME volatiles concentration (µL/kg octanol) in HPT and TT edible films functionalized with TEO, in the beginning of storage (HPT1, TT1) and after ten days of storage (HPT10, TT10) at constant RH and temperature (RH 50%; 25 °C).

Compound	Class	KI	Ions	HPT1	HPT10	TT1	TT10
Tricyclene	MT	935	91;93;77;121	13.96 ± 1.45 [g,A,*]	7.92 ± 0.55 [e,B]	2.09 ± 0.19 [e,C]	1.69 ± 0.11 [e,C]
α-Thujene	MT	946	91;77; 93;65	4.68 ± 0.52 [g,A]	2.61 ± 0.18 [e,B]	0.78 ± 0.06 [e,C]	0.63 ± 0.04 [e,C]
α-Pinene	MT	957	91;77;93;65	26.97 ± 2.13 [f,g,A]	14.91 ± 1.22 [e,B]	5.23 ± 0.44 [d,e,C]	4.38 ± 0.36 [e,C]
Camphene	MT	965	91;93;121;136;77	60.13 ± 5.25 [f,g,A]	24.67 ± 2.31 [d,e,B]	12.10 ± 1.14 [c,d,e,C]	7.92 ± 0.80 [e,C]
1S-α-Pinene	MT	970	91;67;79;93	58.50 ± 4.48 [f,g,A]	29.97 ± 2.71 [d,e,B]	10.70 ± 1.01 [d,e,C]	9.26 ± 0.88 [d,e,C]
α-Phellandrene	MT	974	91;93;77;139;51	18.68 ± 0.95 [g,A]	6.62 ± 0.72 [e,B]	8.15 ± 0.99 [d,e,B]	3.63 ± 0.34 [e,C]
α-Terpinene	MT	983	91;93;77;136	426.51 ± 38.74 [c,A]	213.03 ± 20.19 [c,B]	87.70 ± 8.99 [c,d,e,C]	65.58 ± 6.69 [c,d,e,C]
p-Cymene	MT	991	119;91;134;117	1125.78 ± 121.25 [a,A]	510.79 ± 49.88 [b,B]	377.02 ± 45.20 [b,B,C]	256.25 ± 27.48 [b,C]
α-Copaene	SQT	1038	105;91;119;161	11.38 ± 1.08 [g,B]	26.44 ± 2.14 [d,e,A]	10.82 ± 1.42 [d,e,B]	7.73 ± 0.89 [e,B]
β-Phellandrene	MT	1044	91;93;79;77	34.44 ± 3.29 [f,g,A]	14.18 ± 1.22 [e,C]	26.70 ± 2.57 [c,d,e,B]	12.74 ± 1.56 [c,d,e,C]
γ-Terpinene	MT	1047	67;95;108;193	28.72 ± 2.14 [f,g,A]	1.88 ± 0.09 [e,C]	30.35 ± 3.04 [c,d,e,B]	15.22 ± 1.68 [c,d,e,A]
Thymol methyl ether	AOMT	1057	149;91;164;117	202.55 ± 15.42 [d,A]	171.39 ± 16.12 [c,A]	52.56 ± 7.88 [c,d,e,C]	125.19 ± 11.42 [c,d,B]
Caryophyllene	SQT	1063	91;105;133;77	193.94 ± 14.69 [d,e,A]	189.42 ± 1.56 [c,A]	145.30 ± 15.22 [c,B]	125.33 ± 14.18 [c,d,B]
δ-Cadinene	SQT	1073	93;95;91;121	9.69 ± 0.87 [g,A]	5.33 ± 0.49 [e,B]	10.99 ± 1.25 [d,e,A]	4.25 ± 1.77 [e,B]
γ-Muurolene	SQT	1078	161;105;91;204	88.60 ± 8.36 [e,f,g,A]	59.88 ± 5.74 [d,e,A]	96.86 ± 44.12 [c,d,e,A]	50.19 ± 4.12 [c,d,e,A]
Bicyclogermacrene	SQT	1089	91;105;133;189	41.64 ± 4.25 [f,g,A]	30.97 ± 2.09 [d,e,B]	8.35 ± 0.92 [d,e,D]	16.48 ± 1.99 [c,d,e,C]
γ-Cadinene	SQT	1092	161;105;91;119	24.75 ± 2.21 [f,g,B]	19.65 ± 1.74 [e,B]	43.03 ± 3.39 [c,d,e,A]	25.63 ± 2.12 [c,d,e,B]
α-Calacorene	SQT	1124	91;93;67;79;121	6.27 ± 0.52 [g,A]	5.18 ± 0.49 [e,A]	5.53 ± 0.55 [d,e,A]	5.60 ± 1.13 [d,e,A]
Caryophyllene oxide	OSQT	1243	429;355;430;295	41.04 ± 3.22 [f,g,A]	8.25 ± 0.72 [e,B]	4.76 ± 1.12 [d,e,B]	10.06 ± 1.31 [d,e,B]
α-Guaiene	SQT	1255	185;200;201;204	5.17 ± 0.48 [g,B]	4.66 ± 0.38 [e,B]	10.82 ± 1.22 [d,e,A]	6.16 ± 0.74 [e,B]
γ-Guaiene	SQT	1263	105;133;148;91	17.15 ± 1.62 [g,A]	12.94 ± 1.16 [e,A]	17.20 ± 1.97 [c,d,e,A]	17.16 ± 1.98 [c,d,e,A]
α-Maaliene	SQT	1270	221;213;429;187	11.83 ± 1.05 [g,A,B]	8.66 ± 0.71 [e,B]	12.95 ± 1.42 [c,d,e,A]	11.22 ± 1.64 [d,e,A,B]
Thymol	AOMT	1391	135;150;91;115	1674.78 ± 112.88 [a,A]	1640.55 ± 154.49 [a,A]	1815.69 ± 200.28 [a,A]	1611.06 ± 180.14 [a,A]
Carvacrol	AOMT	1396	135;150;91;115	128.45 ± 13.49 [d,e,f,A]	124.47 ± 11.76 [c,d,A]	136.06 ± 14.12 [c,d,A]	127.65 ± 13.14 [c,A]
γ-Himachalene	SQT	1398	161;91;135;105	2.68 ± 0.19 [g,B]	2.41 ± 0.12 [e,B]	4.24 ± 0.51 [d,e,A]	3.08 ± 0.28 [e,B]

* different letters indicate significant differences ($p < 0.05$) among columns (small caps) and rows (capital letters) by post-hoc Tuckey test; MT—monoterpenes; SQT—sesquiterpenes; AOMT—aromatic monoterpenes; OSQT—oxide sesquiterpenes.

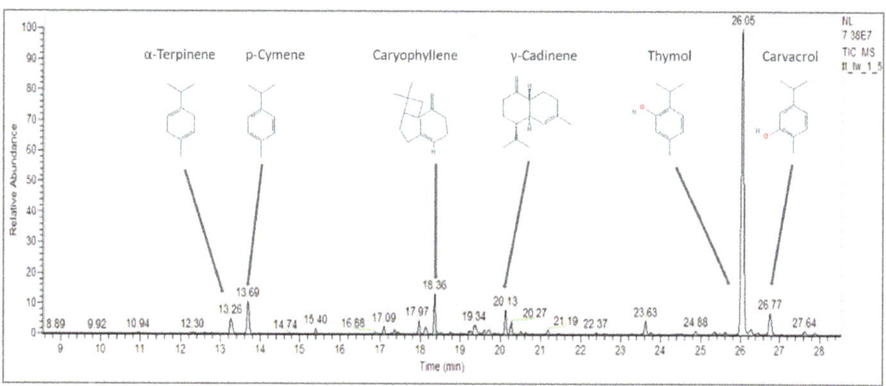

Figure 3. Fingerprint of the main volatiles present in the TT film functionalized with thyme, in the first day of storage.

The combined HPT treatment resulted into a denser film compared with the thermally treated ones and with better defined individual oil droplets inside the film structure as shown by microscopy analysis (Figure 1). This observation could indicate a better entrapment capacity but a weaker linkage of TEO in HPT compared to TT films.

The dried protein films complemented with tween surfactant, glycerol and thyme that went through different preliminary processing methods (HPT and TT), were then assessed in relation with the capacity to withhold the aromatic molecules during ten days of storage. The edible films were kept at constant relative humidity (RH 50%) and environmental temperature (25 °C).

In the SPME GC-Ms analysis the samples were kept in the same equilibrium environment for 10 days and later on, they were tested, basically measuring the remaining VOCs in the edible film matrix.

When evaluating the fingerprints of HPT and TT after 10 days it can be noticed that HPT film lost higher amounts of p-cymene (54.63%) and α-terpinene (50.06%) (HPT10 1) compared the thermally treated ones 32.03% and 25.22%, respectively (TT10_1) (Figure 4).

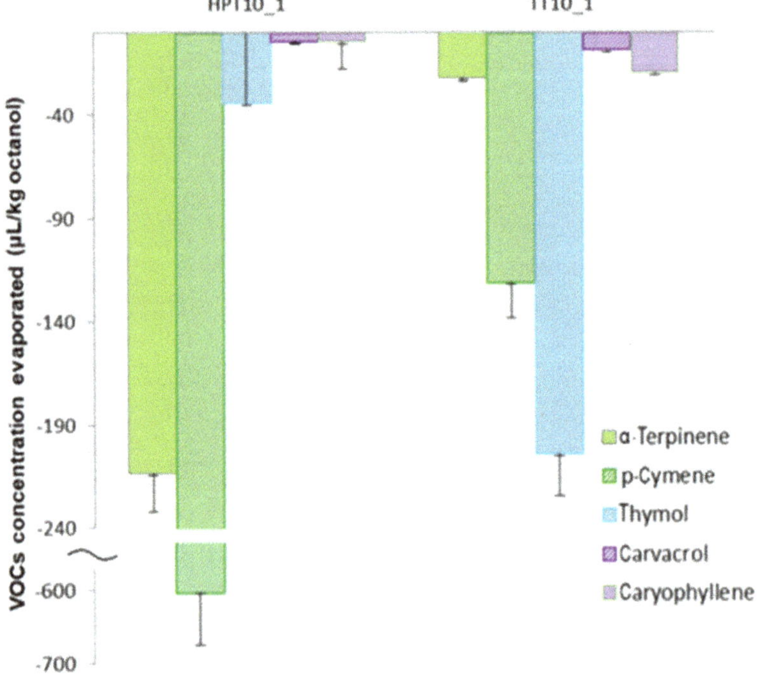

Figure 4. The loss of main volatiles in the HPT and TT edible films during 10 days of storage at 50 ± 3% RH and 25 ± 1 °C.

Another VOC that was consistently reduced by 79.90% after 10 days of storage is caryophyllene oxide in the HPT film. The most desired property of the antimicrobial packaging materials is the controlled release of the antimicrobial agents from the film to the food surface. A burst release of VOCs causes fast consumption of the antimicrobial agent after which the minimum concentration required for the inhibition of microbial growth is not maintained on the food surface [55]. On the other hand, spoilage reactions on the food surface may start if the release rate of the antimicrobial agent from the film is too slow. Thus, the controlled release of the active agent over a long period of time is necessary to extend the shelf life of the packaged food [56].

The edible film structures obtained in this research showed that HPT displayed over time a 2-fold lower capacity to retain the monoterpenes (MTs) with high volatility (KI from 935 to 1044) compared to TT. This finding demonstrates that forces involved in the VOCs entrapment in HPT treatment are weak so these components are more susceptible of fast leaving the films compared to TT. Despite the initially better capacity to retain volatiles the HPT matrix demonstrated a lower capacity to retain over storage especially the MTs with high volatility.

3.7. PCA Analysis

The PCA could explain 94% of the total variation of VOCs in the sample with the highest contribution explained by PC1 (Figure 5). The association of the volatiles and samples given by the PCA analysis shows that the highest contribution in PC1 is made by the TT1 with α-guaiene, cadinene but also by TT10 associated with high concentrations of thymol and carvacrol. Oppositely influencing the PC1, is the HPT structure, from the first day (HPT1), containing camphene, and p-cymene. After 10 days of storage the content in bicylogemacrene and thymol methyl ether in the HPT film could explain most of the variation influencing the PC2.

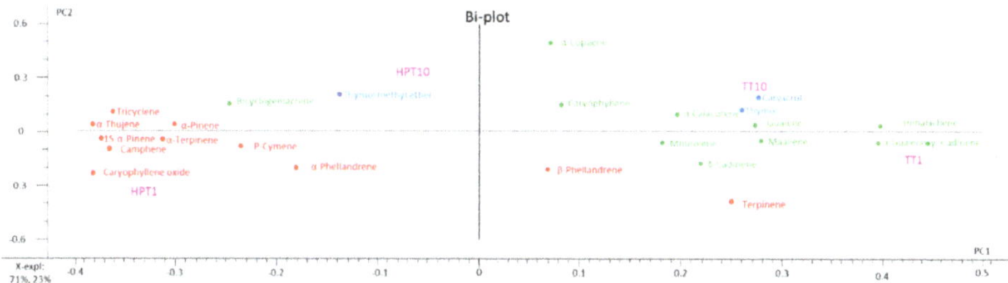

Figure 5. The Bi-plot of the principal component analysis of HPT and TT fingerprint during 10 days of storage at 50 ± 3% RH and 25 ± 1 °C.

4. Conclusions

This study showed that HPT denaturation of whey proteins result in different structures compared to the TT. The HPT films were more prone to swell and presented a lower WVP than TT films. The antimicrobial activity for the films contained in glass Petri dishes were comparable, however a slightly better antimicrobial activity of the vapors was demonstrated by the TT films against *Geotrichum candidum* in the first day and against *Bacillus subtilis* in the 10th day of storage.

The HPT functionalized with TEO film had a better capacity to embed the volatiles after drying, however over time is released more easily the monoterpenes from the film structure showing a weaker capacity to withhold the highly volatile components when compared to TT film when stored in controlled environment (25 °C, 50% RH). The use of EFs in the food industry could require either long time or short-time protection of food depending on its durability, so the selected pretreatment, either thermal of combined pressure thermal pretreatment, could be elected in relation with the type of application EFs are intended for.

The current study can be considered a starting point for future designing of EF with controlled release of thyme antimicrobial components, by understanding the molecular dynamic equilibrium between the protein matrix, TEO and environment.

Supplementary Materials: The following are available online at http://www.mdpi.com/2304-8158/9/7/855/s1. Figure S1: Volatile fingerprint of TT-WPC-EF in the beginning of storage, Figure S2: Volatile fingerprint of HPT-WPC-EF in the beginning of storage, Figure S3: Volatile fingerprint of TT-WPC-EF after 10 days of storage, Figure S4: Volatile fingerprint of HPT-WPC-EF after 10 days of storage.

Author Contributions: I.B.: investigation, methodology, formal analysis, writing, review and editing; E.E.: imagistic investigation, writing; D.B.: GC investigation, software, conceptualization, editing, supervision. All authors have read and agreed to the published version of the manuscript.

Funding: This work was supported by the project "Excellence, performance and competitiveness in the Research, Development and Innovation activities at "Dunarea de Jos" University of Galati", acronym "EXPERT", financed by the Romanian Ministry of Research and Innovation in the framework of Programme 1 – Development of the national research and development system, Sub-programme 1.2—Institutional Performance—Projects for financing excellence in Research, Development and Innovation, Contract no. 14PFE/17.10.2018.

Acknowledgments: The authors wish to thank Re-SPIA project, SMIS code 11377, for the research infrastructure provided for this study, Dima Stefan for providing ultrasound technology and professional support, Kuk Company for the WPC provided and SC Hofigal SA for providing the TEO.

Conflicts of Interest: The authors declare no conflict of interest. The funders had no role in the design of the study; in the collection, analyses, or interpretation of data; in the writing of the manuscript, or in the decision to publish the results.

References

1. U.S. FDA. *Title 21: Food and Drugs Part 182—Substances Generally Recognized As Safe Subpart A—General Provisions*; U.S. FDA: Montgomery/Prince Georges, MD, USA, 2009; pp. 26–28.
2. Gamage, G.R.; Park, H.J.; Kim, K. Effectiveness of antimicrobial coated oriented polypropylene/polyethylene films in sprout packaging. *Food Res. Int.* **2009**, *42*, 832–839. [CrossRef]
3. Atarés, L.; Chiralt, A. Essential oils as additives in biodegradable films and coatings for active food packaging. *Trends Food Sci. Technol.* **2016**, *48*, 51–62. [CrossRef]
4. Hu, W.; Feng, K.; Xiu, Z.; Jiang, A.; Lao, Y. Thyme oil alginate-based edible coatings inhibit growth of pathogenic microorganisms spoiling fresh-cut cantaloupe. *Food Biosci.* **2019**, *32*, 100467.
5. Cuomo, F.; Iacovino, S.; Messia, M.C.; Sacco, P.; Lopez, F. Protective action of lemongrass essential oil on mucilage from chia (Salvia hispanica) seeds. *Food Hydrocoll.* **2020**, *105*, 105860. [CrossRef]
6. Arfat, Y.A.; Benjakul, S.; Prodpran, T.; Sumpavapol, P.; Songtipya, P. Properties and antimicrobial activity of fish protein isolate/fish skin gelatin film containing basil leaf essential oil and zinc oxide nanoparticles. *Food Hydrocoll.* **2014**, *41*, 265–273. [CrossRef]
7. Froiio, F.; Ginot, L.; Paolino, D.; Lebaz, N.; Bentaher, A.; Fessi, H.; Elaissari, A. Essential oils-loaded polymer particles: Preparation, characterization and antimicrobial property. *Polymers* **2019**, *11*, 1017. [CrossRef]
8. Hossain, F.; Follett, P.; Salmieri, S.; Vu, K.D.; Fraschini, C.; Lacroix, M. Antifungal activities of combined treatments of irradiation and essential oils (EOs) encapsulated chitosan nanocomposite films in in vitro and in situ conditions *Int. J. Food Microbiol.* **2019**, *295*, 33–40. [CrossRef]
9. Liu, Q.; Zhang, M.; Bhandari, B.; Xu, J.; Yang, C. Effects of nanoemulsion-based active coatings with composite mixture of star anise essential oil, polylysine, and nisin on the quality and shelf life of ready-to-eat Yao meat products. *Food Control.* **2020**, *107*, 106771. [CrossRef]
10. Seydim, C.; Sarikus, G. Antimicrobial activity of whey protein based edible films incorporated with oregano, rosemary and garlic essential oils. *Food Res. Int.* **2006**, *39*, 639–644. [CrossRef]
11. Zinoviadou, K.G.; Koutsoumanis, K.P.; Biliaderis, C.G. Physico-chemical properties of whey protein isolate films containing oregano oil and their antimicrobial action against spoilage flora of fresh beef. *Meat Sci.* **2009**, *82*, 338–345. [CrossRef]
12. Esmaeili, H.; Cheraghi, N.; Khanjari, A.; Rezaeigolestani, M.; Basti, A.A.; Kamkar, A.; Aghaee, E.M. Incorporation of nanoencapsulated garlic essential oil into edible films: A novel approach for extending shelf life of vacuum-packed sausages. *Meat Sci.* **2020**, *166*, 108135. [CrossRef] [PubMed]
13. Cruz-Diaz, K.; Cobos, Á.; Fernández-Valle, M.E.; Díaz, O.; Cambero, M.I. Characterization of edible films from whey proteins treated with heat, ultrasounds and/or transglutaminase. Application in cheese slices packaging. *Food Packag. Shelf Life* **2019**, *22*, 100397. [CrossRef]
14. Devi, A.F.; Buckow, R.; Hemar, Y.; Kasapis, S. Modification of the structural and rheological properties of whey protein/gelatin mixtures through high pressure processing. *Food Chem.* **2014**, *156*, 243–249. [CrossRef] [PubMed]
15. Ramos, O.L.; Pereira, R.N.; Rodrigues, R.; Teixeira, J.A.; Vicente, A.A.; Malcata, F.X. Physical effects upon whey protein aggregation for nano-coating production. *Food Res. Int.* **2014**, *66*, 344–355. [CrossRef]
16. Gul, O.; Saricaoglu, F.T.; Besir, A.; Atalar, I.; Yazici, F. Effect of ultrasound treatment on the properties of nano-emulsion films obtained from hazelnut meal protein and clove essential oil. *Ultrason. Sonochem.* **2018**, *41*, 466–474. [CrossRef] [PubMed]

17. Bleoanca, I.; Neagu, C.; Dima, S.; Borda, D. Development of multicomponent edible films from milk-serum proteins. *J. Biotechnol.* **2015**, *208*, S47. [CrossRef]

18. Tcholakova, S.; Denkov, N.D.; Ivanov, I.B.; Campbell, B. Coalescence stability of emulsions containing globular milk proteins. *Adv. Colloid Interface Sci.* **2006**, *123*, 259–293. [CrossRef]

19. Bagamboula, C.F.; Uyttendaele, M.; Debevere, J. Inhibitory effect of thyme and basil essential oils, carvacrol, thymol, estragol, linalool and p-cymene towards Shigella sonnei and S. flexneri. *Food Microbiol.* **2004**, *21*, 33–42. [CrossRef]

20. Bleoancă, I.; Saje, K.; Mihalcea, L.; Oniciuc, E.A.; Smole-Mozina, S.; Nicolau, A.I.; Borda, D. Contribution of high pressure and thyme extract to control Listeria monocytogenes in fresh cheese-A hurdle approach. *Innov. Food Sci. Emerg. Technol.* **2016**, *38*, 7–14. [CrossRef]

21. Gounga, M.; Shi-Ying, X.; Wang, Z. Sensory attributes of freshly roasted and roasted freeze dried chinese chestnut coated with wpisolate-pullulan edible coating. *Int. J. Agric. Res.* **2007**, *2*, 959–964.

22. Ramos, O.L.; Silva, S.; Soares, J.C.; Fernandes, J.C.; Poças, M.F.; Pintado, M.E.; Malcata, F.X. Features and performance of edible films, obtained from whey protein isolate formulated with antimicrobial compounds. *Food Res. Int.* **2012**, *45*, 351–361. [CrossRef]

23. ASTM. Standard practice for conditioning plastics for testing. In *Annual Book of ASTM Standards*; ASTM: Montgomery, PA, USA, 2013.

24. ASTM. Standard test method for moisture content of paper and paperboard by oven drying. In *Annual Book of ASTM Standards*; ASTM: Montgomery, PA, USA, 1994; pp. 1–2.

25. Limmatvapirat, S.; Limmatvapirat, C.; Puttipipatkhachorn, S.; Nuntanid, J.; Luangtana-anan, M. Enhanced enteric properties and stability of shellac films through composite salts formation *Eur. J. Pharm. Biopharm.* **2007**, *67*, 690–698. [CrossRef]

26. van den Berg, C.; Bruin, S. Water activity and its estimation in food systems: Theoretical aspects. In *Water Activity: Influences on food Quality*; Rockland, L.B., Stewarts, G.F., Eds.; Academic Press: Cambridge, MA, USA, 1981; pp. 2–61.

27. Tudose, C.; Neagu, C.; Borda, D.; Alexe, P. The impact of water activity on storage stability of a newly reformulated salami: A pilot scale study. *Rev. Chim.* **2017**, *68*, 763–767. [CrossRef]

28. ASTM. *Standard Test. Methods for Water Vapor Transmission of Materials*; ASTM: Montgomery, PA, USA, 1997.

29. Du, W.X.; Avena-Bustillos, R.J.; Sheng, S.; Hua, T.; McHugh, T.H. *Antimicrobial Volatile Essential Oils in Edible Films for Food Safety In Science Against Microbial Pathogens: Communicating Current Research and Technological Advances*; Mendez-Vilas, A., Ed.; Formatex Research Center: Badajoz, Spain, 2011; Volume 2, pp. 1124–1134.

30. Perestrelo, R.; Silva, C.L.; Silva, P.; Medina, S.; Pereira, R.; Câmara, J.S. Untargeted fingerprinting of cider volatiles from different geographical regions by HS-SPME/GC-MS. *Microchem. J.* **2019**, *148*, 643–651. [CrossRef]

31. Krochta, J.; Perez-Gago, M. Formation and properties of whey protein films and coatings. In *Protein-Based Films and Coatings*; Gennadios, A., Ed.; CRC Press: Boca Raton, FL, USA, 2002; pp. 159–180.

32. Le Tien, C.; Letendre, M.; Ispas-Szabo, P.; Mateescu, M.A.; Delmas-Patterson, G.; Yu, H.L.; Lacroix, M. Development of Biodegradable Films from Whey Proteins by Cross-Linking and Entrapment in Cellulose. *J. Agric. Food Chem.* **2000**, *48*, 5566–5575. [CrossRef]

33. Bolumar, T.; Middendorf, D. Structural changes in foods caused by high-pressure processing. In *High Pressure Processing of Food: Principles, Technology and Applications*; Balasubramaniam, L., Barbosa-Canova, V.M., Eds.; Springer: New York City, NY, USA, 2016; pp. 271–294.

34. Kokoszka, S.; Debeaufort, F.; Lenart, A.; Voilley, A. Water vapour permeability, thermal and wetting properties of whey protein isolate based edible films. *Int. Dairy J.* **2010**, *20*, 53–60. [CrossRef]

35. Osés, J.; Fernández-Pan, I.; Mendoza, M.; Maté, J.I. Stability of the mechanical properties of edible films based on whey protein isolate during storage at different relative humidity. *Food Hydrocoll.* **2009**, *23*, 125–131. [CrossRef]

36. Simelane, S.; Ustunol, Z. Mechanical properties of heat-cured whey protein–based edible films compared with collagen casings under sausage manufacturing conditions. *JFS Food Eng. Phys. Prop.* **2005**, *69*, 271–276. [CrossRef]

37. Huntrakul, K.; Harnkarnsujarit, N. Effects of plasticizers on water sorption and aging stability of whey protein/carboxy methyl cellulose films. *J. Food Eng.* **2020**, *272*, 109809. [CrossRef]

38. Mali, S.; Sakanaka, L.S.; Yamashita, F.; Grossmann, M.V.E. Water sorption and mechanical properties of cassava starch films and their relation to plasticizing effect. *Carbohydr. Polym.* **2005**, *60*, 283–289. [CrossRef]

39. Wang, Y.; Liu, L.; Zhou, J.; Ruan, X.; Lin, J.; Fu, L. Effect of chitosan nanoparticle coatings on the quality changes of postharvest whiteleg shrimp, litopenaeus vannamei, during Storage at 4 °C. *Food Bioprocess Technol.* **2014**, *8*, 907–915.

40. Andrade, M.A.; Ribeiro-Santos, R.; Costa Bonito, M.C.; Saraiva, M.; Sanches-Silva, A. Characterization of rosemary and thyme extracts for incorporation into a whey protein based film. *LWT* **2018**, *92*, 497–508. [CrossRef]

41. Gontard, N.; Guilbert, S.; Cuq, J. Edible wheat gluten films: Influence of the main process variables on film properties using response surface methodology. *J. Food Sci.* **1992**, *57*, 190–195. [CrossRef]

42. Bahram, S.; Rezaei, M.; Soltani, M.; Kamali, A.; Ojagh, S.M.; Abdollahi, M. Whey protein concentrate edible film activated with cinnamon essential oil. *J. Food Process. Preserv.* **2014**, *38*, 1251–1258. [CrossRef]

43. Ojagh, S.; Rezaei, M.; Razavi, S.; Hosseini, S. Development and evaluation of a novel biodegradable film made from chitosan and cinnamon essential oil with low affinity toward water. *Food Chem.* **2010**, *122*, 161–166. [CrossRef]

44. Cofelice, M.; Cuomo, F.; Chiralt, A. Alginate films encapsulating lemongrass essential oil as affected by spray calcium application. *Colloids Interfaces* **2019**, *3*, 58. [CrossRef]

45. Laird, K.; Phillips, C. Vapour phase: A potential future use for essential oils as antimicrobials? *Lett. Appl. Microbiol.* **2012**, *54*, 169–174. [CrossRef]

46. Mandras, N.; Nostro, A.; Roana, J.; Scalas, D.; Banche, G.; Ghisetti, V.; Del Re, S.; Fucale, G.; Cuffini, A.M. Liquid and vapour-phase antifungal activities of essential oils against Candida albicans and non-albicans Candida. *BMC Complement. Altern. Med.* **2016**, *16*, 330. [CrossRef]

47. Reyes-Jurado, F.; Navarro-Cruz, A.R.; Ochoa-Velasco, C.E.; Palou, E.; López-Malo, A.; Ávila-Sosa, R. Essential oils in vapor phase as alternative antimicrobials: A review. *Crit. Rev. Food Sci. Nutr.* **2020**, *60*, 1641–1650. [CrossRef]

48. Nadjib, B.M.; Amine, F.M.; Abdelkrim, K.; Fairouz, S.; Maamar, M. Liquid and vapour phase antibacterial activity of eucalyptus globulus essential oil=susceptibility of selected respiratory tract pathogens. *Am. J. Infect. Dis.* **2014**, *10*, 105–117. [CrossRef]

49. Goñi, P.; López, P.; Sánchez, C.; Gómez-Lus, R.; Becerril, R.; Nerín, C. Antimicrobial activity in the vapour phase of a combination of cinnamon and clove essential oils. *Food Chem.* **2009**, *116*, 982–989. [CrossRef]

50. Kohiyama, C.Y.; Ribeiro, M.M.Y.; Mossini, S.A.G.; Bando, É.; Bomfim, N.D.S.; Nerilo, S.B.; Rocha, G.H.O.; Grespan, R.; Mikcha, J.M.G.; Machinski, M., Jr. Antifungal properties and inhibitory effects upon aflatoxin production of Thymus vulgaris L. by Aspergillus flavus. *Food Chem.* **2015**, *173*, 1006–1010. [CrossRef] [PubMed]

51. Chizzola, R.; Michitsch, H.; Franz, C. Antioxidative properties of thymus vulgaris leaves : Chemotypes. *J. Agric. Food Chem.* **2008**, *56*, 6897–6904. [CrossRef] [PubMed]

52. Nuin, M.; Alfaro, B.; Cruz, Z.; Argarate, N.; George, S.; Le Marc, Y.; Olley, J.; Pin, C. Modelling spoilage of fresh turbot and evaluation of a time-temperature integrator (TTI) label under fluctuating temperature. *Intl. J. Food Microbiol.* **2008**, *127*, 193–199. [CrossRef]

53. Walkenström, A.-M.; Hermansson, P. Mixed gels of gelatin and whey proteins, formed by combining temperature and high pressure. *Food Hydrocoll.* **1997**, *11*, 457–470. [CrossRef]

54. Krešić, G.; Lelas, V.; Jambrak, A.R.; Herceg, Z.; Brnčić, S.R. Influence of novel food processing technologies on the rheological and thermophysical properties of whey proteins. *J. Food Eng.* **2008**, *87*, 64–73. [CrossRef]

55. Nguyen Van Long, N.; Joly, C.; Dantigny, P. Active packaging with antifungal activities. *Int. J. Food Microbiol.* **2016**, *220*, 73–90. [CrossRef]

56. Uz, M.; Altinkaya, S.A. Development of mono and multilayer antimicrobial food packaging materials for controlled release of potassium sorbate. *LWT Food Sci. Technol.* **2011**, *44*, 2302–2309. [CrossRef]

Article

Evaluation of the Toxicity of *Satureja intermedia* C. A. Mey Essential Oil to Storage and Greenhouse Insect Pests and a Predator Ladybird

Asgar Ebadollahi [1,*] **and William N. Setzer** [2,3,*]

1   Moghan College of Agriculture and Natural Resources, University of Mohaghegh Ardabili, Ardabil 56199-36514, Iran
2   Department of Chemistry, University of Alabama in Huntsville, Huntsville, AL 35899, USA
3   Aromatic Plant Research Center, 230 N 1200 E, Suite 100, Lehi, UT 84043, USA
*   Correspondence: ebadollahi@uma.ac.ir (A.E.); wsetzer@chemistry.uah.edu (W.N.S.)

Received: 13 May 2020; Accepted: 21 May 2020; Published: 2 June 2020

Abstract: The use of chemical insecticides has had several side-effects, such as environmental contamination, foodborne residues, and human health threats. The utilization of plant-derived essential oils as efficient bio-rational agents has been acknowledged in pest management strategies. In the present study, the fumigant toxicity of essential oil isolated from *Satureja intermedia* was assessed against cosmopolitan stored-product insect pests: *Trogoderma granarium* Everts (khapra beetle), *Rhyzopertha dominica* (Fabricius) (lesser grain borer), *Tribolium castaneum* (Herbst) (red flour beetle), and *Oryzaephilus surinamensis* (L.) (saw-toothed grain beetle). The essential oil had significant fumigant toxicity against tested insects, which positively depended on essential oil concentrations and the exposure times. Comparative contact toxicity of *S. intermedia* essential oil was measured against *Aphis nerii* Boyer de Fonscolombe (oleander aphid) and its predator *Coccinella septempunctata* L. (seven-spot ladybird). Adult females of *A. nerii* were more susceptible to the contact toxicity than the *C. septempunctata* adults. The dominant compounds in the essential oil of *S. intermedia* were thymol (48.1%), carvacrol (11.8%), *p*-cymene (8.1%), and γ-terpinene (8.1%). The high fumigant toxicity against four major stored-product insect pests, the significant aphidicidal effect on *A. nerii*, and relative safety to the general predator *C. septempunctata* make terpene-rich *S. intermedia* essential oil a potential candidate for use as a plant-based alternative to the detrimental synthetic insecticides.

Keywords: *Aphis nerii*; *Coccinella septempunctata*; plant-based insecticide; *Oryzaephius surinamensis*; *Rhyzopertha dominica*; *Tribolium castaneum*; *Trogoderma granarium*

---

1. Introduction

The Khapra Beetle {*Trogoderma granarium* Everts (Coleoptera: Dermestidae)}, lesser grain borer {*Rhyzopertha dominica* (Fabricius) (Coleoptera: Bostrichidae)}, red flour beetle {*Tribolium castaneum* (Herbst) (Coleoptera: Tenebrionidae)}, and saw-toothed grain beetle {*Oryzaephilus surinamensis* (L.) (Coleoptera: Silvanidae)} are among the most well-known and economically-important stored-product pests with world-wide distribution. Along with direct damage due to feeding on various stored products, the quality of products is strictly diminished because of their residues and mechanically associated microbes [1–5].

Oleander aphid {*Aphis nerii* Boyer de Fonscolombe (Hemiptera: Aphididae)}, as a cosmopolitan obligate parthenogenetic aphid, is a common insect pest of many ornamental plants comprising several species of Asclepiadaceae, Apocynaceae, Asteraceae, Convolvulaceae, and Euphorbiaceae, especially in greenhouse conditions. Along with direct damage, *A. nerii* is able to transmit pathogenic viruses to many plants [6–8]. The seven-spot ladybird beetle {*Coccinella septempunctata* L. (Coleoptera:

Coccinellidae)} is a natural enemy of various soft-bodied pests like aphids, thrips, and spider mites, and is considered an important biocontrol agent for greenhouse crops [9–11].

The utilization of chemical insecticides is the main strategy in the management of insect pests. However, there is a global concern about their numerous side effects including environmental pollution, insecticide resistance, resurgence of secondary pests, and toxicity to non-target organisms ranging from soil microorganisms to pollinator, predator and parasitoid insects, fish, and even humans [12–14]. Therefore, the search for eco-friendly and efficient alternative agents for insect pest management is urgent.

Based on the low toxicity to mammals, rapid biodegradation in the environment, and very low chance of insect pest resistance, the use of essential oils extracted from different aromatic plants has been the motivating subject of many researchers in pest management strategies over the past decade [15–18].

Sixteen species of the *Satureja* genus from the Lamiaceae have been reported in the Iranian flora, of which *S. atropatana* Bunge, *S. bachtiarica* Bunge, *S. edmondi* Briquet, *S. intermedia* C. A. Mey, *S. isophylla* Rech., *S. kallarica* Jamzad, *S. khuzistanica* Jamzad, *S. macrosiphonia* Bornm., *S. sahendica* Bornm., and *S. rechingeri* Jamzad are endemic to Iran [19]. *S. intermedia*, as a small delicate perennial plant growing on rock outcrops, is among aromatic plants with considerable amount (1.45% (*w/w*)) of essential oil [20]. The essential oil of *S. intermedia* is rich in terpenes such as 1,8-cineole, *p*-cymene, limonene, γ-terpinene, α-terpinene, thymol, and β-caryophyllene, which are classified in four main groups; monoterpene hydrocarbons, oxygenated monoterpenoids, sesquiterpene hydrocarbons, and oxygenated sesquiterpenoids [20–22]. Some important biological effects of *S. intermedia* essential oil include antifungal, antibacterial, and antioxidant effects, and cytotoxic effects have been reported in previous studies [21–23]. Although the susceptibility of insect pests to the essential oils isolated from some *Satureja* species such as *S. hortensis*, *S. montana* L., *S. parnassica* Heldr. & Sart ex Boiss., *S. spinosa* L., and *S. thymbra* L. was documented in recent years [24–26], the insecticidal effects of *S. intermedia* essential oil have not reported yet.

As part of a screening program for eco-friendly and efficient plant-derived insecticides, the evaluation of the fumigant toxicity against four major Coleopteran stored-product insect pests *O. surinamensis*, *R. dominica*, *T. castaneum* and *T. granarium* and the contact toxicity against a greenhouse insect pest *Aphis nerii* of the essential oil of *S. intermedia* was the main objective of the present study. Because of the importance of studying the effects of insecticides on the natural enemies of insect pests, the toxicity of *S. intermedia* essential oil against *C. septempunctata* was also investigated.

2. Materials and Methods

2.1. Plant Materials and Essential Oil Extraction

Aerial parts (3.0 kg) of *S. intermedia* were gathered from the Heiran regions, Ardebil province, Iran (38°23′ N, 48°35′ E, elevation 907 m). It was identified according to the keys provided by Jamzad [27]. The voucher specimen was deposited in the Department of Plant Production, Moghan College of Agriculture and Natural Resources, Ardabil, Iran. The fresh leaves and flowers were separated and dried under shade within a week. One hundred grams of the specimen were poured into a 2-L round-bottom flask and subjected to hydrodistillation using a Clevenger apparatus for 3 h. The extraction was repeated in triplicate and the obtained essential oil was dried over anhydrous Na_2SO_4 and stored in a refrigerator at 4 °C.

2.2. Essential Oil Characterization

The chemical profile of the *S. intermedia* essential oil was evaluated using gas chromatography (Agilent 7890B) coupled with mass-spectrometer (Agilent 5977A). The analysis was carried out by a HP-5 ms capillary column (30 m × 0.25 mm × 0.25 μm). The temperature of the injector was 280 °C and the column temperature adjusted from 50 to 280 °C using the temperature program: 50 °C (hold for 1 min), increase to 100 °C at 8°/min, increase to 185 °C at 5°/min, increase to 280 °C at 15°/min, and hold at 280 °C for 2 min. The carrier gas was helium (99.999%) with flow rate of 1 mL/min. Essential oil was diluted in methanol, and 1 µL solution was injected (split 1:10 at 0.75 min). The identification of components was performed by comparing mass spectral fragmentation patterns and retention indices with those reported in the databases [28–30].

2.3. Insects

The required colonies of *Oryzaephilus surinamensis* and *Rhyzopertha dominica* were reared on wheat grains for several generations at the Department of Plant Production, Moghan College of Agriculture and Natural Resources, University of Mohaghegh Ardabili (Ardabil province, Iran). *Tribolium castaneum* and *Trogoderma granarium* adults were collected from infested stored wheat grains in Moghan region (Ardabil province, Iran). Insects were identified by Asgar Ebadollahi. Fifty unsexed pairs of adult insects were separately released onto wheat grains and removed from breeding container after 48 h. Wheat grains contaminated with insect eggs were separately kept in an incubator at 25 ± 2 °C, 65 ± 5% relative humidity and a photoperiod of 14:10 (L:D) h. Finally, one to fourteen-day-old adults of *O. surinamensis*, *R. dominica*, *T. castaneum* and *T. granarium* were designated for fumigant bio-assays.

Aphis nerii and its natural predator *Coccinella septempunctata* were used to evaluate the contact toxicity of the *S. intermedia* essential oil. Cohorts of apterous adult females of *A. nerii* and unsexed adults of *C. septempunctata* were taken directly from homegrown oleander (*Nerium oleander* L.) and a chemically untreated alfalfa (*Medicago sativa* L.) field (Moghan region, Ardabil province, Iran), respectively.

2.4. Fumigant Toxicity

The fumigant toxicity of *S. intermedia* essential oil was tested on adults of *O. surinamensis*, *R. dominica*, *T. castaneum*, and *T. granarium*. To determine the fumigant toxicity of the essential oil, filter papers (Whatman No. 1, 2 × 2 cm) were impregnated with essential oil concentrations and were attached to the under surface of the screw cap of glass containers (340-mL) as fumigant chambers. A series of concentrations (4.71–14.71, 7.06–20.88, 20.59–58.82, and 8.82–35.29 µL/L for *O. surinamensis*, *R. dominica*, *T. castaneum*, and *T. granarium*, respectively) was organized to assess the toxicity of *S. intermedia* essential oil after an initial concentration setting experiment for each insect species. Twenty unsexed adults (1–14 days old) of each insect species were separately put into glass containers and their caps were tightly affixed. The same conditions without any essential oil concentration were used for control groups and each treatment was replicated five times. Insects mortality was documented 24, 48 and 72 h after initial exposure to the essential oil. Insects were considered dead when no leg or antennal movements were observed [31].

2.5. Contact Toxicity

The contact toxicity of *S. intermedia* essential oil against the apterous adult females of *A. nerii* and unsexed adults of *C. septempunctata* was tested through filter paper discs (Whatman No. 1), 9 cm diameter, positioned in glass petri dishes (90 × 10 mm). Range-finding experiments were established to find the proper concentrations for each insect. Concentrations ranging from 200 to 750 µg/mL for *A. nerii* and from 500 to 1400 µg/mL for *C. septempunctata* were prepared via 1.00% aqueous Tween-80 as an emulsifying agent. Each solution (200 µL) was applied to the surface of the filter paper. Ten insects were separately released onto each treated disc, the dishes sealed with Parafilm® and kept at 25 ± 2 °C, 65 ± 5% relative humidity and a photoperiod of 16:8 h (light:dark). Except for the addition of essential oil concentrations, all other procedures were unchanged for the control groups. Four replications were made for each treatment and mortality was documented after 24 h. Aphids and ladybirds were considered dead if no leg or antennal movements were detected when softly prodded [32,33].

2.6. Data Analysis

The mortality percentage was corrected using Abbott's formula: $Pt = [(Po - Pc)/(100 - Pc)] \times 100$, in which Pt is the corrected mortality percentage, Po is the mortality (%) caused by essential oil concentrations and Pc is the mortality (%) in the control groups [34].

Analysis of variance (ANOVA) and Tukey's test at $p = 0.05$ were used to statistically identify the effects of independent factors (essential oil concentration and exposure time) on insect mortality and the differences among mean mortality percentage of insects, respectively. Probit analysis was used to estimate LC_{50} and LC_{95} values with 95% fiducial limits, the data heterogeneity and linear regression information using SPSS 24.0 software package (Chicago, IL, USA).

3. Results

3.1. Chemical Composition of Essential Oil

The chemical composition of *S. intermedia* essential oil is presented in Table 1. A total of 47 compounds were identified in the essential oil, in which the phenolic monoterpenoids thymol (48.1%) and carvacrol (11.8%), along with *p*-cymene (8.1%), γ-terpinene (8.1%), carvacryl methyl ether (4.0%), α-pinene (2.7%), and β-caryophyllene (2.4%) were dominants. Terpenoids were the most abundant components (98.6%), especially monoterpene hydrocarbons (20.5%) and oxygenated monoterpenoids (68.4%) with only minor amounts of phenylpropanoids or fatty acid-derived compounds.

Table 1. Chemical composition of the essential oil isolated from aerial parts of *Satureja intermedia*.

RI_{calc}	RI_{db}	Compound	%	RI_{calc}	RI_{db}	Compound	%
929	932	α-Pinene	2.7	1384	1387	β-Bourbonene	0.1
984	974	1-Octen-3-ol	0.3	1389	1379	Geranyl acetate	tr
990	988	Myrcene	0.4	1423	1417	β-Caryophyllene	2.4
1016	1020	*p*-Cymene	8.1	1428	1431	β-Gurjunene	0.1
1034	1024	Limonene	0.5	1432	1442	α-Maaliene	0.1
1037	1026	1,8-Cineole	1.7	1438	1439	Aromadendrene	0.7
1060	1054	γ-Terpinene	8.1	1454	1452	α-Humulene	0.3
1066	1065	*cis*-Sabinene hydrate	0.4	1476	1478	γ-Muurolene	0.5
1083	1086	Terpinolene	0.2	1487	1489	β-Selinene	0.2
1083	1089	*p*-Cymenene	0.2	1496	1496	Viridiflorene	0.7
1092	1095	Linalool	0.2	1500	1500	α-Muurolene	0.2
1094	1098	*trans*-Sabinene hydrate	0.1	1510	1505	β-Bisabolene	1.3
1121	1128	*allo*-Ocimene	0.2	1515	1513	γ-Cadinene	0.3
1164	1165	Borneol	0.4	1523	1522	δ-Cadinene	0.7
1176	1174	Terpinen-4-ol	0.8	1530	1533	*trans*-Cadina-1,4-diene	0.1
1187	1191	Hexyl butyrate	0.1	1535	1537	α-Cadinene	tr
1239	1241	Carvacryl methyl ether	4.0	1540	1544	α-Calacorene	0.3
1284	1282	(*E*)-Anethole	0.7	1557	1553	Thymohydroquinone	0.5
1290	1289	Thymol	48.1	1578	1577	Spathulenol	0.9
1298	1298	Carvacrol	11.8	1581	1582	Caryophyllene oxide	0.8
1340	1340	Piperitenone	tr			Monoterpene hydrocarbons	20.5
1346	1346	α-Terpinyl acetate	0.1			Oxygenated monoterpenoids	68.4
1349	1349	Thymyl acetate	0.2			Sesquiterpene hydrocarbons	8.0
1357	1356	Eugenol	0.1			Oxygenated sesquiterpenoids	1.7
1365	1373	α-Ylangene	0.1			Phenylpropanoids	0.8
1371	1374	α-Copaene	0.2			Others	0.4
1376	1372	Carvacryl acetate	0.1			Total identified	99.8

RI_{calc} = Retention index determined with respect to a homologous series of *n*-alkanes on a HP-5 ms column; RI_{db} = Retention index from the databases [28–30]; tr = trace (<0.05%).

3.2. Fumigant Toxicity

Analysis of variance (ANOVA) revealed that the tested concentrations of *S. intermedia* essential oil ($F = 239.462$ and $p < 0.0001$ for *O. surinamensis*, $F = 223.629$ and $p < 0.0001$ for *R. dominica*, $F = 169.615$ and $p < 0.0001$ for *T. castaneum*, and $F = 89.032$ and $p < 0.0001$ for *T. granarium* with df = 4, 45) and the considered exposure times ($F = 212.855$ and $p < 0.0001$ for *O. surinamensis*, $F = 281.180$ and $p < 0.0001$

for *R. dominica*, $F = 84.705$ and $p < 0.0001$ for *T. castaneum*, and $F = 84.501$ and $p < 0.0001$ for *T. granarium* with df = 2, 45) had significant effects on the mortality of all insect pests. According to Figure 1 and relatively high R^2 values, there is a positive correlation between the fumigation of essential oil concentrations and the mortality of four storage insect pests at all exposure times. Furthermore, the steep slopes indicate a homogenous toxic response among beetles to the essential oil.

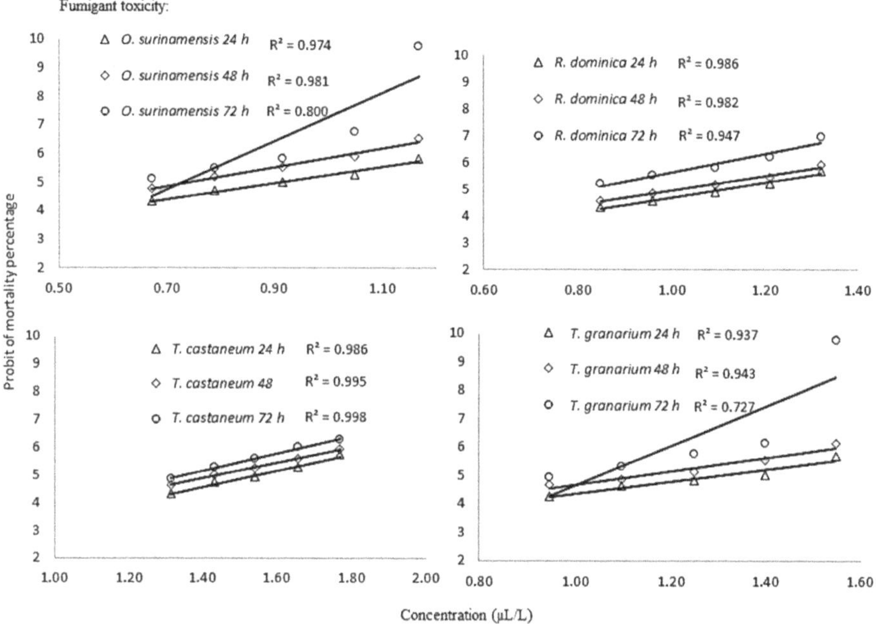

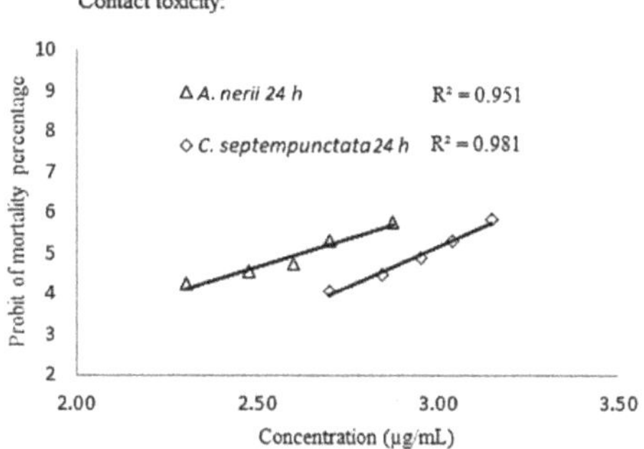

Figure 1. Concentration–response lines of contact and fumigant toxicity of Satureja intermedia essential oil against Aphis nerii and Coccinella septempunctata, and Oryzaephilus surinamensis, Rhyzopertha dominica, Tribolium castaneum, and Trogoderma granarium, respectively.

According to Table 2, an obvious difference in the mean mortality percentage of all tested storage insect pests was detected, as essential oil concentration and exposure time were increased. For example, 25.00% mortality of *O. surinamensis* adults was observed at 4.71 μL/L and 24-h exposure time, which had increased to 80.00% and 100% at 14.71 μL/L after 24 and 72 h, respectively. It is apparent that the essential oil of *S. intermedia* gave at least 90% mortality against all tested stored-product insect pests at 58.82 μL/L after 72 h (Table 2).

Table 2. Mean mortality ± SE of the adults of *Oryzaephilus surinamensis*, *Rhyzopertha dominica*, *Tribolium castaneum*, and *Trogoderma granarium* exposed to the fumigation of *Satureja intermedia* essential oil after 24, 48, and 72 h.

Insect	Time (h)	Concentration (μL/L)				
		4.71	**6.18**	**8.24**	**11.18**	**14.71**
O. surinamensis	24	25.00 ± 0.41 [j]	38.75 ± 0.63 [i]	50.00 ± 0.41 [g]	60.00 ±0.41 [f]	80.00 ± 0.41 [d]
	48	41.25 ± 0.48 [h]	57.50 ± 0.29 [f,g]	70.00 ± 0.41 [e]	81.25 ± 0.48 [d]	93.75 ± 0.48 [c]
	72	53.75 ± 0.48 [g]	68.75 ± 0.48 [e]	80.00 ± 0.58 [d]	96.25 ± 0.48 [b]	100.00 ± 0.00 [a]
		7.06	**9.12**	**12.35**	**16.18**	**20.88**
R. dominica	24	25.00 ± 0.41 [l]	33.75 ± 0.48 [k]	46.25 ± 0.48 [i]	58.75 ± 0.29 [h]	75.00 ± 0.58 [e]
	48	33.75 ± 0.48 [k]	43.75 ± 0.48 [j]	56.25 ± 0.48 [h]	67.50 ± 0.29 [g]	82.50 ± 0.29 [c]
	72	57.50 ± 0.29 [h]	70.00 ± 0.41 [f]	78.75 ± 0.25 [d]	88.75 ± 0.48 [b]	97.50 ± 0.29 [a]
		20.59	**27.06**	**34.71**	**45.29**	**58.82**
T. castaneum	24	23.75 ± 0.48 [k]	38.75 ± 0.48 [i]	46.25 ± 0.48 [g]	60.00 ±0.41 [e]	76.25 ± 0.25 [c]
	48	35.00 ± 0.58 [j]	50.00 ± 0.58 [f]	58.75 ± 0.63 [e]	71.25 ± 0.48 [d]	82.50 ± 0.50 [b]
	72	43.75 ± 0.48 [h]	60.00 ± 0.41 [e]	71.25 ± 0.25 [d]	83.75 ± 0.63 [b]	90.00 ± 0.50 [a]
		8.82	**12.53**	**17.68**	**25.00**	**35.29**
T. granarium	24	22.50 ± 0.48 [j]	35.00 ± 0.29 [i]	42.50 ± 0.25 [h]	50.00 ± 0.41 [g]	75.00 ± 0.29 [c]
	48	37.50 ± 0.25 [i]	45.00 ± 0.29 [h]	55.00 ± 0.29 [f]	70.00 ± 0.41 [d]	87.50 ± 0.48 [b]
	72	47.50 ± 0.25 [g]	62.50 ± 0.48 [e]	77.50 ± 0.48 [c]	87.50 ± 0.48 [b]	100.00 ± 0.00 [a]

Data that do not have the same letters are statistically significant different at $p = 0.05$ based on Tukey's test. Each datum represents mean ± SE of four replicates with eighty adult insects.

Based on lower LC_{50} values of those stored-product insect pests tested, *O. surinamensis* was significantly the most susceptible insect to the essential oil of *S. intermedia* at all time intervals. In contrast, the adults of *T. castaneum* with highest LC_{50} and LC_{95} values were the most tolerant to fumigation with *S. intermedia* essential oil. Furthermore, the susceptibility of insect pests to the fumigation of *S. intermedia* essential oil followed in the order: *O. surinamensis* > *R. dominica* > *T. granarium* > *T. castaneum* (Table 3).

Table 3. Probit analysis of the data obtained from fumigation of *Satureja intermedia* essential oil on the adults of *Oryzaephilus surinamensis*, *Rhyzopertha dominica*, *Tribolium castaneum*, and *Trogoderma granarium*.

Insect	Time (h)	LC_{50} with 95% Confidence Limits (μL/L)	LC_{90} with 95% Confidence Limits (μL/L)	χ^2 (df = 3)	Slope ± SE	Sig. *
O. surinamensis	24	8.151 (7.396–8.970)	23.177 (18.675–32.578)	1.99	2.824 ± 0.344	0.574
	48	5.542 (4.853–6.119)	13.710 (11.971–16.756)	1.288	3.258 ± 0.378	0.732
	72	4.716 (4.143–5.174)	9.200 (8.413–10.405)	5.134	4.415 ± 0.504	0.162
R. dominica	24	12.825 (11.661–14.189)	36.901 (29.147–54.0970)	0.885	2.792 ± 0.356	0.829
	48	10.398 (9.265–11.454)	30.455 (24.687–42.838)	1.056	2.746 ± 0.358	0.788
	72	6.358 (5.126–7.296)	15.970 (14.160–19.138)	2.488	3.204 ± 0.432	0.477
T. granarium	24	20.489 (18.114–23.612)	81.507 (58.604–140.911)	4.233	2.137 ± 0.283	0.237
	48	13.654 (11.811–15.364)	49.192 (38.852–71.499)	3.978	2.302 ± 0.289	0.264
	72	9.785 (6.082–12.258)	24.075 (18.870–42.027)	5.842	3.277 ± 0.360	0.12
T. castaneum	24	35.612 (32.538–39.070)	95.948 (77.352–135.744)	0.967	2.977 ± 0.376	0.809
	48	28.048 (24.747–30.916)	80.251 (65.751–111.454)	0.297	2.807 ± 0.378	0.961
	72	22.861 (19.648–25.415)	57.584 (50.068–71.481)	0.139	3.194 ± 0.405	0.987

* Since the significance level is greater than 0.05, no heterogeneity factor is used in the calculation of confidence limits. The number of insects for calculation of LC_{50} values is 200 for *T. granarium* and 400 for other insects in each time.

3.3. Contact Toxicity

The tested concentrations of *S. intermedia* essential oil demonstrated significant contact toxicity on both *A. nerii* ($F = 27.682$, df = 4, 15; $p < 0.0001$) and *C. septempunctata* ($F = 35.607$, df = 4, 15; $p < 0.0001$). A positive correlation between essential oil concentrations and the mortality of *A. nerii* and *C. septempunctata* in the contact assay is also apparent, based on the high R^2 values (Figure 1). Comparisons of the mean mortality percentage of *A. nerii* and its predator *C. septempunctata* caused by *S. intermedia* essential oil are shown in Table 4. The mortality percentages of both insects increased with increasing essential oil concentrations, but their susceptibility to the essential oil was noticeably different. For example, 62.50% mortality was documented for *A. nerii* at 500 μg/mL essential oil concentration while its predator *C. septempunctata* was more tolerant and exhibited only 17.50% mortality at this concentration (Table 4).

Table 4. Mean mortality ± SE of the adults of *Aphis nerii* and *Coccinella septempunctata* exposed to the different concentration of *Satureja intermedia* essential oil after 24 h.

Insect	Concentration (µg/mL)				
	200	300	400	500	750
A. nerii	22.50 ± 0.25 [e]	32.50 ± 0.25 [d]	40.00 ± 0.41 [c]	62.50 ± 0.25 [b]	77.50 ± 0.75 [a]
	500	700	900	1100	1400
C. septempunctata	17.50 ± 0.48 [e]	30.00 ± 0.41 [d]	45.00 ± 0.29 [c]	62.50 ± 0.48 [b]	80.00 ± 0.41 [a]

Data that do not have the same letters are statistically significant different at $p = 0.05$ based on Tukey's test. Each datum represents mean ± SE of four replicates with eighty adult insects.

The results of the probit analysis for the contact toxicity of *S. intermedia* essential oil against *A. nerii* and *C. septempunctata* adults are shown in Table 5. According to low LC_{50} and LC_{95} values, the adult females of *A. nerii* were more susceptible to contact toxicity of *S. intermedia* essential oil than the adults of *C. septempunctata*.

Table 5. Probit analysis of the data obtained from contact toxicity of *Satureja intermedia* essential oil on the adults of *Aphis nerii* and *Coccinella septempunctata*.

Insect	LC_{50} with 95% Confidence Limits (µg/mL)	LC_{90} with 95% Confidence Limits (µg/mL)	χ^2 (df = 3)	Slope ± SE	Sig. *
A. nerii	418.379 (379.586–464.130)	1224.788 (975.704–1738.840)	4.363	2.747 ± 0.318	0.225
C. septempunctata	913.722 (853.739–980.799)	1908.099 (1652.748–2352.473)	1.932	4.008 ± 0.413	0.587

* Since the significance level is greater than 0.05, no heterogeneity factor is used in the calculation of confidence limits. The number of insects for calculation of LC_{50} values is 240 for each insect.

4. Discussion

The susceptibility of *O. surinamensis*, *R. dominica*, *T. castaneum* and *T. granarium* adults to the essential oil of *S. intermedia* with 24-h LC_{50} values of 8.151, 12.825, 20.489, and 35.612 µL/L, respectively, was distinguished in the present study. The fumigant toxicity of some plant-derived essential oils against *O. surinamensis*, *R. dominica*, *T. castaneum* and *T. granarium* has been documented in previous studies; it was found that the essential oils of *Agastache foeniculum* (Pursh) Kuntze, *Achillea filipendulina* Lam., and *Achillea millefolium* L. with respective 24-h LC_{50} values of 18.781, 12.121, and 17.977 µL/L, had high toxicity on the adults of *O. surinamensis* [31,34–36]. The adults of *R. dominica* were also susceptible to the fumigation of essential oils extracted from *Eucalyptus globulus* Labill (24-h LC_{50} = 3.529 µL/L), *Lavandula stoechas* L. (24-h LC_{50} = 5.660 µL/L), and *Apium graveolens* L. (24-h LC_{50} = 53.506 µL/L) [37,38]. The fumigation of the essential oils of *Lippia citriodora* Kunth (24-h LC_{50} = 37.349 µL/L), *Melissa officinalis* L. (24-h LC_{50} = 19.418 µL/L), and *Teucrium polium* L. (24-h LC_{50} = 20.749 µL/L) resulted in significant mortality in *T. castaneum* [39–41]. The essential oils of *Schinus molle* L. (48-h LC_{50} = 806.50 µL/L) and *Artemisia sieberi* Besser (24-h LC_{50} = 33.80 µL/L) also had notable fumigant toxicity against the adults of *T. granarium* [42,43]. The toxicity of all the above-mentioned essential oils was augmented when the exposure time was prolonged. These findings support the results regarding the time-dependent susceptibility of *O. surinamensis*, *R. dominica*, *T. castaneum* and *T. granarium* to plant essential oils. The differences in observed LC_{50} values are likely due to the differences in the essential oil compositions from the different plant species and possibly to differences in the experimental conditions. Furthermore, the *S. intermedia* essential oil with low 24-h LC_{50} value was more toxic on *O. surinamensis* than *A. foeniculum*, *A. filipendulina*, and *A. millefolium* essential oils, on *R. dominica* than *A. graveolens* essential oil, on *T. castaneum* than *Lippia citriodora* essential oil, and on *T. granarium* than *S. molle* essential oil.

The terpenes, especially thymol, carvacrol, *p*-cymene and γ-terpinene, were recognized as the main components of *S. intermedia* essential oil in the present study. In the study of Sefidkon and Jamzad, thymol (32.3%), γ-terpinene (29.3%), *p*-cymene (14.7%), elemicin (4.8%), limonene (3.3%), and α-terpinene (3.3%) were the main components of *S. intermedia* essential oil [20]. In another study, thymol (34.5%), γ-terpinene (18.2%), *p*-cymene (10.5%), limonene (7.3%), α-terpinene (7.1%), carvacrol (6.9%), and elemicin (5.3%) were found to be major components in the essential oil of *S. intermedia* [23]. In the present study, however, limonene was a minor component (0.5%), and neither elemicin nor α-terpinene were detected. Ghorbanpour et al. reported the terpenes thymol (32.3%), *p*-cymene (14.7%), γ-terpinene (3.3%), and carvacrol (1.0%), and the phenylpropanoid elemicin (4.8%) as the main components in the essential oil of *S. intermedia* [22], while the concentrations of γ-terpinene and carvacrol were much lower compared to the present findings. The differences in the chemical profile of the plant essential oils are likely due to the internal and external factors such as seasonal variation, geographical features, plant growth stage, and different extraction conditions [19,44,45]. The insecticidal properties of several terpenes, especially monoterpene hydrocarbons and monoterpenoids, which accounted for 88.9% of the *S. intermedia* essential oil in the present study, have been documented in recent investigations. For example, insecticidal activities of *p*-cymene, α-pinene, γ-terpinene, 1,8-cineole, and limonene have been demonstrated against several detrimental insect pests [46–50]. Previous studies have also indicated that the monoterpenoids thymol and carvacrol had significant toxicity against insect pests [46,51,52]. Accordingly, the insecticidal efficiency of *S. intermedia* essential oil can be attributed to such components.

The contact toxicity of the essential oil of *Eucalyptus globulus* Labill. against *A. nerii* has been reported by Russo et al. [53]. Although this is the only previous study to investigate the susceptibility of *A. nerii* to a plant essential oil, its findings confirm the results of the present study about the possibility of *A. nerii* management through plant essential oils. Indeed, the toxicity of *S. intermedia* essential oil was evaluated for the first time in the present study against *A. nerii* and its natural enemy *C. septempunctata*. The essential oil of *S. intermedia* was more toxic on *A. nerii* (LC_{50}: 418 µg/mL) than the predator ladybird *C. septempunctata* (LC_{50}: 914 µg/mL), suggesting that the predator was more tolerant than the aphid to *S. intermedia* essential oil, which is very valuable in terms of predator protection. Similar results were obtained for controlling aphids [54,55] and some other insect pests [56–58] using plant-derived essential oils along with protecting their predators. However, the destructive side-effects of some essential oils on parasitoids have been reported [59–61]. Therefore, it is important to select efficient pesticides with lower side effects on natural enemies at operative concentrations to the pests, which has been achieved in the current study.

5. Conclusions

In conclusion, the terpene-rich essential oil of *S. intermedia* has significant fumigant toxicity against the adults of *O. surinamensis*, *R. dominica*, *T. castaneum*, and *T. granarium*, and may be considered as a natural effective fumigant on stored products. This bio-rational agent also has significant contact toxicity on the adult females of *A. nerii*, one of the cosmopolitan insect pests of ornamental plants. Furthermore, the predator ladybird *C. septempunctata* was more tolerant to the essential oil than the aphid. Accordingly, *S. intermedia* essential oil can be nominated as an eco-friendly efficient insecticide by decreasing the risks associated with the application of synthetic chemicals. However, the exploration of any side-effects of the essential oil on other useful insects such as parasitoids and pollinators, its phytotoxicity on the treated plants and crops, any adverse tastes or odors on stored products, and the preparation of novel formulations to increase its stability in the environment for practical utilization are needed.

Author Contributions: Conceptualization, A.E.; methodology, A.E. and W.N.S.; validation, A.E. and W.N.S.; formal analysis, A.E. and W.N.S.; investigation, A.E.; resources, A.E.; data curation, A.E.; writing—original draft preparation, A.E.; writing—review and editing, A.E. and W.N.S.; project administration, A.E.; funding acquisition, A.E. All authors have read and agreed to the published version of the manuscript.

Funding: This research was funded by the University of Mohaghegh Ardabili.

Acknowledgments: W.N.S. participated in this work as part of the activities of the Aromatic Plant Research Center (APRC, https://aromaticplant.org/). This study received financial support from the University of Mohaghegh Ardabili, which is greatly appreciated.

Conflicts of Interest: The authors declare no conflict of interest.

References

1. Edde, P.A. A review of the biology and control of *Rhyzopertha dominica* (F.) the lesser grain borer. *J. Stored Prod. Res.* **2012**, *48*, 1–18. [CrossRef]

2. Bosly, H.A.; Kawanna, M.A. Fungi species and red flour beetle in stored wheat flour under Jazan region conditions. *Toxicol. Ind. Health* **2014**, *30*, 304–310. [CrossRef] [PubMed]

3. Garcia, D.; Girardi, N.S.; Passone, M.A.; Nesci, A.; Etcheverry, M. Harmful effects on *Oryzaephilus surinamensis* (L.) and *Tribolium castaneum* by food grade antioxidants and their formulations in peanut kernel. *J. Food Chem. Nanotechnol.* **2017**, *3*, 86–92. [CrossRef]

4. Özberk, F. Impacts of khapra beetle (*T. granarium* Everts) onto marketing price and relevant traits in bread wheat (*T. aestivum* L.). *Appl. Ecol. Environ. Res.* **2018**, *16*, 6143–6153. [CrossRef]

5. Athanassiou, C.G.; Phillips, T.W.; Wakil, W. Biology and control of the khapra beetle, *Trogoderma granarium*, a major quarantine threat to global food security. *Annu. Rev. Entomol.* **2019**, *64*, 131–148. [CrossRef]

6. Blackman, R.L.; Eastop, V.F. *Aphids on the World's Crops*, 2nd ed.; John Wiley & Sons Ltd.: Chichester, UK, 2000; ISBN 978-0471851912.

7. Zehnder, C.B.; Hunter, M.D. A comparison of maternal effects and current environment on vital rates of *Aphis nerii*, the milkweed-oleander aphid. *Ecol. Entomol.* **2007**, *32*, 172–180. [CrossRef]

8. Colvin, S.M.; Yeargan, K.V. Predator fauna associated with oleander aphids on four milkweed species and the effect of those host plants on the development and fecundity of *Cycloneda munda* and *Harmonia axyridis*. *J. Kans. Entomol. Soc.* **2014**, *87*, 280–298. [CrossRef]

9. Gupta, G.; Kumar, N.R. Growth and development of ladybird beetle *Coccinella septempunctata* L. (Coleoptera: Coccinellidae), on plant and animal based protein diets. *J. Asia Pac. Entomol.* **2017**, *20*, 959–963. [CrossRef]

10. Cheng, Y.; Zhi, J.; Li, F.; Jin, J.; Zhou, Y. An artificial diet for continuous maintenance of *Coccinella septempunctata* adults (Coleoptera: Coccinellidae). *Biocontrol Sci. Technol.* **2018**, *28*, 242–252. [CrossRef]

11. Liu, T.; Wang, Y.; Zhang, L.; Xu, Y.; Zhang, Z.; Mu, W. Sublethal effects of four insecticides on the seven-spotted lady beetle (Coleoptera: Coccinellidae). *J. Econ. Entomol.* **2019**, *112*, 2177–2185.

12. Damalas, C.A.; Eleftherohorinos, I.G. Pesticide exposure, safety issues, and risk assessment indicators. *Int. J. Environ. Res. Public Health* **2011**, *8*, 1402–1419. [CrossRef] [PubMed]

13. Mulé, R.; Sabella, G.; Robba, L.; Manachini, B. Systematic review of the effects of chemical insecticides on four common butterfly families. *Front. Environ. Sci.* **2017**, *5*, 32. [CrossRef]

14. Zikankuba, V.L.; Mwanyika, G.; Ntwenya, J.E.; James, A. Pesticide regulations and their malpractice implications on food and environment safety. *Cogent Food Agric.* **2019**, *5*, 1601544. [CrossRef]

15. Isman, M.B.; Grieneisen, M.L. Botanical insecticide research: Many publications, limited useful data. *Trends Plant Sci.* **2014**, *19*, 140–145. [CrossRef] [PubMed]

16. Chau, D.T.M.; Chung, N.T.; Huong, L.T.; Hung, N.H.; Ogunwande, I.A.; Dai, D.N.; Setzer, W.N. Chemical compositions, mosquito larvicidal and antimicrobial activities of leaf essential oils of eleven species of Lauraceae from Vietnam. *Plants* **2020**, *9*, 606. [CrossRef]

17. Pavela, R.; Benelli, G. Essential oils as ecofriendly biopesticides? Challenges and constraints. *Trends Plant Sci.* **2016**, *21*, 1000–1007. [CrossRef]

18. Spochacz, M.; Chowański, S.; Walkowiak-Nowicka, K.; Szymczak, M.; Adamski, Z. Plant-derived substances used against beetles–Pests of stored crops and food–and their mode of action: A review. *Compr. Rev. Food Sci. Food Saf.* **2018**, *17*, 1339–1366. [CrossRef]

19. Alizadeh, A. Essential oil composition, phenolic content, antioxidant, and antimicrobial activity of cultivated *Satureja rechingeri* Jamzad at different phenological stages. *Zeitschrift für Naturforschung C* **2015**, *70*, 51–58. [CrossRef]

20. Sefidkon, F.; Jamzad, Z. Chemical composition of the essential oil of three Iranian *Satureja* species (*S. mutica, S. macrantha* and *S. intermedia*). *Food Chem.* **2005**, *91*, 1–4. [CrossRef]

21. Shahnazi, S.; Khalighi-Sigaroodi, F.; Ajani, Y.; Yazdani, D.; Taghizad-Farid, R.; Ahvazi, M.; Abdoli, M. The chemical composition and antimicrobial activity of essential oil of *Satureja intermedia* CA Mey. *J. Med. Plants* **2008**, *1*, 85–92.

22. Ghorbanpour, M.; Hadian, J.; Hatami, M.; Salehi-Arjomand, H.; Aliahmadi, A. Comparison of chemical compounds and antioxidant and antibacterial properties of various *Satureja* species growing wild in Iran. *J. Med. Plants* **2016**, *3*, 58–72.

23. Sadeghi, I.; Yousefzadi, M.; Behmanesh, M.; Sharifi, M.; Moradi, A. In vitro cytotoxic and antimicrobial activity of essential oil from *Satureja intermedia*. *Iran. Red Crescent Med. J.* **2013**, *15*, 70–74. [CrossRef] [PubMed]

24. Michaelakis, A.; Theotokatos, S.A.; Koliopoulos, G.; Chorianopoulos, N.G. Essential oils of *Satureja* species: Insecticidal effect on *Culex pipiens* larvae (Diptera: Culicidae). *Molecules* **2007**, *12*, 2567–2578. [CrossRef] [PubMed]

25. Tozlu, E.; Cakir, A.; Kordali, S.; Tozlu, G.; Ozer, H.; Akcin, T.A. Chemical compositions and insecticidal effects of essential oils isolated from *Achillea gypsicola*, *Satureja hortensis*, *Origanum acutidens* and *Hypericum scabrum* against broadbean weevil (*Bruchus dentipes*). *Sci. Hortic.* **2011**, *130*, 9–17. [CrossRef]

26. Magierowicz, K.; Górska-Drabik, E.; Sempruch, C. The insecticidal activity of *Satureja hortensis* essential oil and its active ingredient carvacrol against *Acrobasis advenella* (Zinck.) (Lepidoptera, Pyralidae). *Pestic. Biochem. Physiol.* **2019**, *153*, 122–128. [CrossRef]

27. Jamzad, Z. *Thymus and Satureja Species of Iran*, 1st ed.; Research Institute of Forests and Rangelands: Tehran, Iran, 2009.

28. Adams, R.P. *Identification of Essential Oil Components by Gas Chromatography/Mass Spectrometry*, 4th ed.; Allured Publishing: Carol Stream, IL, USA, 2007.

29. Mondello, L. *FFNSC 3*; Shimadzu Scientific Instruments: Columbia, MD, USA, 2016.

30. NIST. *NIST17*; National Institute of Standards and Technology: Gaithersburg, MD, USA, 2017.

31. Alkan, M. Chemical composition and insecticidal potential of different *Origanum* spp. (Lamiaceae) essential oils against four stored product pests. *Turk. Entomol. Derg.* **2020**, *44*, 149–163. [CrossRef]

32. Behi, F.; Bachrouch, O.; Boukhris-Bouhachem, S. Insecticidal activities of *Mentha pulegium* L., and *Pistacia lentiscus* L., essential oils against two citrus aphids *Aphis spiraecola* Patch and *Aphis gossypii* Glover. *J. Essent. Oil Bear. Plants* **2019**, *22*, 516–525. [CrossRef]

33. Ikbal, C.; Pavela, R. Essential oils as active ingredients of botanical insecticides against aphids. *J. Pest Sci.* **2019**, *92*, 971–986. [CrossRef]

34. Islam, R.; Khan, R.I.; Al-Reza, S.M.; Jeong, Y.T.; Song, C.H.; Khalequzzaman, M. Chemical composition and insecticidal properties of *Cinnamomum aromaticum* (Nees) essential oil against the stored product beetle *Callosobruchus maculatus* (F.). *J. Sci. Food Agric.* **2009**, *89*, 1241–1246. [CrossRef]

35. Ebadollahi, A.; Safaralizadeh, M.H.; Pourmirza, A.A.; Gheibi, S.A. Toxicity of essential oil of *Agastache foeniculum* (Pursh) Kuntze to *Oryzaephilus surinamensis* L. and *Lasioderma serricorne* F. *J. Plant Prot. Res.* **2010**, *50*, 215–219. [CrossRef]

36. Ebadollahi, A. Essential oil isolated from Iranian yarrow as a bio-rational agent to the management of saw-toothed grain beetle, *Oryzaephilus surinamensis* (L.). *Korean J. Appl. Entomol.* **2017**, *56*, 395–402.

37. Ebadollahi, A.; Safaralizadeh, M.H.; Pourmirza, A.A. Fumigant toxicity of essential oils of *Eucalyptus globulus* Labill and *Lavandula stoechas* L., grown in Iran, against the two coleopteran insect pests; *Lasioderma serricorne* F. and *Rhyzopertha dominica* F. *Egypt. J. Biol. Pest Control* **2010**, *20*, 1–5.

38. Ebadollahi, A. Fumigant toxicity and repellent effect of seed essential oil of celery against lesser grain borer, *Rhyzopertha dominica*. *J. Essent. Oil Bear. Plants* **2018**, *21*, 146–154. [CrossRef]

39. Ebadollahi, A.; Ashrafi Parchin, R.; Farjaminezhad, M. Phytochemistry, toxity and feeding inhibitory activity of *Melissa officinalis* L. essential oil against a cosmopolitan insect pest; *Tribolium castaneum* Herbst. *Toxin Rev.* **2016**, *35*, 77–82. [CrossRef]

40. Ebadollahi, A.; Razmjou, J. Chemical composition and toxicity of the essential oils of *Lippia citriodora* from two different locations against *Rhyzopertha dominica* and *Tribolium castaneum*. *Agric. For.* **2019**, *65*, 135–146. [CrossRef]

41. Ebadollahi, A.; Taghinezhad, E. Modeling and optimization of the insecticidal effects of *Teucrium polium* L. essential oil against red flour beetle (*Tribolium castaneum* Herbst) using response surface methodology. *Inf. Process. Agric.* **2019**, in press. [CrossRef]

42. Abdel-Sattar, E.; Zaitoun, A.A.; Farag, M.A.; El Gayed, S.H.; Harraz, F.M.H. Chemical composition, insecticidal and insect repellent activity of *Schinus molle* L. leaf and fruit essential oils against *Trogoderma granarium* and *Tribolium castaneum*. *Nat. Prod. Res.* **2010**, *24*, 226–235. [CrossRef]

43. Nouri-Ganbalani, G.; Borzoui, E. Acute toxicity and sublethal effects of *Artemisia sieberi* Besser on digestive physiology, cold tolerance and reproduction of *Trogoderma granarium* Everts (Col.: Dermestidae). *J. Asia Pac. Entomol.* **2017**, *20*, 285–292. [CrossRef]

44. Sefidkon, F.; Abbasi, K.; Khaniki, G.B. Influence of drying and extraction methods on yield and chemical composition of the essential oil of *Satureja hortensis*. *Food Chem.* **2006**, *99*, 19–23. [CrossRef]

45. Hadian, J.; Ebrahimi, S.N.; Salehi, P. Variability of morphological and phytochemical characteristics among *Satureja hortensis* L. accessions of Iran. *Ind. Crop. Prod.* **2010**, *32*, 62–69. [CrossRef]

46. Lee, B.-H.; Choi, W.-S.; Lee, S.-E.; Park, B.-S. Fumigant toxicity of essential oils and their constituent compounds towards the rice weevil, *Sitophilus oryzae* (L.). *Crop Prot.* **2001**, *20*, 317–320. [CrossRef]

47. Lee, B.-H.; Lee, S.-E.; Annis, P.C.; Pratt, S.J.; Park, B.-S.; Tumaalii, F. Fumigant toxicity of essential oils and monoterpenes against the red flour beetle, *Tribolium castaneum* Herbst. *J. Asia Pac. Entomol.* **2002**, *5*, 237–240. [CrossRef]

48. Yildirim, E.; Emsen, B.; Kordali, S. Insecticidal effects of monoterpenes on *Sitophilus zeamais* Motschulsky (Coleoptera: Curculionidae). *J. Appl. Bot. Food Qual.* **2013**, *86*, 198–204.

49. Saad, M.M.G.; Abou-Taleb, H.K.; Abdelgaleil, S.A.M. Insecticidal activities of monoterpenes and phenylpropenes against *Sitophilus oryzae* and their inhibitory effects on acetylcholinesterase and adenosine triphosphatases. *Appl. Entomol. Zool.* **2018**, *53*, 173–181. [CrossRef]

50. Liu, T.-T.; Chao, L.K.-P.; Hong, K.-S.; Huang, Y.-J.; Yang, T.-S. Composition and insecticidal activity of essential oil of *Bacopa caroliniana* and interactive effects of individual compounds on the activity. *Insects* **2020**, *11*, 23. [CrossRef]

51. Youssefi, M.R.; Tabari, M.A.; Esfandiari, A.; Kazemi, S.; Moghadamnia, A.A.; Sut, S.; Acqua, S.D.; Benelli, G.; Maggi, F. Efficacy of two monoterpenoids, carvacrol and thymol, and their combinations against eggs and larvae of the West Nile vector *Culex pipiens*. *Molecules* **2019**, *24*, 1867. [CrossRef]

52. Lu, X.; Weng, H.; Li, C.; He, J.; Zhang, X.; Ma, Z. Efficacy of essential oil from *Mosla chinensis* Maxim. cv. Jiangxiangru and its three main components against insect pests. *Ind. Crop. Prod.* **2020**, *147*, 112237. [CrossRef]

53. Russo, S.; Yaber Grass, M.A.; Fontana, H.C.; Leonelli, E. Insecticidal activity of essential oil from *Eucalyptus globulus* against *Aphis nerii* (Boyer) and *Gynaikothrips ficorum* (Marchal). *AgriScientia* **2018**, *35*, 63–67. [CrossRef]

54. Abramson, C.I.; Wanderley, P.A.; Wanderley, M.J.A.; Miná, A.J.S.; de Souza, O.B. Effect of essential oil from citronella and alfazema on fennel aphids *Hyadaphis foeniculi* Passerini (Hemiptera: Aphididae) and its predator *Cycloneda sanguinea* L. (Coleoptera: Coccinelidae). *Am. J. Environ. Sci.* **2007**, *3*, 9–10. [CrossRef]

55. Kimbaris, A.C.; Papachristos, D.P.; Michaelakis, A.; Martinou, A.F.; Polissiou, M.G. Toxicity of plant essential oil vapours to aphid pests and their coccinellid predators. *Biocontrol Sci. Technol.* **2010**, *20*, 411–422. [CrossRef]

56. Faraji, N.; Seraj, A.A.; Yarahmadi, F.; Rajabpour, A. Contact and fumigant toxicity of *Foeniculum vulgare* and *Citrus limon* essential oils against *Tetranychus turkestani* and its predator *Orius albidipennis*. *J. Crop Prot.* **2016**, *5*, 283–292. [CrossRef]

57. Ribeiro, N.; Camara, C.; Ramos, C. Toxicity of essential oils of *Piper marginatum* Jacq. against *Tetranychus urticae* Koch and *Neoseiulus californicus* (McGregor). *Chil. J. Agric. Res.* **2016**, *76*, 71–76. [CrossRef]

58. Seixas, P.T.L.; Demuner, A.J.; Alvarenga, E.S.; Barbosa, L.C.A.; Marques, A.; Farias, E.D.S.; Picanço, M.C. Bioactivity of essential oils from *Artemisia* against *Diaphania hyalinata* and its selectivity to beneficial insects. *Sci. Agric.* **2018**, *75*, 519–525. [CrossRef]

59. Titouhi, F.; Amri, M.; Messaoud, C.; Haouel, S.; Youssfi, S.; Cherif, A.; Mediouni Ben Jemâa, J. Protective effects of three *Artemisia* essential oils against *Callosobruchus maculatus* and *Bruchus rufimanus* (Coleoptera: Chrysomelidae) and the extended side-effects on their natural enemies. *J. Stored Prod. Res.* **2017**, *72*, 11–20. [CrossRef]

60. Haouel-Hamdi, S.; Abdelkader, N. Combined use of *Eucalyptus salmonophloia* essential oils and the parasitoid *Dinarmus basalis* for the control of the cowpea seed beetle *Callosobruchus maculatus*. Tunis. J. Plant Prot. **2018**, *13*, 123–137.
61. De Souza, M.T.; de Souza, M.T.; Bernardi, D.; Krinski, D.; de Melo, D.J.; Oliveira, D.C.; Rakes, M.G.; Zarbin, P.H.; de Noronha Sales Maia, B.H.L.; Zawadneak, M.A.C. Chemical composition of essential oils of selected species of *Piper* and their insecticidal activity against *Drosophila suzukii* and *Trichopria anastrephae*. Environ. Sci. Pollut. Res. **2020**, *27*, 13056–13065. [CrossRef]

Article

Common Plant-Derived Terpenoids Present Increased Anti-Biofilm Potential against *Staphylococcus* Bacteria Compared to a Quaternary Ammonium Biocide

Dimitra Kostoglou, Ioannis Protopappas and Efstathios Giaouris *

Laboratory of Biology, Microbiology and Biotechnology of Foods, Department of Food Science and Nutrition, School of the Environment, University of the Aegean, GR-81 400 Myrina, Lemnos, Greece; dimitra_kostoglou@outlook.com.gr (D.K.); jproto2009@yahoo.gr (I.P.)
* Correspondence: stagiaouris@aegean.gr; Tel.: +30-22540-83115

Received: 7 May 2020; Accepted: 20 May 2020; Published: 1 June 2020

Abstract: The antimicrobial actions of three common plant-derived terpenoids (i.e., carvacrol, thymol and eugenol) were compared to those of a typical quaternary ammonium biocide (i.e., benzalkonium chloride; BAC), against both planktonic and biofilm cells of two widespread *Staphylococcus* species (i.e., *S. aureus* and *S. epidermidis*). The minimum inhibitory and bactericidal concentrations (MICs, MBCs) of each compound against the planktonic cells of each species were initially determined, together with their minimum biofilm eradication concentrations (MBECs). Various concentrations of each compound were subsequently applied, for 6 min, against each type of cell, and survivors were enumerated by agar plating to calculate log reductions and determine the resistance coefficients (Rc) for each compound, as anti-biofilm effectiveness indicators. Sessile communities were always more resistant than planktonic ones, depending on the biocide and species. Although lower BAC concentrations were always needed to kill a specified population of either cell type compared to the terpenoids, for the latter, the required increases in their concentrations, to be equally effective against the biofilm cells with respect to the planktonic ones, were not as intense as those observed in the case of BAC, presenting thus significantly lower Rc. This indicates their significant anti-biofilm potential and advocate for their further promising use as anti-biofilm agents.

Keywords: *Staphylococcus aureus*; *S. epidermidis*; carvacrol; thymol; eugenol; benzalkonium chloride; biofilms; planktonic; disinfection; natural products

---

1. Introduction

Staphylococcus aureus is a common facultative anaerobic Gram-positive bacterial pathogen associated with a wide spectrum of minor to serious community and hospital-acquired infections. This non-motile, catalase and coagulase positive coccus is equipped with a tremendous range of virulence factors which allow its survival within the living host [1]. In addition, its ability to produce various heat stable enterotoxins in foodstuffs, makes staphylococcal foodborne intoxication one of the most common foodborne diseases worldwide [2]. Foods are usually contaminated through infected food handlers (via manual contact or their respiratory tract activity), while animal origin contamination is also frequent in products such as raw milk and cheeses [3]. *S. epidermidis* is usually a harmless commensal bacterium highly abundant on the human skin playing an important role in balancing the normal microflora. Nevertheless, this can still switch to an invasive lifestyle under certain predetermined conditions. Compared to *S. aureus*, this has, however, a more limited repertoire of virulence factors resulting in lower pathogenicity [1]. Nevertheless, this has still emerged as the most frequent cause of nosocomial infections primarily in patients with indwelling medical devices [4].

Both of these two species display a great ability to attach to various surfaces and create robust biofilms [5,6]. These surface-attached aggregated microbial communities are surrounded by a self-produced matrix of extracellular polymeric substances (EPS), allowing them to cope with many stresses and survive in inhospitable environments [7]. Indeed, biofilm formation is one of the most critical features that contributes to the success of these bacteria and is considered essential for the emergence of their pathogenesis and persistence [8]. Inside a biofilm, *Staphylococcus* bacteria (as well as other microbial human pathogens) can evade the host immune system and are in parallel protected against antibiotic treatment, making infections hard to eradicate [9]. In addition, pathogenic biofilms, formed on abiotic food-contact surfaces encountered within the food industry, including those being created by/containing staphylococci, allow embedded microorganisms to withstand killing action of common sanitizers, used at their recommended or even much higher concentrations, resulting in survival, cross contamination (through the ultimate dispersal of the remaining viable cells) and diseases transmission [10].

Therefore, there is currently an urgent demand to develop alternatives to conventional treatments (such as antibiotics and chemical sanitizers) to control unwanted biofilms in both healthcare and industrial environments [11]. In addition, due to the potential hazards of several synthetic biocides for both public health and the environment, novel eco-friendly approaches are nowadays preferred [12]. In this respect, numerous plant extracts and phytochemicals have been successfully evaluated as anti-biofilm agents in different model systems [13,14]. Besides their green status, these may present different modes of action from classical biocides, making them more efficient and probably helping to overcome the problem of resistance [15]. For instance, some phytocompounds have even been found to be capable of inhibiting biofilm formation in much lower concentrations than those required to inhibit planktonic growth, mainly through their interference with quorum sensing (QS) signaling pathways, something that seems to reduce the selective pressure exerted on the target microorganisms, in comparison with other antimicrobials, such as the antibiotics [16,17].

Carvacrol (CAR), thymol (THY) and eugenol (EUG) are natural terpenoids included in the most bioactive phytochemicals isolated from essential oils (EOs), all well-recognized for their wide spectrum of antimicrobial action, mainly due to their considerable deleterious actions on the cytoplasmic membranes [18]. Thus, CAR and THY are the main components occurring in EOs isolated from plants of the Lamiaceae family (e.g., oregano, thyme), which are commonly used as flavouring and preservative agents by the food industry processors, in commercial mosquito repellents, in aromatherapy, and in traditional medicine [19,20]. On the other hand, EUG is found in high concentrations in the EO of clove and has till now been applied in the agricultural, food, cosmetic and pharmaceutical industries [21]. All three of these plant metabolites are authorized as food flavourings across Europe [22], while EUG is also a permitted food additive by the U.S. Food and Drug Administration [23].

Benzalkonium chloride (BAC) is a synthetic quaternary ammonium compound (QAC) widely used as preservative, sanitizer and surface disinfectant in households, healthcare, agricultural and industrial settings, due to its broad antimicrobial spectrum against bacteria, fungi, and viruses [24]. In general, QACs, including BAC, exert their action by disrupting the bilayer and charge distribution of the cellular membranes, through the alkyl chains and charged nitrogen these are containing, respectively. Alarmingly, long-term low-dose microbial exposure to BAC might confer selective pressure and results in increased resistance both towards this compound, as well as other distinct chemicals, such as clinically relevant antibiotics, through cross-resistance mechanisms [25–27]. These last include changes in membrane composition, overexpression or modification of efflux pumps, downregulation of porins, horizontal transfer of stress response genes, biodegradation, and biofilm formation [24]. Not surprisingly, BAC-resistant staphylococci have been isolated from a variety of (seemingly distant) samples, such as environmental, hospital-acquired, animals, and foods, with several QAC resistance genes to have till now been identified, mainly and alarmingly easily transferable plasmid-borne ones encoding for efflux proteins [28–31]. Besides this great antimicrobial resistance problem, safety concerns regarding the use of BAC have also been emerged [32], with some countries to have already prohibited its use for some applications [24].

Considering all the above, it is evident that new antimicrobial agents that will be safe, cost-effective and in parallel exhibit as low as possible possibilities for resistance development are urgently required, especially to get rid of the most resistant biofilm-enclosed pathogenic microorganisms. For the effective development and application of such novel agents, it is, however, important to have previously compared their efficiency with the classically applied ones. Although several studies have been published in recent years related to the anti-biofilm action of many plant compounds, including CAR, THY and EUG, against various bacteria [33–37], including staphylococci [38–41], very few of them have compared their actions with those of standard chemical antimicrobials [42–46]. In addition, and to the best of our knowledge, no other study has been published comparing in parallel the efficiency of these three common plant-derived terpenoids (i.e., CAR, THY, and EUG) and of BAC against both planktonic and biofilm *Staphylococcus* bacteria or of other species.

Thus, the main objective of the present study was to compare the disinfection efficiencies of all these compounds (i.e., CAR, THY, EUG, and BAC) against both planktonic and biofilm cells of both *S. aureus* and *S. epidermidis*. For this, the minimum inhibitory and bactericidal concentrations (MICs, MBCs) of each compound against the planktonic cells of each bacterial species were initially determined, together with their minimum biofilm eradication concentrations (MBECs), by applying standard protocols for these purposes. Subsequently, both planktonic and biofilm cells of each species were exposed for 6 min to various concentrations ($n = 3$–4) of each compound, based on the previous determination of MBCs and MBECs, and the remaining viable cells were then enumerated by agar plating to calculate log reductions for each compound and at each tested concentration. This last made it possible to create the linear regression plots correlating these two parameters (log reductions vs. concentrations). These plots (for each compound, bacterial species, and cell type; $n = 16$) were finally used to accurately determine the resistance coefficients (Rc) of each compound against the biofilm cells of each species compared to its planktonic ones, as indicators for its anti-biofilm effectiveness. Results revealed the significant anti-biofilm potential of all three natural terpenoids (i.e., CAR, THY, and EUG) over the synthetic biocide (i.e., BAC), advocating for their further promising exploitation as anti-biofilm agents.

2. Materials and Methods

2.1. Chemicals and Stock Solutions

Carvacrol (CAR), thymol (THY), eugenol (EUG) and benzalkonium chloride (BAC) were purchased from Sigma-Aldrich (liquid, ≥98%, molar mass: 150.22 g/mol, density: 0.976 g/mL; product code: W224502), Penta Chemicals (powder, >99.0%, molar mass: 150.22 g/mol; product code: 27450-30100), Alfa Aesar (liquid, ≥98.5%, molar mass: 164.21 g/mol, density: 1.068 g/mL; product code: A14332), and Acros Organics (liquid, alkyl distribution from C8H17 to C16H33, density: 0.98 g/mL; product code: 215411000), respectively. With respect to the terpenoids (i.e., CAR, THY, and EUG), two stock solutions for each one were prepared in absolute ethanol at 10% and 40% (*v/v* for CAR and EUG; *w/v* for THY), for subsequent use against planktonic and biofilm cells, respectively, following appropriate dilutions (see below), while the stock solution of BAC (1% *v/v*) was prepared in sterile distilled water. All stock solutions were maintained at −20 °C for up to 1 month. The chemical formulas of the four tested compounds are presented in Figure 1, while Table 1 summarizes their main physical and chemical properties, together with the correlations in the concentrations (for each compound) expressed in either as ppm or molarity (M), using the 0.1% (*v/v* or *w/v*) as a reference concentration.

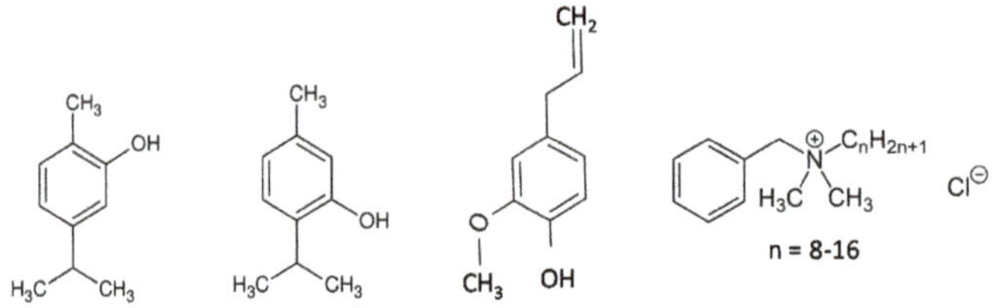

carvacrol   thymol   eugenol   benzalkonium chloride

Figure 1. Chemical formulas of the four tested compounds.

Table 1. Main physical and chemical properties of the four tested compounds, together with the correlations in concentrations (for each compound) expressed in either as ppm or molarity (M), using the 0.1% (*v/v* or *w/v*) as a reference concentration.

Compound	Physical Form (20 °C)	Molar Mass (g/mol)	Density (g/mL)	Concentration		
				%	ppm	M (mol/L)
carvacrol	liquid	150.22	0.976	0.1 (*v/v*)	1000	0.00650
thymol	powder	150.22	unknown [1]	0.1 (*w/v*)	1000	0.00666
eugenol	liquid	164.21	1.068	0.1 (*v/v*)	1000	0.00650
BAC	liquid	unknown [2]	0.98	0.1 (*v/v*)	1000	unknown [2]

[1] Not provided by the manufacturer; [2] BAC was provided a mixture of QACs with different lengths for the alkyl chain (ranging from C8 to C16).

2.2. Bacterial Strains and Preparation of the Working Saline Suspensions

The two bacterial strains used in this research were the *S. aureus* DFSN_B26, isolated in our lab from non-pasteurized milk cheese and the *S. epidermidis* DFSN_B4 (C5M6), originally isolated from fermenting grape juice and kindly provided by Professor G.-J. Nychas (Agricultural University of Athens, Greece). Before their use in the subsequent experiments, both strains were stored frozen (at −80 °C) in Tryptone Soy Broth (TSB; Lab M, Heywood, Lancashire, UK) containing 15% glycerol in cryovials and was then each one revivified by streaking a loopful of its frozen culture on to the surface of Tryptone Soy Agar (TSA; Lab M) and incubating at 37 °C for 24 h (precultures). Working cultures were prepared by inoculating, using a microbiological loop, cells of a district and well isolated colony from each preculture into 10 mL of fresh TSB and incubating at 37 °C for 18 h. Bacteria from each final working culture were collected by centrifugation (4000× *g* for 10 min at RT), washed twice with quarter-strength Ringer's solution (Lab M), and finally suspended in the same solution, so as to display an absorbance at 600 nm (A$_{600\,nm}$) equal to 0.1 (*ca.* 10^7 CFU/mL).

2.3. Determination of Minimum Inhibitory and Bactericidal Concentrations (MIC, MBC) of Each Compound against Planktonic Bacteria

The MIC of each compound (i.e., CAR, THY, EUG, and BAC) against the planktonic cells of each *Staphylococcus* species was determined using the broth microdilution method, as previously described [46]. Briefly, on the day of application, ten different concentrations for each compound were prepared by appropriately diluting its stock solution (i.e., 10% and 1%, for terpenoids and BAC, respectively) in fresh TSB. For terpenoids, the tested concentrations ranged from 19.5 to 10,000 ppm (two-fold dilutions), while for BAC those ranged from 1 to 10 ppm. Subsequently, 180 μL of each dilution were transferred to a well (in duplicate) of a sterile flat-bottomed 96-well polystyrene (PS) microtiter plate (transparent, hydrophobic, Ref 655101; Greiner bio-one GmbH, Frickenhausen, Germany) and 20 μL of a 10-fold dilution of the appropriate bacterial suspension (A$_{600\,nm}$ = 0.1) in quarter-strength Ringer's solution were then added, so as to have an initial bacterial concentration in each well of ca. 10^5 CFU/mL. Wells without bacteria and wells without any added compound served as negative and positive growth controls (for bacterial growth), respectively. The plates were sealed with parafilm and statically incubated at 37 °C for 24 h. The growth in each well was finally turbidimetrically assessed by naked eye observation and confirmed by measuring absorbances at 620 nm using a computer-controlled microplate reader (Halo Led 96; Dynamica Scientific Ltd., Livingston, UK). The MIC value was considered as the lowest concentration of each compound that totally inhibited the visible bacterial growth. To calculate MBCs, from all the wells showing no visible growth, 10 µL were aspirated and spotted on TSA and the number of colonies was counted following incubation at 37 °C for 48 h. MBC for each compound was defined as its lowest concentration, reducing the initial inoculum by at least three logs (i.e., no appearance of colonies).

2.4. Determination of Minimum Biofilm Eradication Concentration (MBEC) of Each Compound against Biofilm Bacteria

The MBEC of each compound (i.e., CAR, THY, EUG, and BAC) against the biofilm cells of each *Staphylococcus* species was determined following a previously described protocol, with some modifications [47]. Briefly, 200 µL of each bacterial suspension ($A_{600 \text{ nm}}$ = 0.1) were transferred into a well (in quadruplicate) of a sterile 96-well PS microtiter plate, and the plate was then statically incubated at 37 °C for 2 h, in order to allow bacteria to adhere to its surface. Following this adhesion step, the planktonic bacterial suspension was removed from each well, this was then washed with quarter-strength Ringer's solution (to remove the loosely attached cells), and 200 µL of TSB containing 5% NaCl were added. The plate was then statically incubated at 37 °C for 48 h to allow biofilm growth. Following biofilm formation, the planktonic suspensions were removed, and each well was twice washed with quarter-strength Ringer's solution (to remove the loosely attached cells). 200 µL of the appropriate antimicrobial solution were then added and left in contact for 6 min at 20 °C. Each compound was tested in five different concentrations, ranging from 8 to 128 × MBC (two-fold dilutions), which were all prepared in sterile distilled water starting from each stock solution (i.e., 40% and 1% for terpenoids and BAC, respectively). Sterile distilled water (also containing 6% *v/v* ethanol when CAR/THY were tested, or 24% *v/v* ethanol when EUG was tested) was used as the negative disinfection control. Those ethanol concentrations were included in the negative controls since were the maximum ones existing in the highest tested concentration for the terpenoids (i.e., 128 × MBC). Following disinfection, the antimicrobial solution was carefully removed from each well and this was then washed with quarter-strength Ringer's solution, to remove any disinfectant residues. Subsequently, 200 µL of quarter-strength Ringer's solution were added, and the strongly attached/biofilm bacteria were removed from the PS surface by thoroughly scratching with a plastic pipette tip, vortexed, serially diluted and finally enumerated by counting colonies on spot inoculated (10 µL) TSA plates following their incubation at 37 °C for 48 h. The MBEC for each compound was determined as its lowest concentration reducing biofilm cells by at least five logs (i.e., no appearance of colonies) with respect to the negative disinfection control.

2.5. Disinfection of Planktonic Bacteria

The disinfection of planktonic bacteria was carried out as previously described [46]. Briefly, 1 mL of each bacterial suspension ($A_{600 \text{ nm}}$ = 0.1) was centrifuged at 5000× *g* for 10 min at 20 °C, supernatant was discarded, and each pellet (ca. 10^7 cells) was then suspended in 1 mL of the appropriate antimicrobial solution and left in contact for 6 min at 20 °C. Four different concentrations for each compound were tested (based on the previous MBC determination) and were all prepared in sterile distilled water by appropriately diluting its stock solution (i.e., 10% and 1% for terpenoids and BAC, respectively). Following disinfection, the antimicrobial action was interrupted by transferring a volume (1:9) to Dey-Engley neutralizing broth (Lab M) and leaving there for 10 min at 20 °C. Serial decimal dilutions were then prepared in quarter-strength Ringer's solution, TSA plates were spot inoculated (10 µL) and colonies were counted following incubation at 37 °C for 48 h. Sterile distilled water (also containing 2.25% *v/v* ethanol when the terpenoids were tested) was used as the negative disinfection control. This ethanol concentration was included in the negative control since this was the maximum one with the highest preliminary tested concentrations for the terpenoids (i.e., 2500 ppm). For each compound and tested concentration, the logarithmic reduction ($\log_{10}$ CFU/mL) of cells following disinfection was calculated by subtracting the $\log_{10}$ of the survivors from that counted following disinfection with water (negative control).

2.6. Disinfection of Biofilm Bacteria

The disinfection of biofilm bacteria was carried out as previously described for the determination of the MBECs (Section 2.4), but this time, each terpenoid was tested in three different concentrations, while BAC was applied at four different concentrations (based on the previous MBEC determination). All these concentrations were lower than the MBECs, since the aim of this specific disinfection protocol was not to completely kill the cells, but to leave survivors for calculating log reductions at each tested concentration, so as to later be able to accurately calculate the resistance coefficients for each compound (Section 2.7). Sterile distilled water (also containing 0.4% *v/v* ethanol when the terpenoids were tested) was used as the negative disinfection control. This ethanol concentration was included in the negative control since was the maximum one existing in the highest tested concentration for the terpenoids (i.e., 2500 ppm). Survivors were again enumerated by counting colonies on spot inoculated (10 µL) TSA plates, while plate counts were converted to $\log_{10}$ CFU/cm^2 before the calculation of log reductions ($\log_{10}$ CFU/cm^2).

2.7. Calculation of Resistance Coefficients (Rc) of Each Compound against Biofilm Cells Compared to Planktonic Ones

To compare the antimicrobial action of each compound between the two cell types (i.e., planktonic, biofilm), its resistance coefficient was determined as the ratio of concentrations (Rc) required to achieve the same log reductions in both populations ($C_{biofilm}/C_{planktonic}$) [48]. Thus, for instance, a Rc equal to 10 means that a ten-fold more concentrated compound is needed to kill the same level of biofilm cells as planktonic. To accurately calculate Rc for each compound and against each bacterial species, a linear regression plot (standard curve) was constructed by plotting the log reductions achieved (for each cell type) at each tested compound's concentration (based on the results of disinfection protocols presented in Sections 2.5 and 2.6). The mathematical equations of each regression plot ($y = a \cdot x + b$; 16 equations in total i.e., 4 compounds $\times$ 2 bacterial species $\times$ 2 cell types; Figures 2 and 3) were then used to calculate those concentrations required (x) to achieve prespecified log reductions (y). For this, at least 100 different log reduction values were considered for each linear regression equation (based on the total range of those covered by each standard curve). For each of those calculated log reduction—concentration combinations between the two cell types, the Rc value was obtained (by dividing the concentrations corresponding to the same log reduction: $C_{biofilm}/C_{planktonic}$) and finally the average Rc was determined for each compound and bacterial species. All calculations were done using the Excel® module of the Microsoft® Office 365 suite (Redmond, Washington, DC, USA).

2.8. Statistics

Each experiment was repeated at least three times using independent bacterial cultures. Plate counts were always transformed to logarithms before means and standard deviations were computed. All the disinfection data obtained for each compound (i.e., CAR, THY, EUG, and BAC), tested concentration (ppm), bacterial species (i.e., *S. aureus*, *S. epidermidis*), and cell type (i.e., planktonic, biofilm) were analysed by analysis of variance (ANOVA) to check for any significant effects of compound's type, concentration and bacterial species on disinfection efficiency (expressed as log reduction), using the statistical software STATISTICA® (StatSoft Inc.; Tulsa, OK 74104, USA). Following this analysis, least square means of log reductions were separated by Fisher's least significant difference (LSD) test. The same test was also used to check for significant differences between the Rc values for each compound and bacterial species. Pearson correlation analysis was also applied to determine the significance of the correlations between log reductions ($\log_{10}$ CFU/mL or cm^2) and tested concentrations (ppm) for each compound, bacterial species and cell type. All differences are reported at a significance level of 0.05.

3. Results

3.1. Determination of MICs, MBCs and MBECs of Each Compound

The MICs, MBCs and MBECs of each compound against each bacterial species are presented in Table 2. Thus, both CAR and THY presented an MIC against both species equal to 156.3 ppm, while eugenol was four times less efficient, presenting an MIC against both species equal to 625 ppm. As expected, BAC was capable of inhibiting bacterial growth at much lower concentrations, presenting an MIC against both species equal to just 3 ppm. At all cases, MBCs were two times more the respective MICs, confirming the bactericidal nature of all the compounds. With respect to the efficiency of the terpenoids (i.e., CAR, THY, and EUG) against the biofilm cells, someone observes that the MBECs against *S. aureus* were always two-fold lower compared to those observed against *S. epidermidis*, something that implies that *S. aureus* biofilm was less hard to eradicate using those compounds compared to that formed by *S. epidermidis*. On the contrary, the MBEC of BAC against *S. aureus* was two times more than that observed against *S. epidermidis*, indicating that *S. aureus* biofilm was less susceptible to BAC compared to *S. epidermidis* one. Similarly, to the antimicrobial efficiencies of each compound against the planktonic cells, BAC was again the most effective compound also against the biofilm cells, followed by CAR and THY (both these terpenoids present equal MBECs), while EUG was the least effective, needed for both species to be used in the highest concentration to eradicate their biofilm cells. However, it should be noted that the required increases in the compounds' concentrations to be able to eradicate biofilm cells with regard the planktonic ones were always much lower for the terpenoids compared to BAC and for both bacterial species. This indicates that although terpenoids were always needed to be used at higher concentrations compared to BAC to kill the cells (either planktonic or biofilm), these still presented a better efficiency for destroying the biofilm cells than BAC when considering their "inherent" antimicrobial efficiencies against the planktonic bacteria. This last was more evident for EUG, than for the other two terpenoids (i.e., CAR, THY). Thus, EUG was capable of eradicating *S. aureus* biofilm population at just eight times more than its MBC (i.e., 10,000 ppm), whereas for the same to happen, BAC was needed to be used at 128 times more than its MBC (i.e., 768 ppm).

Table 2. MICs, MBCs and MBECs of each compound against each bacterial species.

Compound	MIC [1]		MBC [1]		MBEC [1]	
	S. aureus	*S. epidermidis*	*S. aureus*	*S. epidermidis*	*S. aureus*	*S. epidermidis*
carvacrol	156.3	156.3	312.5 (2 × MIC)	312.5 (2 × MIC)	5000 (16 × MBC)	10,000 (32 × MBC)
thymol	156.3	156.3	312.5 (2 × MIC)	312.5 (2 × MIC)	5000 (16 × MBC)	10,000 (32 × MBC)
eugenol	625	625	1250 (2 × MIC)	1250 (2 × MIC)	10,000 (8 × MBC)	20,000 (16 × MBC)
BAC	3	3	6 (2 × MIC)	6 (2 × MIC)	768 (128 × MBC)	384 (64 × MBC)

[1] All concentrations are expressed as ppm (1000 ppm = 0.1% *v/v*).

3.2. Comparative Evaluation of Disinfection Efficiencies of Each Compound against Planktonic and Biofilm Bacteria

The log reductions of planktonic ($\log_{10}$ CFU/mL) and biofilm ($\log_{10}$ CFU/cm^2) cells of each species, following the 6 min exposure to each compound (i.e., CAR, THY, EUG, and BAC) being applied at different concentrations (ppm) are presented in Figures 2 and 3, respectively. By observing these results, the following general remarks can be formulated. Firstly, log reductions always increased as the compounds' concentrations increased. This means that more cells died when increasing a compound's concentration; something that was rather expected (at least for the planktonic populations). However, it is worth noting that under the range of concentrations tested, the killing rates increased significantly faster for planktonic cells than for biofilm ones, highlighting the greater recalcitrance of the later. This is also clear when observing the concentrations needed for each compound to kill the same level of biofilm cells as planktonic. For instance, to kill 99% of planktonic *S. epidermidis* cells (i.e., to cause a 2-logs reduction), 20 ppm of BAC were enough (Figure 2), whereas this compound needed to be applied at 200 ppm (i.e., ten-fold more highly concentrated) to kill the same number of biofilm cells (Figure 3). Similarly, thymol at 450 ppm reduced planktonic *S. aureus* population by 99.9% (i.e., 3 logs), while this needed to be applied at 2500 ppm (i.e., more than five times more) to kill the same level of biofilm cells. Secondly, and in accordance to MBC and MBEC results previously presented (Table 2), EUG was the least effective compound, whereas BAC was the most effective one for both species and cell types (i.e., planktonic, biofilm). Thus, for instance, 1450 ppm of EUG were needed to reduce planktonic *S. aureus* population by 4 logs, whereas 30 ppm of BAC were enough for the same effect (Figure 2). Similarly, biofilm population of the same species was reduced by 1.5 log upon applying 200 ppm of BAC, whereas for the same log reduction EUG needed to be applied at ten times higher concentration (2000 ppm) (Figure 3). Thirdly, the resistance of biofilm cells seems to be significantly influenced by the forming species and compound tested. Thus, *S. epidermidis* biofilm was always more resistant (i.e., presenting lower log reductions) to both THY and EUG compared to the *S. aureus* one. However, the opposite occurred when these biofilms were exposed to BAC, with *S. aureus* always presenting lower log reductions than *S. epidermidis*. This last observation is in full accordance with the MBEC results previously presented (Table 2).

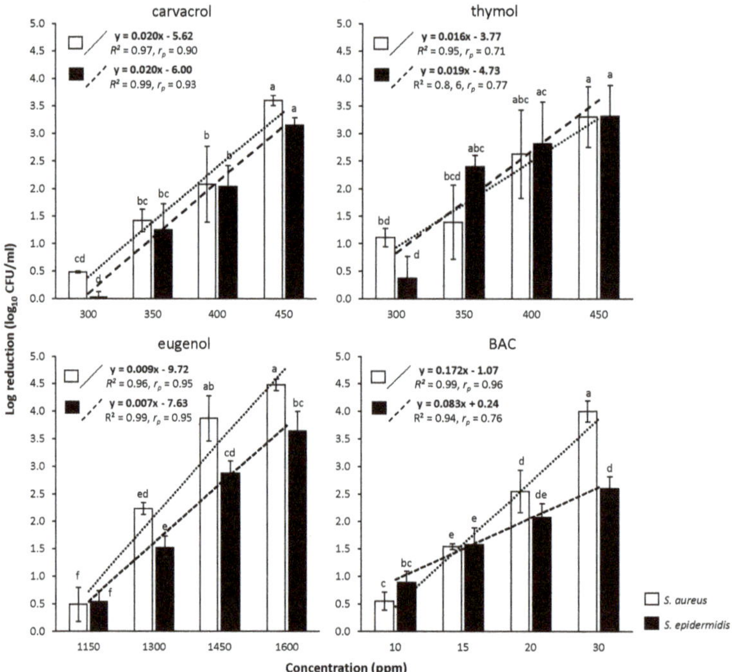

Figure 2. Log reductions (log$_{10}$ CFU/mL) of planktonic cells for each bacterial species (□ *S. aureus*; ■ *S. epidermidis*) following 6 min exposure to each compound (i.e., CAR, THY, EUG, and BAC) applied at four different concentrations (ppm). The bars represent the mean values ± standard deviations. For each separate graph, mean values sharing at least one common letter shown above the bars are not significantly different ($p > 0.05$). Dotted lines illustrate linear regression correlations between the log reductions achieved (for each species) at each tested compound's concentration. The mathematical equations of these regression plots, together with their regression coefficients (R^2) and Pearson's correlation coefficients (r_p), are also shown.

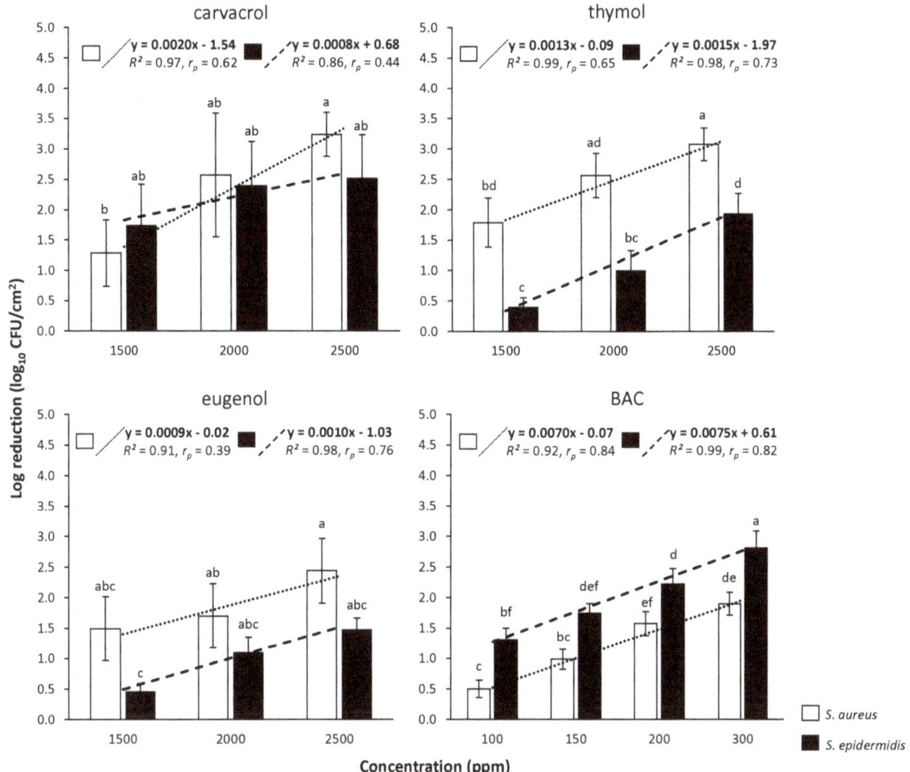

Figure 3. Log reductions ($\log_{10}$ CFU/cm^2) of biofilm cells for each bacterial species (□ *S. aureus*; ■ *S. epidermidis*) following 6 min exposure to each compound (i.e., CAR, THY, EUG, and BAC) applied at different concentrations (ppm). The bars represent the mean values ± standard deviations. For each separate graph, mean values sharing at least one common letter shown above the bars are not significantly different ($p > 0.05$). Dotted lines illustrate linear regression correlations between the log reductions achieved (for each species) at each tested compound's concentration. The mathematical equations of these regression plots, together with their regression coefficients (R^2) and Pearson's correlation coefficients (r_p) are also shown.

To accurately compare and easily perceive the efficiency of each compound against each cell type (i.e., planktonic vs. biofilm), its resistance coefficient (Rc) was determined, based on the results of log reductions for each cell type following disinfection and the respective regression plots (Figures 2 and 3). The calculated Rc values are presented in Figure 4. Thus, the quaternary ammonium compound BAC was found to exhibit the highest Rc values equal to 13.6 and 8.5 against *S. aureus* and *S. epidermidis*, respectively. This means that this compound needed to be applied at concentrations 13.6 and 8.5 times higher to kill the same numbers of biofilm cells as the planktonic ones. On the contrary, EUG exhibited the lowest Rc values (i.e., 1.6 against both bacterial species), highlight its almost similar efficiency against both cell types. The other two terpenoids (i.e., CAR, THY) presented Rc values near to 4 (with some minor differences between them and depending on the bacterial species), meaning that these needed to be applied in concentrations approximately four times greater against biofilm cells to achieve similar log reductions with respect to planktonic ones. This remarkable potential of all three terpenoids against the biofilm cells was also previously noticed upon presenting the MBEC results (Table 2).

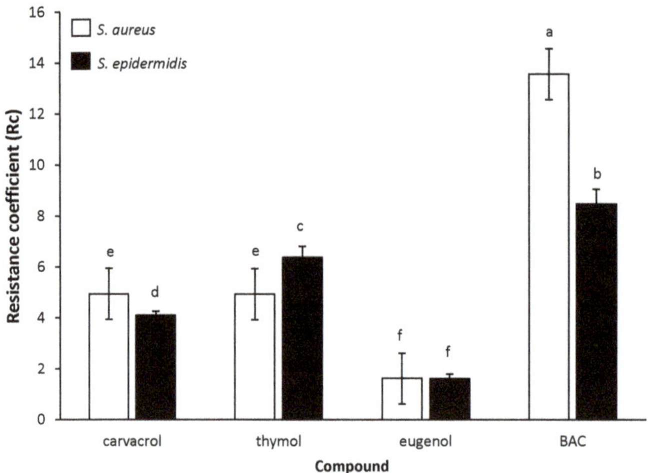

Figure 4. Resistance coefficients (Rc) of each compound for each bacterial species (□ *S. aureus*; ■ *S. epidermidis*). The bars represent the mean values ± standard deviations. Mean values sharing at least one common letter shown above the bars are not significantly different ($p > 0.05$).

4. Discussion

To comparatively evaluate the disinfection efficiencies of each compound (i.e., CAR, THY, EUG, and BAC) against each cell type (i.e., planktonic, biofilm) and for each bacterial species (i.e., *S. aureus*, *S. epidermidis*), their MICs, MBCs and MBECs were initially determined following some standard protocols (Table 2). It was revealed that for both cell types and species, the synthetic biocide BAC was quite a bit more efficient than the three plant-derived terpenoids, presenting the lowest MICs, MBCs and MBECs. The identical MIC value for both staphylococci (i.e., 3 ppm) reveals their intermediate planktonic resistance, according to the Clinical and Laboratory Standards Institute guidelines [49], which define staphylococci as being resistant to BAC upon presenting an MIC greater than 3 ppm. On the contrary, the least effective compound was EUG, presenting the highest MICs, MBCs and MBECs. Compared to those, CAR and THY displayed intermediate and equal efficiencies. These results were rather expected based on the rich available literature concerning the antimicrobial actions of these compounds. Thus, like our results, the MIC of EUG was found to be four times greater than that of CAR against an *S. aureus* strain (1000 and 250 ppm, respectively), previously determined with a broth liquid method where sterile filter papers impregnated with each compound had been placed into inoculated broth tube cultures [50]. The slight differences between those MIC values and ours could just be due to the different strain and method followed to determine these values.

More generally, the lower efficiency of EUG compared to either CAR or THY should be attributed to its lower hydrophobicity with respect to the latter compounds, considering that the most hydrophobic cyclic hydrocarbons are generally reported to present more toxic effects and as such be more antimicrobial [51]. In addition, CAR and THY are isomeric compounds that only differ in the position of their free hydroxyl group, and they can both release the proton of this group more easily than EUG, which also presents a methoxyl group in ortho position (Figure 1). This better proton exchange activity is believed to allow CAR and THY to more easily collapse the proton gradient (motive force) across the cytoplasmic membrane [50]. Relatively close to the present results, the MIC of EUG against *S. aureus* strains recovered from the milk of cows with subclinical mastitis was 392 ppm [52].

In a previous similar study evaluating the susceptibility of 26 methicillin-susceptible (MSS) and 21 methicillin-resistant staphylococci (MRS) to CAR and THY using an agar dilution method,

MIC values of 150–300 ppm and 300–600 ppm were reported for CAR and THY, respectively, with no significant differences between MSS and MRS regarding their susceptibility [53]. Another study also found that the MICs of THY against 6 *S. aureus* strains (ATCC29213 and 5 MRSA strains) ranged from 250 to 375 ppm, with the MBECs also found to be two- to three-fold higher than those (530–1070 ppm) [54]. In another previous study evaluating the effect of CAR and THY on biofilm-grown *S. aureus* and *S. epidermidis* strains (6 strains per each species), as well as their effects on biofilm formation, it was found that for most of the strains tested, the biofilm eradication concentrations (i.e., 1250–5000 ppm) were two- to four-fold greater than the concentrations required to inhibit planktonic growth [55]. However, it should be noted that in all those previous studies, the protocol used to form biofilms (i.e., in TSB containing 0.25–1% *v/v* glucose at 37 °C for 24 h, with no initial attachment step) was quite different from the one here applied, while in addition and more importantly the terpenoids had been left to act for 24 h, whereas a short 6-min exposure was applied here, thus making any attempted comparison risky. Thus, in the present study, biofilms of both species were left to be formed in a general purpose medium (i.e., TSB) and in the presence of high salt concentration (i.e., 5% *v/v* NaCl), at 37 °C for 48 h (following a 2-h initial attachment step in saline), since it is known that high osmolarity usually induces the biofilm-forming potential of staphylococci, mainly through the increase in the expression of several biofilm-associated genes this can provoke [56,57]. Preliminary experiments by our group have also confirmed this positive influence of NaCl on biofilm formation by the two staphylococci strains applied here (results not shown). In addition, the short exposure time (i.e., 6 min) was selected here to imitate conditions that could be applied within the food industry or for surface disinfection in other environments, such as the clinical ones, where a short disinfection period is usually desired. In this direction, standard protocols approved for the evaluation of the bactericidal activity of chemical disinfectants also propose exposure times ranging from 1 to 60 min (e.g., EN 1276) [58].

All three terpenoids tested here (i.e., CAR, THY, and EUG), being phenolic compounds with both hydrophilic and hydrophobic properties, are known to be capable of interacting with the lipid bilayer of the cytoplasmic membranes, provoking the loss of their integrity, disruption of the proton's motive force, impairment of intracellular pH homeostasis, and leakage of cellular material including ATP [50]. In addition, their relative hydrophilic nature conferred by the free hydroxyl group these are all containing, is believed to further allow their ease diffusion through the polar polysaccharide biofilm matrix, and as such the efficient killing of the enclosed bacteria [50]. Interestingly, time-lapse confocal laser scanning microscopy (CLSM) has previously revealed the significant advantage of another plant mixture rich in CAR and also containing both THY and EUG (i.e., the hydrosol of the Mediterranean spice *Thymbra capitata*) for easily penetrating into the three-dimensional (3D) biofilm structure of *Salmonella* Typhimurium and quickly killing the cells, when compared to BAC [42]. In that study, the Rc value for that hydrosol mixture was found to be quite low (1.6), a value equal to that found in our study for EUG. On the other hand, in that previous study BAC was found to present an Rc value equal to 208.3, whereas an average Rc value of 11.1 (i.e., 13.6 and 8.5 for *S. aureus* and *S. epidermidis*, respectively) was determined here for this compound (Figure 4). In the literature, the Rc values for the BAC range significantly from 10 to 1000, but in most cases, these surpass 50 [48]. It is surely difficult, if not impossible, for someone to compare results obtained in different studies, due to the large variations in the experimental setup (e.g., different bacterial strains, support materials, growth media, biofilm forming procedure, incubation temperatures and times), which can drastically influence the phenotypic behaviour (including resistance) of the formed biofilms. Disinfection exposure times also vary greatly between the different studies.

The lower Rc values of EUG found here against both bacterial species compared to either CAR or THY, and as thus its relative better anti-biofilm efficiency when also considering the "inherent" antimicrobial action of all these terpenoids against planktonic cells, may be attributed to its lower hydrophobicity, and as thus its better solubility and diffusion in the water containing EPS biofilm matrixes [50]. This is surely something that deserves to be further investigated and verified through microscopy. In a planktonic system, however, where EPS are either absent or encountered in low amounts, the higher hydrophobicity of both CAR and THY, together with their better proton release abilities, seems to increase their toxic effects against the freely accessible bacteria, as previously reported [50]. It is also worth noting that the Rc values determined here for both CAR and THY (i.e., from 4.1 to 6.4, depending on the compound and bacterial species; Figure 4), are close enough to those previously reported in the literature for these two compounds [48]. It should still be noted that the approach we here followed to calculate Rc took into account a large range of different log reduction-concentration combinations (>100), through the previous construction of the regression plots significantly correlating these two interrelated parameters (Figures 2 and 3), whereas in all the previous studies, the Rc values were usually calculated based on either a limited number of tested concentrations or solely through the comparison of MBC and MBEC results. Our more sophisticated approach not only confirmed the MBC and MBEC results determined here (Table 2), but also seems to more accurately calculate the Rc values for each compound (as reliable anti-biofilm effectiveness indicators).

In another study comparing the antimicrobial action of CAR to that of a peroxide-based commercial sanitizer at various stages of dual-species biofilm development by *S. aureus* and *S. Typhimurium* (in a constant-depth film fermenter system for up to 21 days), it was found that the commercial sanitizer was more biocidal than CAR only during early biofilm development (<3 days), whereas the natural terpenoid outmatched it when the biofilm had reached a quasi-steady state [44]. This last point undoubtedly further highlights the importance of biofilm maturation stage when someone evaluates the effectiveness of antimicrobial treatments. In our study, biofilms were left to be formed for 48 h under static conditions, resulting in both species achieving biofilm populations of over 10^7 CFU/cm^2 by the end of incubation (just before disinfection; results not shown). Such cell-concentration levels are considered adequate for sufficient (mature) biofilm formation (and not just individual cells attachment), with many other previous studies having left staphylococci to form biofilms on PS microtiter plates for just 24 h before further experimentation [59–61]. However, we still do not have any other further info regarding the structure and composition of the extracellular material of the biofilms formed here or whether and in which way these characteristics, together with the variation in biofilm incubation time (or many other parameters that could potentially influence biofilm growth), could affect the resistance of the enclosed bacteria to the tested antimicrobials. Nevertheless, the higher resistance of *S. epidermidis* biofilms to all three terpenoids tested here compared to those formed by *S. aureus*, together with the increased resistance of the latter to BAC (Table 2 and Figure 3), should probably imply a different matrix structure and/or composition of the biofilms formed by these two distinct species, given their similar planktonic resistance (Table 2 and Figure 2). The important roles of biofilm matrix on the overall physiology of the enclosed microorganisms and their interactions with the environment (including disinfectants) have also been well documented in the literature [62]. Not only does its synthesis depend on the involved microbial species, but in addition, its exact composition and conformation can considerably vary even within the same species, depending on the strain and the prevailing environmental conditions [63].

Obviously, the high heterogeneity that biofilms may present, even those formed by the same microorganism under different environmental conditions, is something that should be always considered when studying biofilms and their resistance, since it could drastically influence the results obtained. Future studies also employing different strains of various species, being left to develop mixed-culture sessile communities, could also further increase our knowledge of the efficiency of novel anti-biofilms approaches, and their superiority (if any) over the traditional ones. We should not forget that in nature and in several other habitats as well (e.g., food industry, healthcare), biofilms may be composed of a variety of different microorganisms interacting in quite complex ways with each other [64]. All these interactions could ultimately leave their notorious imprint on biofilm robustness and resistance.

5. Conclusions

The three plant-derived terpenoids (i.e., CAR, THY, and EUG) were found to present increased anti-biofilm potential against staphylococci, when compared to BAC. Thus, the required increases in their concentrations to be equally effective against biofilm cells as they are against the planktonic ones were always much lower compared to the synthetic biocide. This was more evident for EUG, which was found to present a very low Rc (i.e., 1.6), revealing almost similar effectiveness against both cell types, quite probably due to its good diffusion through the biofilm matrix. These results confirm and increase our knowledge of the significant bactericidal and in parallel anti-biofilm actions of all these three terpenoids, advocating for their further use as promising alternatives or supplementary agents (e.g., application together with antibiotics or other sanitizers) for dealing with biofilm-enclosed resistant microorganisms and as thus improve the quality of modern human life.

Author Contributions: Conceptualization, E.G.; methodology, D.K., I.P. and E.G.; validation, D.K. and I.P.; formal analysis, E.G.; investigation, D.K. and I.P.; resources, E.G.; data curation, D.K., I.P. and E.G.; writing—original draft preparation, E.G.; writing—review and editing, E.G.; visualization, D.K. and E.G.; supervision, E.G.; project administration, E.G., All authors have read and agreed to the published version of the manuscript.

Funding: This research received no external funding.

Conflicts of Interest: The authors declare no conflict of interest.

References

1. Rosenstein, R.; Götz, F. What distinguishes highly pathogenic staphylococci from medium- and non-pathogenic? *Curr. Top. Microbiol. Immunol.* **2013**, *358*, 33–89. [PubMed]
2. Kadariya, J.; Smith, T.C.; Thapaliya, D. Staphylococcus aureus and Staphylococcal Food-Borne Disease: An Ongoing Challenge in Public Health. *BioMed Res. Int.* **2014**, *2014*, 1–9. [CrossRef] [PubMed]
3. Le Loir, Y.; Baron, F.; Gautier, M. Staphylococcus aureus and food poisoning. *Genet. Mol. Res.* **2003**, *2*, 63–76. [PubMed]
4. Gomes, F.; Teixeira, P.; Oliveira, R. Mini-review: Staphylococcus epidermidis as the most frequent cause of nosocomial infections: Old and new fighting strategies. *Biofouling* **2013**, *30*, 131–141. [CrossRef]
5. Archer, N.K.; Mazaitis, M.J.; Costerton, J.W.; Leid, J.G.; Powers, M.E.; Shirtliff, M. Staphylococcus aureus biofilms. *Virulence* **2011**, *2*, 445–459. [CrossRef]
6. Büttner, H.; Mack, D.; Rohde, H. Structural basis of Staphylococcus epidermidis biofilm formation: Mechanisms and molecular interactions. *Front. Microbiol.* **2015**, *5*. [CrossRef]
7. Abdallah, M.; Benoliel, C.; Drider, D.; Dhulster, P.; Chihib, N.-E. Biofilm formation and persistence on abiotic surfaces in the context of food and medical environments. *Arch. Microbiol.* **2014**, *196*, 453–472. [CrossRef]
8. Figueiredo, A.M.S.; Ferreira, F.; Beltrame, C.O.; Côrtes, M.F. The role of biofilms in persistent infections and factors involved in ica -independent biofilm development and gene regulation in Staphylococcus aureus. *Crit. Rev. Microbiol.* **2017**, *2*, 1–19. [CrossRef]
9. Høiby, N.; Bjarnsholt, T.; Givskov, M.; Molin, S.; Ciofu, O. Antibiotic resistance of bacterial biofilms. *Int. J. Antimicrob. Agents* **2010**, *35*, 322–332. [CrossRef]
10. Bridier, A.; Sanchez-Vizuete, P.; Guilbaud, M.; Piard, J.-C.; Naïtali, M.; Briandet, R. Biofilm-associated persistence of food-borne pathogens. *Food Microbiol.* **2015**, *45*, 167–178. [CrossRef]
11. Coughlan, L.M.; Cotter, P.D.; Hill, C.; Alvarez-Ordóñez, A. New Weapons to Fight Old Enemies: Novel Strategies for the (Bio)control of Bacterial Biofilms in the Food Industry. *Front. Microbiol.* **2016**, *7*, 172. [CrossRef] [PubMed]
12. Ashraf, M.A.; Ullah, S.; Ahmad, I.; Qureshi, A.K.; Balkhair, K.; Rehman, M.A. Green biocides, a promising technology: Current and future applications to industry and industrial processes. *J. Sci. Food Agric.* **2013**, *94*, 388–403. [CrossRef] [PubMed]
13. Borges, A.; Saavedra, M.J.; Simões, M. Insights on antimicrobial resistance, biofilms and the use of phytochemicals as new antimicrobial agents. *Curr. Med. Chem.* **2015**, *22*, 2590–2614. [CrossRef] [PubMed]
14. Sakarikou, C.; Kostoglou, D.; Simões, M.; Giaouris, E. Exploitation of plant extracts and phytochemicals against resistant Salmonella spp. in biofilms. *Food Res. Int.* **2019**, *128*, 108806. [CrossRef]

15. Apolónio, J.D.; Miguel, M.G.; Faleiro, M.; Neto, L. No induction of antimicrobial resistance in Staphylococcus aureus and Listeria monocytogenes during continuous exposure to eugenol and citral. *FEMS Microbiol. Lett.* **2014**, *354*, 92–101. [CrossRef]

16. Nazzaro, F.; Fratianni, F.; Coppola, R. Quorum Sensing and Phytochemicals. *Int. J. Mol. Sci.* **2013**, *14*, 12607–12619. [CrossRef]

17. Ta, C.A.K.; Arnason, J.T. Mini Review of Phytochemicals and Plant Taxa with Activity as Microbial Biofilm and Quorum Sensing Inhibitors. *Molecules* **2015**, *21*, 29. [CrossRef]

18. Vergis, J.; Gokulakrishnan, P.; Agarwal, R.K.; Kumar, A. Essential Oils as Natural Food Antimicrobial Agents: A Review. *Crit. Rev. Food Sci. Nutr.* **2013**, *55*, 1320–1323. [CrossRef]

19. Sharifi-Rad, M.; Varoni, E.M.; Iriti, M.; Martorell, M.; Setzer, W.N.; Contreras, M.D.M.; Salehi, B.; Soltani-Nejad, A.; Rajabi, S.; Tajbakhsh, M.; et al. Carvacrol and human health: A comprehensive review. *Phytother. Res.* **2018**, *32*, 1675–1687. [CrossRef]

20. Marchese, A.; Orhan, I.E.; Daglia, M.; Barbieri, R.; Di Lorenzo, A.; Nabavi, S.F.; Gortzi, O.; Izadi, M.; Nabavi, S.M. Antibacterial and antifungal activities of thymol: A brief review of the literature. *Food Chem.* **2016**, *210*, 402–414. [CrossRef]

21. Kamatou, G.P.; Vermaak, I.; Viljoen, A.M. Eugenol—from the remote Maluku islands to the international market place: A review of a remarkable and versatile molecule. *Molecules* **2012**, *17*, 6953–6981. [CrossRef] [PubMed]

22. EC (European Commission). 2020. Available online: https://ec.europa.eu/food/safety/food_improvement_agents/flavourings_en (accessed on 21 May 2020).

23. FDA (Food and Drug Administration). 2020. Available online: https://www.fda.gov/food/food-additives-petitions/food-additive-status-list (accessed on 21 May 2020).

24. Pereira, B.M.P.; Tagkopoulos, I. Benzalkonium Chlorides: Uses, Regulatory Status, and Microbial Resistance. *Appl. Environ. Microbiol.* **2019**, *85*. [CrossRef]

25. Kampf, G. Adaptive microbial response to low-level benzalkonium chloride exposure. *J. Hosp. Infect.* **2018**, *100*, e1–e22. [CrossRef] [PubMed]

26. Kim, M.; Weigand, M.R.; Oh, S.; Hatt, J.K.; Krishnan, R.; Tezel, U.; Pavlostathis, S.G.; Konstantinidis, K.T. Widely Used Benzalkonium Chloride Disinfectants Can Promote Antibiotic Resistance. *Appl. Environ. Microbiol.* **2018**, *84*, AEM.01201-18. [CrossRef]

27. Tandukar, M.; Oh, S.; Tezel, U.; Konstantinidis, K.T.; Pavlostathis, S.G. Long-Term Exposure to Benzalkonium Chloride Disinfectants Results in Change of Microbial Community Structure and Increased Antimicrobial Resistance. *Environ. Sci. Technol.* **2013**, *47*, 9730–9738. [CrossRef] [PubMed]

28. Bjorland, J.; Steinum, T.; Kvitle, B.; Waage, S.; Sunde, M.; Heir, E. Widespread Distribution of Disinfectant Resistance Genes among Staphylococci of Bovine and Caprine Origin in Norway. *J. Clin. Microbiol.* **2005**, *43*, 4363–4368. [CrossRef]

29. He, G.-X.; Landry, M.; Chen, H.; Thorpe, C.; Walsh, D.; Varela, M.F.; Pan, H. Detection of benzalkonium chloride resistance in community environmental isolates of staphylococci. *J. Med. Microbiol.* **2014**, *63*, 735–741. [CrossRef]

30. Sidhu, M.S.; Heir, E.; Sørum, H.; Holck, A. Genetic linkage between resistance to quaternary ammonium compounds and beta-lactam antibiotics in food-related *Staphylococcus* spp. *Microb. Drug Resist.* **2001**, *7*, 363–371. [CrossRef]

31. Zhang, M.; O'Donoghue, M.M.; Ito, T.; Hiramatsu, K.; Boost, M. Prevalence of antiseptic-resistance genes in Staphylococcus aureus and coagulase-negative staphylococci colonising nurses and the general population in Hong Kong. *J. Hosp. Infect.* **2011**, *78*, 113–117. [CrossRef]

32. Ferk, F.; Mišík, M.; Hoelzl, C.; Parzefall, W.; Nersesyan, A.; Grummt, T.; Ehrlich, V.; Uhl, M.; Fuerhacker, M.; Grillitsch, B.; et al. Benzalkonium chloride (BAC) and dimethyldioctadecyl-ammonium bromide (DDAB), two common quaternary ammonium compounds, cause genotoxic effects in mammalian and plant cells at environmentally relevant concentrations. *Mutagenesis* **2007**, *22*, 363–370. [CrossRef]

33. Junior, G.B.; Sutili, F.; Gressler, L.; Ely, V.; Da Silveira, B.P.; Tasca, C.; Reghelin, M.; Matter, L.; Vargas, A.; Baldisserotto, B. Antibacterial potential of phytochemicals alone or in combination with antimicrobials against fish pathogenic bacteria. *J. Appl. Microbiol.* **2018**, *125*, 655–665. [CrossRef] [PubMed]

34. Gutiérrez, S.; Morán, A.; Blanco, H.M.; Ferrero, M.A.; Rodríguez-Aparicio, L.B. The Usefulness of Non-Toxic Plant Metabolites in the Control of Bacterial Proliferation. *Probiotics Antimicrob. Proteins* **2017**, *9*, 323–333. [CrossRef] [PubMed]

35. Miladi, H.; Zmantar, T.; Kouidhi, B.; Chaabouni, Y.; Mahdouani, K.; Bakhrouf, A.; Chaieb, K. Use of carvacrol, thymol, and eugenol for biofilm eradication and resistance modifying susceptibility of Salmonella enterica serovar Typhimurium strains to nalidixic acid. *Microb. Pathog.* **2017**, *104*, 56–63. [CrossRef] [PubMed]

36. Neyret, C.; Herry, J.-M.; Meylheuc, T.; Dubois-Brissonnet, F. Plant-derived compounds as natural antimicrobials to control paper mill biofilms. *J. Ind. Microbiol. Biotechnol.* **2013**, *41*, 87–96. [CrossRef] [PubMed]

37. Upadhyay, A.; Upadhyaya, I.; Johny, A.K.; Venkitanarayanan, K. Antibiofilm effect of plant derived antimicrobials on Listeria monocytogenes. *Food Microbiol.* **2013**, *36*, 79–89. [CrossRef]

38. Miladi, H.; Zmantar, T.; Kouidhi, B.; Al Qurashi, Y.M.A.; Bakhrouf, A.; Chaabouni, Y.; Mahdouani, K.; Chaieb, K. Synergistic effect of eugenol, carvacrol, thymol, p-cymene and γ-terpinene on inhibition of drug resistance and biofilm formation of oral bacteria. *Microb. Pathog.* **2017**, *112*, 156–163. [CrossRef]

39. Moran, A.; Gutierrez, S.; Blanco, H.M.; Ferrero, M.; Monteagudo-Mera, A.; Rodríguez-Aparicio, L.B. Non-toxic plant metabolites regulate Staphylococcus viability and biofilm formation: A natural therapeutic strategy useful in the treatment and prevention of skin infections. *Biofouling* **2014**, *30*, 1175–1182. [CrossRef]

40. Walsh, D.J.; Livinghouse, T.; Durling, G.M.; Chase-Bayless, Y.; Arnold, A.D.; Stewart, P.S. Sulfenate Esters of Simple Phenols Exhibit Enhanced Activity against Biofilms. *ACS Omega* **2020**, *5*, 6010–6020. [CrossRef]

41. Walsh, D.J.; Livinghouse, T.; Goeres, D.M.; Mettler, M.; Stewart, P.S. Antimicrobial Activity of Naturally Occurring Phenols and Derivatives Against Biofilm and Planktonic Bacteria. *Front. Chem.* **2019**, *7*, 653. [CrossRef]

42. Karampoula, F.; Giaouris, E.; Deschamps, J.; Doulgeraki, A.I.; Nychas, G.-J.E.; Dubois-Brissonnet, F. Hydrosol of Thymbra capitata Is a Highly Efficient Biocide against Salmonella enterica Serovar Typhimurium Biofilms. *Appl. Environ. Microbiol.* **2016**, *82*, 5309–5319. [CrossRef]

43. Karpanen, T.J.; Worthington, T.; Hendry, E.R.; Conway, B.R.; Lambert, P.A. Antimicrobial efficacy of chlorhexidine digluconate alone and in combination with eucalyptus oil, tea tree oil and thymol against planktonic and biofilm cultures of Staphylococcus epidermidis. *J. Antimicrob. Chemother.* **2008**, *62*, 1031–1036. [CrossRef] [PubMed]

44. Knowles, J.R.; Roller, S.; Murray, D.B.; Naidu, A.S. Antimicrobial Action of Carvacrol at Different Stages of Dual-Species Biofilm Development by Staphylococcus aureus and Salmonella enterica Serovar Typhimurium. *Appl. Environ. Microbiol.* **2005**, *71*, 797–803. [CrossRef] [PubMed]

45. Vázquez-Sánchez, D.; Galvão, J.A.; Mazine, M.R.; Gloria, E.M.; Oetterer, M. Control of Staphylococcus aureus biofilms by the application of single and combined treatments based in plant essential oils. *Int. J. Food Microbiol.* **2018**, *286*, 128–138. [CrossRef]

46. Vetas, D.; Dimitropoulou, E.; Mitropoulou, G.; Kourkoutas, Y.; Giaouris, E. Disinfection efficiencies of sage and spearmint essential oils against planktonic and biofilm Staphylococcus aureus cells in comparison with sodium hypochlorite. *Int. J. Food Microbiol.* **2017**, *257*, 19–25. [CrossRef] [PubMed]

47. Poimenidou, S.V.; Chrysadakou, M.; Tzakoniati, A.; Bikouli, V.C.; Nychas, G.-J.E.; Skandamis, P.N. Variability of Listeria monocytogenes strains in biofilm formation on stainless steel and polystyrene materials and resistance to peracetic acid and quaternary ammonium compounds. *Int. J. Food Microbiol.* **2016**, *237*, 164–171. [CrossRef]

48. Bridier, A.; Briandet, R.; Thomas, V.; Dubois-Brissonnet, F. Resistance of bacterial biofilms to disinfectants: A review. *Biofouling* **2011**, *27*, 1017–1032. [CrossRef]

49. CLSI (Clinical and Laboratory Standards Institute). *Methods for Dilution Antimicrobial Susceptibility Tests for Bacteria That Grow Aerobically; Approved Standard*, 7th ed.; M7-A7; Clinical and Laboratory Standards Institute: Wayne, PA, USA, 2006.

50. Ben Arfa, A.; Combes, S.; Preziosi-Belloy, L.; Gontard, N.; Chalier, P. Antimicrobial activity of carvacrol related to its chemical structure. *Lett. Appl. Microbiol.* **2006**, *43*, 149–154. [CrossRef]

51. Sikkema, J.; A De Bont, J.; Poolman, B. Mechanisms of membrane toxicity of hydrocarbons. *Microbiol. Rev.* **1995**, *59*, 201–222. [CrossRef]

52. Budri, P.; Silva, N.C.C.; Bonsaglia, E.C.R.; Fernandes, A.; Araujo, J.; Doyama, J.; Gonçalves, J.L.; Dos Santos, M.V.; Fitzgerald-Hughes, D.; Rall, V.L.M. Effect of essential oils of Syzygium aromaticum and Cinnamomum zeylanicum and their major components on biofilm production in Staphylococcus aureus strains isolated from milk of cows with mastitis. *J. Dairy Sci.* **2015**, *98*, 5899–5904. [CrossRef]

53. Nostro, A.; Blanco, A.R.; A Cannatelli, M.; Enea, V.; Flamini, G.; Morelli, I.; Roccaro, A.S.; Alonzo, V. Susceptibility of methicillin-resistant staphylococci to oregano essential oil, carvacrol and thymol. *FEMS Microbiol. Lett.* **2004**, *230*, 191–195. [CrossRef]

54. Kifer, D.; Mužinić, V.; Klarić, M. Šegvić; Vedran Mužinić; Maja Šegvić Klarić Antimicrobial potency of single and combined mupirocin and monoterpenes, thymol, menthol and 1,8-cineole against Staphylococcus aureus planktonic and biofilm growth. *J. Antibiot.* **2016**, *69*, 689–696. [CrossRef]

55. Nostro, A.; Roccaro, A.S.; Bisignano, G.; Marino, A.; Cannatelli, M.A.; Pizzimenti, F.C.; Cioni, P.L.; Procopio, F.; Blanco, A.R. Effects of oregano, carvacrol and thymol on Staphylococcus aureus and Staphylococcus epidermidis biofilms. *J. Med. Microbiol.* **2007**, *56*, 519–523. [CrossRef]

56. Ferreira, R.B.R.; Ferreira, M.C.; Glatthardt, T.; Silvério, M.P.; Chamon, R.C.; Salgueiro, V.C.; Guimarães, L.C.; Alves, E.S.; Dos Santos, K.R.N. Osmotic stress induces biofilm production by Staphylococcus epidermidis isolates from neonates. *Diagn. Microbiol. Infect. Dis.* **2019**, *94*, 337–341. [CrossRef]

57. Rode, T.M.; Langsrud, S.; Holck, A.; Møretrø, T. Different patterns of biofilm formation in Staphylococcus aureus under food-related stress conditions. *Int. J. Food Microbiol.* **2007**, *116*, 372–383. [CrossRef]

58. EN 1276. European Standard. *Chemical Disinfectants and Antiseptics—Quantitative Suspension Test for the Evaluation of Bactericidal Activity of Chemical Disinfectants and Antiseptics Used in Food, Industrial, Domestic, and Institutional Areas—Test Methods and Requirements (Phase 2/Step 1)*; European Standard: Brussels, Belgium, 2009.

59. Chaieb, K.; Zmantar, T.; Souiden, Y.; Mahdouani, K.; Bakhrouf, A. XTT assay for evaluating the effect of alcohols, hydrogen peroxide and benzalkonium chloride on biofilm formation of Staphylococcus epidermidis. *Microb. Pathog.* **2011**, *50*, 1–5. [CrossRef] [PubMed]

60. Yadav, M.K.; Chae, S.-W.; Im, G.J.; Chung, J.-W.; Song, J.-J. Eugenol: A Phyto-Compound Effective against Methicillin-Resistant and Methicillin-Sensitive Staphylococcus aureus Clinical Strain Biofilms. *PLoS ONE* **2015**, *10*, e0119564. [CrossRef] [PubMed]

61. Yuan, Z.; Dai, Y.; Ouyang, P.; Rehman, T.; Hussain, S.; Zhang, T.; Yin, Z.; Fu, H.-L.; Lin, J.; He, C.; et al. Thymol Inhibits Biofilm Formation, Eliminates Pre-Existing Biofilms, and Enhances Clearance of Methicillin-Resistant Staphylococcus aureus (MRSA) in a Mouse Peritoneal Implant Infection Model. *Microorganisms* **2020**, *8*, 99. [CrossRef] [PubMed]

62. Hobley, L.; Harkins, C.; Macphee, C.E.; Stanley-Wall, N.R. Giving structure to the biofilm matrix: An overview of individual strategies and emerging common themes. *FEMS Microbiol. Rev.* **2015**, *39*, 649–669. [CrossRef] [PubMed]

63. Kavanaugh, J.S.; Horswill, A.R. Impact of Environmental Cues on Staphylococcal Quorum Sensing and Biofilm Development. *J. Boil. Chem.* **2016**, *291*, 12556–12564. [CrossRef] [PubMed]

64. Orazi, G.; O'Toole, G.A. 'It takes a village': Mechanisms underlying antimicrobial recalcitrance of polymicrobial biofilms. *J. Bacteriol.* **2019**, *202*. [CrossRef]

Article

Seasonal Effect on the Chemical Composition, Insecticidal Properties and Other Biological Activities of *Zanthoxylum leprieurii* Guill. & Perr. Essential Oils

Evelyne Amenan Tanoh [1,2,*], Guy Blanchard Boué [1], Fatimata Nea [1,2], Manon Genva [2], Esse Leon Wognin [3], Allison Ledoux [4], Henri Martin [2], Zanahi Felix Tonzibo [1], Michel Frederich [4] and Marie-Laure Fauconnier [2]

[1] Laboratory of Biological Organic Chemistry, UFR-SSMT, University Felix Houphouet-Boigny, 01 BP 582 Abidjan 01, Ivory Coast; blancardo2000@yahoo.fr (G.B.B.); neafatima@gmail.com (F.N.); tonzibz@yahoo.fr (Z.F.T.)

[2] Laboratory of Chemistry of Natural Molecules, University of Liège, Gembloux Agro-Bio Tech, 2, Passage des Déportés, 5030 Gembloux, Belgium; m.genva@uliege.be (M.G.); henri.martin@uliege.be (H.M.); marie-laure.fauconnier@uliege.be (M.-L.F.)

[3] Laboratory of Instrumentation Image and Spectroscopy, National Polytechnic Institute Felix Houphouët-Boigny, BP 1093 Yamoussoukro, Ivory Coast; esse.wognin@yahoo.fr

[4] Laboratory of Pharmacognosy, Center for Interdisciplinary Research on Medicines (CIRM), University of Liège, Avenue Hippocrate 15, 4000 Liège, Belgium; allison.ledoux@uliege.be (A.L.); m.frederich@uliege.be (M.F.)

[*] Correspondence: evelynetanoh5@gmail.com; Tel.: +32-(0)4-6566-3587

Received: 27 March 2020; Accepted: 19 April 2020; Published: 1 May 2020

Abstract: This study focused, for the first time, on the evaluation of the seasonal effect on the chemical composition and biological activities of essential oils hydrodistillated from leaves, trunk bark and fruits of *Zanthoxylum leprieurii* (*Z. leprieurii*), a traditional medicinal wild plant growing in Côte d'Ivoire. The essential oils were obtained by hydrodistillation from fresh organs of *Z. leprieurii* growing on the same site over several months using a Clevenger-type apparatus and analyzed by gas chromatography-mass spectrometry (GC/MS). Leaf essential oils were dominated by tridecan-2-one (9.00 ± 0.02–36.80 ± 0.06%), (*E*)-β-ocimene (1.30 ± 0.50–23.57 ± 0.47%), β-caryophyllene (7.00 ± 1.02–19.85 ± 0.48%), dendrolasin (1.79 ± 0.08–16.40 ± 0.85%) and undecan-2-one (1.20 ± 0.03–8.51 ± 0.35%). Fruit essential oils were rich in β-myrcene (16.40 ± 0.91–48.27 ± 0.26%), citronellol (1.90 ± 0.02–28.24 ± 0.10%) and geranial (5.30 ± 0.53–12.50 ± 0.47%). Tridecan-2-one (45.26 ± 0.96–78.80 ± 0.55%), β-caryophyllene (1.80 ± 0.23–13.20 ± 0.33%), α-humulene (4.30 ± 1.09–12.73 ± 1.41%) and tridecan-2-ol (2.23 ± 0.17–10.10 ± 0.61%) were identified as major components of trunk bark oils. Statistical analyses of essential oil compositions showed that the variability mainly comes from the organs. Indeed, principal component analysis (PCA) and hierarchical cluster analysis (HCA) allowed us to cluster the samples into three groups, each one consisting of one different *Z. leprieurii* organ, showing that essential oils hydrodistillated from the different organs do not display the same chemical composition. However, significant differences in essential oil compositions for the same organ were highlighted during the studied period, showing the impact of the seasonal effect on essential oil compositions. Biological activities of the produced essential oils were also investigated. Essential oils exhibited high insecticidal activities against *Sitophilus granarius*, as well as antioxidant, anti-inflammatory and moderate anti-plasmodial properties.

Keywords: *Zanthoxylum leprieurii*; essential oils; *Sitophilus granarius*; tridecan-2-one; β-myrcene; (*E*)-β-ocimene; dendrolasin; antioxidant; anti-inflammatory; insecticidal; anti-plasmodial; Côte d'Ivoire

1. Introduction

Zanthoxylum leprieurii Guill and Perr. (syn. *Fagara leprieurii* Engl and *Fagara Angolensis*) is a plant species belonging to the Genus *Zanthoxylum* of the Rutaceae family, which contains approximatively 150 Genus and 900 species. *Z. leprieurii* is distributed in rain forests and wooded savannahs in Africa, from Senegal (Western Africa), Ethiopia (Eastern Africa), to Angola, Zimbabwe and Mozambique (Southern Africa) [1]. Known as a multipurpose species, *Z. leprieurii* has a wide spectrum of applications, as leaves, trunk bark and roots are used in traditional medicine to cure rheumatism and for the treatment of tuberculosis and generalized body pains in Central and Western Africa [2–4]. Roots are used as chewing sticks to clean the mouth [5]. Moreover, this plant is also used for canoes, boxes, plywood, general carpentry, domestic utensils, beehives and water pots; the pale yellow wood is tough, medium coarse-grained and light [6]. Dried fruits of *Z. leprieurii* are used as spices by local populations in many regions of Africa [7]. The plant has shown antioxidant, antimicrobial, anticancer, cytotoxic, schistosomidal and antibacterial properties [8–12]. From a chemical point of view, a large variety of compounds from different chemical classes were reported in *Z. leprieurii* solvent extracts: acridone alkaloids, benzophenanthridine [13], aliphatic amide [14], coumarins [15] and kaurane diterpenes [11]. Essential oils produced from *Z. leprieurii* revealed that the main constituents in fruit essential oils were (E)-β-ocimene (29.40%) and β-citronellol (17.37%) [15–17]. Our previous study showed the predominance of tridecan-2-one (47.50%) in leaf essential oils from Côte d'Ivoire [18]. Limonene (94.90%) and terpinolene (50.00%) were described as major components in essential oils from Nigeria and Cameroon, respectively [19,20]. The composition of trunk bark essential oils was predominated by sesquiterpenoids in Nigeria and methyl ketones in Côte d'Ivoire [18,19].

All these studies highlighted significant differences in the composition of essential oils extracted from different organs of *Z. leprieurii* from the same growing site, as well as significant differences in the composition of essential oils extracted from *Z. leprieurii* growing in different places. The latter can be explained by the fact that many factors affect essential oil amounts as well as their chemical compositions [21–23]. These include plant genetic differences, as well as environmental factors such as temperature, soil, precipitation, wind speed, rainfall and pests [24].

The first aim of this study was to explore climate-related variations of the composition of essential oils hydrodistillated from different *Z. leprieurii* organs (leaves, trunk bark and fruits). The variation over time of essential oil chemical compositions was then studied at one site in Côte d'Ivoire. GC/MS was used to identify essential oil chemical profiles and statistical analyses were performed on the different chemical compositions. In addition, this study also aimed to evaluate the impact of *Z. leprieurii* essential oil composition variations on their in vitro biological activities, such as antioxidant, anti-inflammatory, insecticidal and anti-plasmodial activities. To our knowledge, such studies have not yet been carried out on this species in Côte d'Ivoire.

2. Materials and Methods

2.1. Plant Material and Hydrodistillation Procedure

Z. leprieurii organs were collected at Adzope (6°06′25″ N, 3°51′36″ W), in south-eastern Côte d'Ivoire, between May and November 2017 for leaves and trunk bark, respectively, and between July and November 2017 for fruits. At each harvest, leaf, fruit and trunk bark samples were taken from 15 randomly selected trees, in the geographical area described before, by taking the same amount of plant material from each tree and pooling it before distillation. The total amount of plant material was between 700 and 1500 g. The same tree was only sampled once to prevent the previous sampling from influencing the next one (e.g., trunk injury). Plants were identified by the Centre Suisse of Research (Adiopodoumé, Abidjan, Côte d'Ivoire) and by the National Flora Center (CNF; Abidjan, Côte d'Ivoire). The vouchers of the specimen (UCJ016132) have been deposited at the CNF Herbarium. Fresh organ material was hydrodistillated for 3 h using a Clevenger-type apparatus. The pale yellow essential oils were treated with anhydrous sodium sulphate (Na_2SO_4) as a drying agent, stored in sealed amber vials, and conserved at 4 °C before analysis. The essential oil yields (*w/w*) were calculated as the rapport between the mass of essential oils obtained compared to the mass of fresh organs.

2.2. GC/MS Chemical Analysis of Essential Oils

Essential oils hydrodistillated from *Z. leprieurii* organs were analyzed by GC/MS. An Agilent GC system 7890B (Agilent, Santa Clara, CA, USA) equipped with a split/splitless injector and an Agilent MSD 5977B detector was used. The experience was repeated three times for each essential oil. One µL of essential oil dilutions (0.01% in hexane; *w/v*) was injected in splitless mode at 300 °C on an HP-5MS capillary column (30 m × 0.25 mm, *df* = 0.25 µm). The temperature was maintained one min at 50 °C, and then increased at a rate of 5 °C/min until 300 °C. The final temperature was maintained for 5 min. The sources and quadrupole temperatures were fixed at 230 °C and 150 °C, respectively. The scan range was 40–400 m/z, and the carrier gas was helium at a flow rate of 1.2 mL/min. The component identification was performed on the basis of chromatographic retention indices (RI) and by comparison of the recorded spectra with a computed data library (Pal 600K®) [25–27]. RI values were measured on an HP-5MS column (Agilent, Santa Clara, CA, USA). RI calculations were performed in temperature program mode according to a mixture of homologues n-alkanes (C7–C30), which were analyzed under the same chromatographic conditions. The main components were confirmed by comparison of their retention and MS spectrum data with co-injected pure references (Sigma, Darmstadt, Germany) when commercially available.

2.3. Biological Activities

2.3.1. Antioxidant Assay

2,2-Diphenyl-1-Picrylhydrazyl (DPPH) Radical Scavenging Capacity

The hydrogen atom- or electron-donating ability of essential oils and Trolox was determined from the bleaching of the purple-colored methanol DPPH solution. Briefly, the samples were tested at 25, 50, 75 and 100 µg/mL. Ten microliters of various concentrations (1 to 5 mg/mL) of each essential oil in methanol were added to 1990 µL of a 10 mg/mL DPPH methanol solution (0.06 mM). Free radical scavenging activities of leaf, trunk bark and fruit essential oils hydrodistillated from *Z. leprieurii* were determined spectrophotometrically [28]. The mixture was vortexed for about 1 min and then incubated at room temperature in the dark for 30 min; absorbance was measured at 517 nm with an Ultrospec UV-visible, dual beam spectrophotometer (GE Healthcare, Cambridge, UK). The same sample procedure was followed for Trolox (Sigma, Darmstadt, Germany) used as standard; methanol (Sigma, Darmstadt, Germany) with DPPH was used as control; and all the samples were tested in triplicate. The optical density was recorded, and the inhibition percentage was calculated using the formula given below:

$$\text{Inhibition percentage of DPPH activity (\%)} = (\text{Abs Blank-Abs sample})/(\text{Abs blank}) \times 100 \qquad (1)$$

Abs Blank = absorbance of the blank sample, Abs sample = absorbance of the test sample

Ferric-Reducing Power Assay

The ferric-reducing antioxidant power (FRAP) of essential oils hydrodistillated from *Z. leprieurii* was determined here. Briefly, four dilutions of essential oils and Trolox were prepared in methanol (25, 50, 75 and 100 µg/mL). Trolox was used as the standard reference. One mL of those methanol solutions were melded with one mL of a phosphate buffer (0.2 M, pH = 6.6) and with one mL of a potassium ferricyanide solution (1%; $K_3Fe(CN)_6$). After 20 min at 50 °C and the addition of one mL of trichloroacetic acid (TCA; 10% *v/v*), the solution was centrifuged at 3000 rpm for 10 min [27–29]. Next, 1.5 mL of the upper phase was recovered and melded with 1.5 mL of distilled water and 150 µL of $FeCl_3$ (0.1% *v/v*). Each concentration was realized as triplicated. Finally, the absorbance of the prepared sample was measured at 700 nm. In comparison with the blank, a higher absorbance shows a high reducing power. For all concentrations, absorbance due to essential oil samples were removed from each measurement.

2.3.2. Anti-Inflammatory Activity

Inhibition Lipoxygenase Assay

The anti-inflammatory activities of *Z. leprieurii* essential oils were determined by the method previously described by Nikhila [30]. In brief, the reaction mixture containing essential oils in various concentrations (100, 75, 50 and 25 µg/mL of methanol) (in triplicate for each concentration), lipoxygenase (Sigma, Darmstadt, Germany) and 35 µL (0.1 mg/mL) of a 0.2 M borate buffer solution (pH = 9.0) was incubated for 15 min at 25 °C. The reaction was then initiated by the addition of 35 µL of a substrate solution (linoleic acid 250 µM), and the absorbance was measured at 234 nm. Quercetin (Sigma, Darmstadt, Germany) was used as a standard inhibitor at the same concentration as the essential oils. The inhibition percentage of lipoxygenase activity was calculated as follows:

$$\text{Inhibition percentage \%} = (\text{Abs Blank-Abs sample})/(\text{Abs blank}) \times 100 \tag{2}$$

where Abs blank is the Absorbance (Abs) of the reaction media without the essential oil, and Absorbance sample is the Abs of the reaction media with the essential oil minus the Abs value of the diluted essential oil (to compensate for absorbance due to the essential oils themselves).

Inhibition of Albumin Denaturation Assay

This test was conducted as described by Kar [31] with some slight modifications. The reaction mixture consisted of 1 mL essential oil samples and diclofenac (standard) at 100, 75, 50 and 25 µg/mL in methanol, 0.5 mL bovine serum albumin (BSA) at 2% in water and 2.5 mL phosphate-buffered saline adjusted with hydrochloric acid (HCl) to pH 6.3. The tubes were incubated for 20 min at room temperature, then heated to 70 °C for 5 min and subsequently cooled for 10 min [32]. The absorbance of these solutions was determined using a spectrophotometer at 660 nm. The experiment was performed in triplicate. The inhibition percentage of albumin denaturation was calculated on a percentage basis relative to the control using the formula:

$$\text{Inhibition percentage of denaturation \%} = (\text{Abs Blank-Abs sample})/(\text{Abs blank}) \times 100 \tag{3}$$

where Abs blank corresponds to the Absorbance (Abs) of the reaction media without the addition of essential oil. Absorbance sample is the Abs of the reaction media with addition of essential oil, subtracted by the Abs value of the diluted essential oil (to compensate for absorbance due to the essential oils themselves).

IC_{50} (half inhibitory concentration), which corresponds to the essential oil concentration needed to inhibit 50% of the activity, was used to express antioxidant and anti-inflammatory properties of essential oils.

2.3.3. Insecticidal Activity

Determination of Mortality Values

Essential oil dilutions (10, 14, 18, 22, 26 and 30 µL/mL) were prepared in acetone. Talisma UL (Biosix, Hermalle-sous-Huy, Belgium), a classical chemical insecticide used for the protection of stored grains against insects, was also used at the same concentrations. For each test, 500 µL of essential oil or standard solution were homogeneously dispersed in tubes containing 20 g of organic wheat grains. The solvent was allowed to evaporate from grains for 20 min before infesting them by 12 adult insects. The granary weevil, *Sitophilus granarius*, was chosen for this study because it is one of the most damaging cereal pests in the world. Moreover, this insect is a primary pest, which means it is able to drill holes in grains, laying its eggs inside them and allowing secondary pests to develop in the grains [33,34]. Acetone was used as a negative control. Six replicates were created for all treatments and controls, and they were incubated at 30 °C. The mortality was recorded after 24 h of incubation. Results from all replicates were subjected to Probit Analysis using Python 3.7 program to determine LC_{50}, LC_{90} and LC_{95} values.

Repulsive Assay

This test has been conducted to evaluate the repulsive effect of essential oils against insects (*Sitophilus granarius*). This experiment was carried out by cutting an 8 cm diameter filter paper in half. The six concentrations (10, 14, 18, 22, 26 and 30 µL/mL) of essential oils were prepared in acetone. Each half disk was treated with 100 µL of the solution, and the other half with acetone. After evaporation of acetone, the two treated parts were joined together by an adhesive tape and placed in a petri dish. Ten insects were placed in the center of each petri dish and were incubated at 30 °C. After two hours of incubation, the number of insects present in the part treated with essential oil and the number of insects present in the part treated only with acetone were counted, as described by Mc Donald [35].

The percentage of repulsively was calculated as follows:

$$Pr = (Nc\text{-}NT)/(Nc\text{+}NT) \times 100 \tag{4}$$

NC: Number of insects present on the disc part treated with acetone; NT: Number of insects present on the part of the disc treated with the essential oil dilution in acetone

2.3.4. Anti-Plasmodial Activity

Ledoux et al. method [36] was used to determine anti-plasmodial activity. To do so, asexual erythrocyte stages of *P. falciparum*, chloroquine-sensitive strain 3D7 were continuously cultivated in vitro using the procedure of Trager and Jensen. The erythrocyte had been initially obtained from a patient from Schipol in the Netherlands (BEI Reagent Search) [37]. ATCC, Bei Ressources provides us with the strains. Red blood cells of A+ group were used as human host cell. The culture medium was RPMI 1640 from Gibco, Fisher Scientific (Loughborough, UK) composed of $NaHCO_3$ (32 mM), HEPES (25 mM) and L-glutamine. Glucose (1.76 g/L) from Sigma-Aldrich (Machelen, Belgium), hypoxanthine (44 mg/mL) from Sigma-Aldrich (Machelen, Belgium), gentamin (100 mg/mL) from Gibco Fisher Scientific (Loughborough, UK) and human pooled serum from A+ group (10%) were added to the medium according to [36,38]. DMSO solutions of essential oils were directly diluted in the medium. The dilutions were performed in triplicate by successive two-fold dilutions in a 96-well plate. The essential oil concentrations are expressed in term of µg/mL of essential oil. As interaction between volatile compounds between samples could occur, we decided to alternate one test line with two lines filled with culture media. The growth of the parasite was recorded after 48 h of incubation using lactate dehydrogenase (pLDH) activity as parameter according to Makker method [39]. A positive control was used in all the repetitions. This positive control was composed of Artemisinin from Sigma-Aldrich (Machelen, Belgium) at a concentration of 100 µg/mL. Sigmoidal curves allowed the determination of half inhibitory concentration (IC_{50}).

2.4. Statistical Analysis

2.4.1. Data Analysis

Hierarchical cluster analysis (HCA) and principal component analysis (PCA) (Ward's method) were used to investigate the seasonal effect on essential oil composition. Analyses on the 19 samples collected during a specific period were performed with Xlstat (Adinsoft, Paris, France).

2.4.2. Biological Activities Analysis

Data were analyzed using IBM SPSS version 20. Results were presented in terms of means. Multiple comparisons of mean values were set up using one-way parametric analysis of variance (ANOVA). The DUNCAN test was used to appreciate the differences between the means at p-value < 0.05. The relationship between the different parameters was studied using Pearson correlation.

3. Results and Discussion

3.1. Chemical Composition of Essential oils and Yields

Z. leprieurii organs were collected over a period of seven months for the leaves and trunk barks and five months for the fruits within one single year. Meteorological data were recorded during the collection period (Table 1). The highest rainfall values were recorded in May, June, October and November, with 164.50 mm, 205.70 mm, 310.80 mm and 206.40 mm. In July and August, the rainfall was moderate, with 71.30 mm and 61.70 mm, respectively. The same trend was observed for the temperature. However, temperature variations were low during the collection period, with temperature ranging from 28.40 °C to 25.00 °C. The trends observed for these two variables in addition to those of relative humidity and daylight confirm that the months of May, June, October and November represent rainy season months; while those of July, August and September were dry season months. Results (Tables 2–4, in bold) showed that essential oil yields obtained in this study (0.02 to 0.04% (*w/w*) for leaves, 0.86 to 1.20% (*w/w*) for trunk bark and 1.13 to 1.51% (*w/w*) for fruits) were consistent with those found in the literature [18,40]. Essential oil yields seem dependent on meteorological variations, as, for each organ, the highest yields were observed in July and August, the collecting moment when the lowest precipitations and temperatures were recorded. When precipitations increased and temperatures were higher, the lowest essential oil yields were obtained. However, as temperature and yields variations were low in this study, those results should to be confirmed with a longer experiment. Results obtained here, though, are supported by previous studies showing that lower precipitation induces higher essential oil yields [41].

Table 1. Meteorological parameters recorded during the collection period of *Zanthoxylum leprieurii* organs.

Months	Rainfall (mm)	Relative Humidity (%)	Daylight (h)	Temperature (°C)
May	164.50	80.00	196.30	28.40
June	205.70	85.00	101.60	26.80
July	71.30	84.00	121.00	25.80
August	61.70	84.00	134.00	25.00
September	102.40	81.00	124.20	25.90
October	310.80	84.00	180.40	27.10
November	206.40	78.00	214.40	27.50

Source: SODEXAM (Société d'Exploitation et de Développement Aéronautique, Aéroportuaire et Météorologique), 2017.

Essential oils hydrodistillated from *Z. leprieurii* organs were analyzed by GC/MS. Representative chromatograms for essential oils hydrodistillated from each organ are presented in Figure 1. Compounds accounting for 97.70–99.50% of global essential oil compositions were identified in the samples. Essential oils hydrodistillated from leaves were dominated by hydrocarbon sesquiterpenes, while methyl ketones were mainly present in trunk bark essential oils. Oxygenated and hydrocarbon monoterpenes were dominant in fruit essential oils. The major compounds identified in these oils were tridecan-2-one and β-caryophyllene in the leaf oils, tridecan-2-one in the trunk bark oils and β-myrcene in the fruit oils.

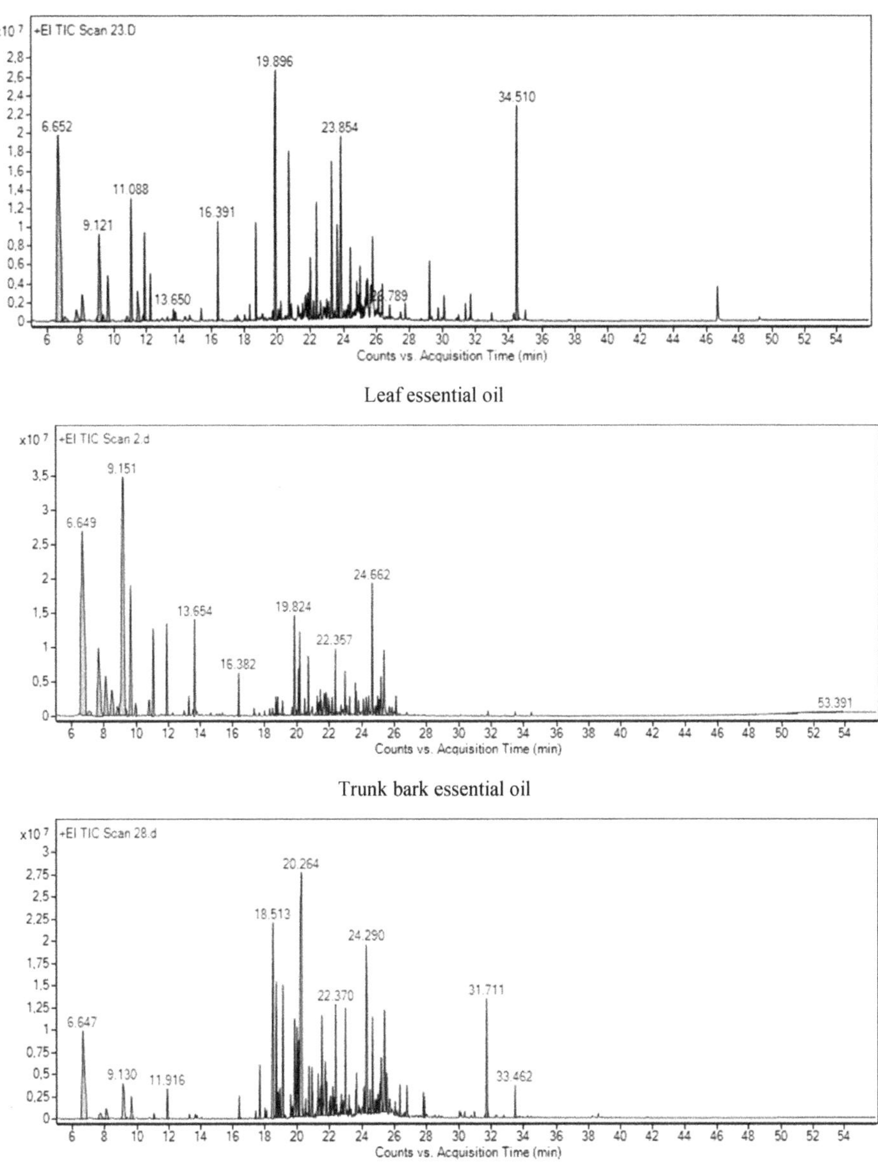

Leaf essential oil

Trunk bark essential oil

Fruit essential oil

Figure 1. Representative chromatograms for essential oils hydrodistillated from each *Z. leprieurii* organ.

3.1.1. Leaf Essential Oils

The analysis of essential oils hydrodistillated from leaves allowed for the identification of 42 compounds ranging from 97.70% to 99.50% of the total composition (Table 2). Sesquiterpenes (34.89–70.8%), methylketones (13.10–42.40%) and monoterpenes (4.5–36.18%) were the main components of these essential oils, which were dominated by tridecan-2-one (9.00% to 36.80%) and β-caryophyllene (7.00% to 19.85%). However, the composition of leaf essential oils was not constant over the collecting period, as some compounds that were present in only a minority in some samples were found in higher quantities during certain months. As a first example, there is a drop in tridean-2-one production in June, which is tricky to explain. This is also the case with (*E*)-β-ocimene, whose content was less than 4% from June to November, while in May, it was found to be at 23.57%. Caryophyllene oxide, which represented 5.7–6% of the total oil compositions from June to July, was only found in trace amounts during the other months. Undecan-2-one was also exceptionally present at 8% in May and August. Dendrolasin was present in significant amounts (4–16.4%) in all months except in May. This last molecule has well-known antimicrobial and antibacterial properties, and is also used in the treatment of cancer [42–44]. We also noticed the presence of thymol, an oxygenated monoterpene, at 13.30% in August. The chemical compositions of essential oils previously reported from two different Côte d'Ivoire locations [18] collected in February and November 2016 were different to those described in this study. *Z. leprieurii* leave essential oils thus exhibiting various chemotypes: for example, we describe here a chemotype with high proportions of dendrolasin. Moreover, the essential oil composition reported from Nigeria and Cameroon was dominated by limonene (94.90%) [19] and ocimene (91.5%) [40], showing that environmental or genetic factors impact essential oil compositions. Most of the major compounds that were detected in leaf essential oils are already known for their beneficial biological activities, such as insecticidal, antioxidant and anti-inflammatory activities. For example, the β-caryophyllene is a molecule characterized by high antioxidant and anti-inflammatory activities [45].

3.1.2. Trunk Bark Essential Oils

In total, 29 compounds were identified in the seven trunk bark oil samples, accounting for 98.30–99.40% of the whole composition (Table 3). Essential oils hydrodistillated from trunk bark were dominated by tridecan-2-one (45.26–78.80%) and α-humulene, which was also present in a significant content (4.3–12.73%). Hydrocarbon monoterpenes were only present as traces. As for leaf essential oils, the composition of essential oils hydrodistillated from the trunk bark was not consistent during the studied period. Indeed, some sesquiterpenes were only present in high a content during a given period: β-caryophyllene (8.1–13.20%) from May to July and September; tridecan-2-ol was found up to 10.10% in June; and (*E,E*)-farnesol (12.5% and 11.1%) in May and July, respectively. This chemical profile shows differences with those reported during our previous work in Côte d'Ivoire. Those differences may be due to the harvesting season and to the harvesting sites, which were not the same in those studies [18]. Moreover, these described compositions are different from those described in Nigeria, in which caryophyllene oxide (23.00%) and humulenol (17.50%) were the major components of trunk bark essential oil [18], showing that the essential oil composition is largely dependent on the plant localization.

In view of the use of tridecan-2-one in the food, pharmaceutical and cosmetic industry [46], trunk bark essential oils of Ivorian *Z. leprieurii* has a high potential. In addition to the major compounds, other minor molecules such as β-caryophyllene and α-humulene were also found in this essential oil, those having interesting antioxidant, anti-inflammatory, antibacterial and insecticidal effects, enhancing the potential use of this essential oil in the pharmaceutical industry [47,48].

3.1.3. Fruit Essential Oils

GC/MS analysis resulted in the identification of 43 constituents of the essential oils hydrodistillated from *Z. leprieurii* fruits (Table 4), accounting for 98.27–99.30% of the total essential oil compositions. This oil was dominated by β-myrcene (16.4–48.27%) but methyl nerate was also present in significant amounts (4.4–6.7%). Moreover, some minor compounds of certain months were present in high quantities in other samples. In particular, citronellol was present at 28.24% in November, but was lower than 6.6% the other months. Furthermore, geranial, which was present in traces in July, saw its content increase in the other months (5.3–6.10%). Some compounds were present in remarkable contents in July: (*E*)-β-ocimene (8.3%); perillene (6.5%); decanal (8.3%); spathulenol (5.2%); and caryophyllene oxide (9.6%).

Table 2. Chemical composition of essential oils hydrodistillated from *Zanthoxylum leprieurii* leaves. Each analysis was realized in triplicate.

N°	Compounds	Cas Number	Identification	Ria	Rib	May	June	July	Aug	Sep	Oct	Nov
								Leaves				
1	α-pinene	80-56-8	MS, RI, STD	931	929	2.48 ± 0.13	8.45 ± 0.57	0.30 ± 0.11	0.60 ± 0.35	0.70 ± 0.15	0.50 ± 0.27	0.30 ± 0.17
2	ß-myrcene	123-35-3	MS, RI, STD	987	988	0.71 ± 0.03	-	5.20 ± 0.21	0.90 ± 0.15	3.40 ± 0.37	0.4 ± 0.31	0.70 ± 0.09
3	p-cymene	25155-15-1	MS, RI, STD	1022	1022	1.48 ± 0.03	-	0.10 ± 0.01	0.30 ± 0.03	-	0.40 ± 0.57	-
4	limonene	138-86-3	MS, RI, STD	1023	1027	-	0.90 ± 0.29	0.40 ± 0.28	-	-	-	-
5	(E)-ß-ocimene	13877-91-3	MS, RI, STD	1041	1046	23.57 ± 047	2.60 ± 0.99	3.90 ± 0.84	2.50 ± 0.04	1.30 ± 0.50	2.30 ± 1.91	-
6	γ-terpinene	99-85-4	MS, RI, STD	1060	1057	0.28 ± 0.03	-	-	-	-	-	-
7	linalool	78-70-6	MS, RI, STD	1094	1098	6.44 ± 0.13	0.50 ± 0.34	1.50 ± 0.55	8.50 ± 0.09	1.70 ± 0.23	4.30 ± 0.95	3.20 ± 0.09
8	alloocimene	673-84-7	MS, RI	1128	1129	1.22 ± 0.04	0.40 ± 0.13	0.20 ± 0.10	-	-	-	-
9	terpineol	98-55-5	MS, RI, STD	1190	1191	-	-	0.20 ± 0.03	-	-	-	-
10	decanal	112-31-2	MS, RI, STD	1191	1204	-	-	1.50 ± 0.52	-	-	-	-
11	citronellol	106-22-9	MS, RI, STD	1225	1227	-	-	2.00 ± 0.48	-	1.50 ± 0.27	-	0.30 ± 0.02
12	geraniol	106-24-1	MS, RI, STD	1250	1253	-	-	1.40 ± 0.53	-	1.00 ± 0.74	-	-
13	decyl alcohol	112-30-1	MS, RI, STD	1262	1263	-	-	1.80 ± 0.92	-	-	-	-
14	thymol	89-83-8	MS, RI, STD	1286	1291	-	2.20 ± 0.96	3.10 ± 0.43	13.30 ± 0.31	4.10 ± 0.22	-	-
15	undecan-2-one	112-12-9	MS, RI, STD	1288	1293	8.51 ± 0.35	1.90 ± 0.50	1.20 ± 0.03	8.40 ± 0.12	2.70 ± 0.56	3.20 ± 1.06	2.10 ± 0.98
16	undecan-2-ol	1653-30-1	MS, RI, STD	1298	1300	-	-	0.20 ± 0.04	0.40 ± 0.04	-	-	-
17	methyl nerate	1862-61-9	MS, RI, STD	1319	1323	-	-	2.20 ± 0.74	-	0.80 ± 0.44	-	-
18	δ-elemene	20307-84-0	MS, RI	1334	1339	-	1.20 ± 0.69	0.20 ± 0.00	-	-	-	-
19	α-copaene	3856-25-5	MS, RI, STD	1376	1378	-	0.60 ± 0.23	0.30 ± 0.08	-	0.60 ± 0.11	0.20 ± 0.07	0.40 ± 0.04
20	ß-elemene	515-13-9	MS, RI	1388	1394	3.92 ± 0.06	-	2.70 ± 0.83	4.20 ± 0.07	5.90 ± 0.26	2.30 ± 0.10	2.90 ± 0.20
21	β-caryophyllene	87-44-5	MS, RI, STD	1419	1423	13.51 ± 0.11	18.90 ± 0.48	15.60 ± 0.54	13.70 ± 0.1	19.85 ± 0.82	7.00 ± 1.02	8.60 ± 0.62
22	cadina-4(14),5-diene	54324-03-7	MS, RI	1430	1433	1.83 ± 0.02	0.90 ± 0.26	0.70 ± 0.16	0.30 ± 0.02	2.40 ± 0.81	1.70 ± 0.26	1.50 ± 0.19
23	γ-elemene	3242-08-8	MS, RI	1432	1435	1.19 ± 0.01	4.40 ± 0.45	1.10 ± 0.21	1.10 ± 0.04	3.10 ± 0.06	0.8 ± 0.12	1.10 ± 0.01
24	α-humulene	6753-98-6	MS, RI, STD	1456	1457	3.92 ± 0.02	6.70 ± 0.62	4.50 ± 0.98	3.60 ± 0.04	6.10 ± 0.27	4.10 ± 0.58	4.20 ± 0.32
25	alloaromadendrene	25246-27-9	MS, RI	1457	1465	-	0.50 ± 0.21	0.30 ± 0.06	-	0.50 ± 0.11	0.40 ± 0.06	0.20 ± 0.01
26	germacrene D	23986-74-5	MS, RI, STD	1482	1485	1.96 ± 0.17	0.90 ± 0.50	0.40 ± 0.10	-	1.60 ± 0.40	1.40 ± 0.21	1.00 ± 0.14

Table 2. *Cont.*

N°	Compounds	Cas Number	Identification	Ria	Rib	Leaves						
						May	June	July	Aug	Sep	Oct	Nov
27	ß-ionone	14901-07-6	MS, RI, STD	1482	1488	-	-	0.40 ± 0.12	-	-	-	0.20 ± 0.05
28	ß-selinene	17066-67-0	MS, RI	1483	1490	0.32 ± 0.01	1.10 ± 0.13	0.70 ± 0.14	0.80 ± 0.01	1.60 ± 0.30	0.20 ± 0.21	0.50 ± 0.05
29	tridecan-2-one	593-08-8	MS, RI, STD	1487	1495	18.74 ± 0.57	9.00 ± 0.02	22.50 ± 0.98	30.20 ± 0.39	15.80 ± 0.84	36.80 ± 0.06	33.70 ± 0.36
30	selina-4(14),7(11)-diene	515-17-3	MS, RI	1495	1498	-	1.30 ± 0.31	0.80 ± 0.15	1.50 ± 0.90	2.90 ± 0.86	-	-
31	tridecan-2-ol	1653-31-2	MS, RI, STD	1495	1501	-	2.00 ± 0.82	1.70 ± 1.17	-	2.30 ± 0.29	2.40 ± 0.80	3.20 ± 0.33
32	(3E,6E)-α-farnesene	502-61-4	MS, RI	1499	1509	3.22 ± 0.29	0.60 ± 0.21	1.20 ± 0.34	0.90 ± 0.07	2.50 ± 0.97	9.10 ± 0.82	4.60 ± 0.55
33	γ-cadinene	39029-4-9	MS, RI	1513	1517	0.42 ± 0.06	1.10 ± 0.36	0.60 ± 0.08	0.20 ± 0.04	1.00 ± 0.21	1.20 ± 0.26	1.30 ± 0.02
34	δ-cadinene	483-76-1	MS, RI	1524	1526	2.25 ± 0.09	4.20 ± 1.01	2.20 ± 0.09	1.00 ± 0.02	4.30 ± 0.13	3.40 ± 0.49	4.30 ± 0.33
35	elemol	639-99-6	MS, RI, STD	1547	1552	0.28 ± 0.02	0.60 ± 0.22	0.20 ± 0.01	0.30 ± 0.01	0.30 ± 0.05	0.40 ± 0.03	1.20 ± 0.09
36	nerolidol	7212-44-4	MS, RI, STD	1557	1564	-	1.50 ± 0.30	1.10 ± 0.39	1.00 ± 0.03	1.60 ± 0.49	4.80 ± 0.22	3.40 ± 0.32
37	dendrolasin	23262-34-2	MS, RI	1576	1580	1.79 ± 0.08	8.60 ± 0.95	9.40 ± 0.90	4.00 ± 0.06	7.60 ± 0.09	10.60 ± 0.44	16.40 ± 0.85
38	spathulenol	6750-60-3	MS, RI	1578	1582	-	1.50 ± 0.46	1.10 ± 0.03	-	-	-	0.40 ± 0.00
39	caryophyllene oxide	1139-30-6	MS, RI, STD	1583	1588	-	6.00 ± 0.83	5.70 ± 0.79	-	-	-	-
40	ζ-cadinol	5937-11-1	MS, RI	1639	1645	-	1.10 ± 0.08	0.60 ± 0.03	-	0.70 ± 0.21	0.60 ± 0.15	0.80 ± 0.04
41	α-cadinol	481-34-5	MS, RI	1650	1659	0.28 ± 0.03	1.30 ± 0.11	0.70 ± 0.01	-	1.10 ± 0.44	0.80 ± 0.23	0.90 ± 0.09
42	pentadecanal	2765-11-9	MS, RI	1713	1715	-	2.10 ± 0.28	-	-	-	-	1.60 ± 0.79
	Monoterpene hydrocarbons (%)					29.74	12.85	10.10	4.30	5.40	3.60	1.00
	Oxygenated monoterpene (%)					6.44	2.70	11.50	21.80	8.20	4.30	3.50
	Sesquiterpene hydrocarbons (%)					32.54	48.10	30.90	27.30	52.35	31.80	30.60
	Oxygenated sesquiterpenes (%)					2.35	22.70	18.10	5.30	12.10	17.20	25.40
	Others (%)					27.25	13.10	28.90	39.00	20.80	42.40	38.50
	Identified compounds (%)					98.32	99.45	99.50	97.70	98.85	99.30	99.00
	Yield (%)					0.02	0.02	0.03	0.04	0.02	0.02	0.02

Identification methods: RIa, theoretical kovats indices (Pubchem and NIST); RIb, calculated kovats indices; MS, mass spectra comparison with PAL 600®libraries; STD, retention time and mass spectra comparison with commercially available standards; RI, retention index comparison with the literature; CAS number; -, Under perception threshold.

Table 3. Chemical composition of essential oils hydrodistillated from *Zanthoxylum leprieurii* trunk bark. Each analysis was realized in triplicate.

N°	Compounds	Cas Number	Identification	Ria	Rib	Trunk Bark						
						May	June	July	Aug	Sep	Oct	Nov
1	α-pinene	80-56-8	MS, RI, STD	931	929	0.52 ± 0.03	-	-	-	-	-	-
2	ß-myrcene	123-35-3	MS, RI, STD	987	988	-	-	-	-	1.00 ± 0.14	-	-
3	citronellal	106-23-0	MS, RI, STD	1153	1153	-	-	-	-	1.50 ± 0.08	-	-
4	citronellol	106-22-9	MS, RI, STD	1225	1227	-	-	-	-	1.60 ± 0.08	-	-
5	geraniol	106-24-1	MS, RI, STD	1250	1253	-	-	-	-	1.50 ± 0.14	-	-
6	thymol	89-83-8	MS, RI, STD	1286	1291	-	5.10 ± 0.36	0.50 ± 0.08	-	1.30 ± 0.05	-	4.10 ± 0.50
7	ß-elemene	515-13-9	MS, RI	1388	1394	-	0.60 ± 0.15	-	-	1.30 ± 0.03	-	-
8	α-bergamotene	17699-05-7	MS, RI	1415	1417	4.43 ± 0.10	0.50 ± 0.27	3.20 ± 0.69	1.60 ± 1.12	0.70 ± 0.01	1.50 ± 0.56	4.80 ± 0.60
9	β-caryophyllene	87-44-5	MS, RI, STD	1419	1423	9.51 ± 0.37	13.20 ± 0.33	8.50 ± 0.35	2.10 ± 0.13	8.10 ± 0.08	4.10 ± 0.79	1.80 ± 0.23
10	γ-elemene	3242-08-8	MS, RI	1432	1435	-	0.60 ± 0.21	0.10 ± 0.01	-	-	-	-
11	geranylacetone	3796-70-1	MS, RI	1455	1453	-	1.10 ± 0.51	0.50 ± 0.19	-	0.40 ± 0.02	0.30 ± 0.06	-
12	α-humulene	6753-98-6	MS, RI, STD	1456	1457	12.73 ± 1.41	4.30 ± 1.09	7.70 ± 0.62	6.20 ± 0.94	7.40 ± 0.11	8.10 ± 1.01	6.30 ± 0.03
13	α-curcumene	644-30-4	MS, RI	1482	1484	0.83 ± 0.18	-	0.70 ± 0.10	0.30 ± 0.11	-	0.40 ± 0.13	1.30 ± 0.08
14	ß-selinene	17066-67-0	MS, RI	1483	1490	-	0.30 ± 0.14	0.10 ± 0.01	-	0.30 ± 0.01	-	0.25 ± 0.33
15	tridecan-2-one	593-08-8	MS, RI, STD	1487	1495	45.26 ± 0.96	56.30 ± 0.31	51.4 ± 1.15	78.80 ± 0.55	54.40 ± 0.56	70.2 ± 0.95	71.36 ± 0.70
16	tridecan-2-ol	1653-31-2	MS, RI, STD	1495	1501	2.23 ± 0.17	10.10 ± 0.61	6.25 ± 0.17	4.30 ± 1.43	5.70 ± 0.20	6.40 ± 0.04	4.20 ± 0.40
17	(3E,6E)-α-farnesene	502-61-4	MS, RI	1503	1509	3.07 ± 0.28	1.20 ± 0.03	1.70 ± 0.04	2.70 ± 0.12	2.50 ± 0.09	2.30 ± 0.21	1.70 ± 0.03
18	ß-bisabolene	495-61-4	MS, RI	1505	1511	1.30 ± 0.05	-	1.20 ± 0.24	0.60 ± 0.37	-	0.50 ± 0.06	1.37 ± 0.20
19	γ-cadinene	39029-4-9	MS, RI	1513	1517	-	-	0.20 ± 0.02	-	0.60 ± 0.08	-	0.2 ± 0.08
20	δ-cadinene	483-76-1	MS, RI	1524	1526	-	0.20 ± 0.04	0.30 ± 0.04	0.10 ± 0.03	1.20 ± 0.05	-	0.38 ± 0.23
21	elemol	639-99-6	MS, RI, STD	1547	1552	-	0.30 ± 0.15	-	-	2.90 ± 0.07	-	0.31 ± 0.30
22	nerolidol	7212-44-4	MS, RI, STD	1557	1564	3.13 ± 0.27	0.40 ± 0.16	1.80 ± 0.63	0.40 ± 0.11	0.70 ± 0.03	1.20 ± 0.35	0.60 ± 0.03
23	caryophyllene oxide	1139-30-6	MS, RI, STD	1583	1588	-	0.20 ± 0.11	0.30 ± 0.00	-	-	-	-
24	ζ-cadinol	5937-11-1	MS, RI	1639	1645	-	-	-	-	0.30 ± 0.02	-	-
25	α-cadinol	481-34-5	MS, RI	1650	1659	-	-	-	-	0.50 ± 0.03	-	-
26	pentadecan-2-one	2345-28-0	MS, RI, STD	1696	1697	0.48 ± 0.02	0.50 ± 0.23	0.70 ± 0.02	-	-	-	-

Table 3. *Cont.*

N°	Compounds	Cas Number	Identification	Ria	Rib	Trunk Bark						
						May	June	July	Aug	Sep	Oct	Nov
27	(*E,E*)-farnesol	106-28-5	MS, RI	1722	1723	12.5 ± 0.85	3.00 ± 0.96	11.1 ± 0.15	1.40 ± 0.90	2.20 ± 0.45	1.90 ± 0.28	-
28	farnesal	19317-11-4	MS, RI	1738	1744	1.36 ± 0.02	0.40 ± 0.15	1.20 ± 0.05	-	0.30 ± 0.07	0.20 ± 0.04	-
29	methyl farnesoate	3675-00-1	MS, RI	1779	1785	1.90 ± 0.10	1.00 ± 0.35	1.60 ± 0.17	-	0.90 ± 0.08	0.70 ± 0.14	-
	Monoterpene hydrocarbons (%)					0.52	0.00	0.00	0.00	1.00	0.00	0.00
	Oxygenated monoterpene (%)					0.00	5.30	0.50	0.00	5.90	0.00	4.10
	Sesquiterpene hydrocarbons (%)					31.87	21.80	24.20	13.60	22.20	16.90	18.20
	Oxygenated sesquiterpenes (%)					19.37	5.80	16.70	1.80	8.20	4.30	1.10
	Others (%)					47.46	66.40	57.65	83.10	62.10	77.10	75.60
	Identified compounds (%)					99.22	99.30	99.05	98.50	99.40	98.30	99.00
	Yield (%)					**0.88**	**0.91**	**1.18**	**1.20**	**0.86**	**0.89**	**0.86**

Identification methods: RIa, theoretical kovats indices (Pubchem and NIST); RIb, calculated kovats indices; MS, mass spectra comparison with PAL 600®libraries; STD, retention time and mass spectra comparison with commercially available standards; RI, retention index comparison with the literature; CAS number; -, Under perception threshold.

Table 4. Chemical composition of essential oils hydrodistillated from *Zanthoxylum leprieurii* fruits. Each analysis was realized in triplicate.

N°	Compounds	Cas Number	Identification	Ria	Rib	Fruits				
						July	Aug	Sep	Oct	Nov
1	α-pinene	80-56-8	MS, RI, STD	931	929	2.30 ± 0.90	0.90 ± 0.12	0.80 ± 0.04	1.90 ± 0.66	0.60 ± 0.13
2	ß-myrcene	123-35-3	MS, RI, STD	987	988	16.40 ± 0.91	44.80 ± 0.13	46.30 ± 0.10	48.27 ± 0.26	24.7 ± 0.97
3	α-terpinene	99-86-5	MS, RI, STD	1008	1017	-	-	-	-	0.30 ± 0.06
4	p-cymene	25155-15-1	MS, RI, STD	1022	1022	0.20 ± 0.08	0.70 ± 0.09	0.40 ± 0.04	0.30 ± 0.05	0.80 ± 0.04
5	limonene	138-86-3	MS, RI, STD	1023	1027	1.10 ± 0.21	2.60 ± 0.29	2.00 ± 0.07	2.20 ± 0.16	6.00 ± 0.42
6	(*E*)-ß-ocimene	13877-91-3	MS, RI, STD	1041	1046	8.30 ± 0.29	2.00 ± 0.12	1.50 ± 0.08	1.50 ± 0.06	2.0 ± 0.15
7	γ-terpinene	99-85-4	MS, RI, STD	1060	1057	-	-	-	-	0.20 ± 0.01
8	linalool	78-70-6	MS, RI, STD	1094	1098	-	2.50 ± 0.27	3.40 ± 0.12	2.10 ± 0.10	4.20 ± 0.6
9	perillene	539-52-6	MS, RI, STD	1094	1100	6.50 ± 0.12	-	-	-	-

Table 4. *Cont.*

N°	Compounds	Cas Number	Identification	Ria	Rib	Fruits				
						July	Aug	Sep	Oct	Nov
10	alloocimene	673-84-7	MS, RI	1128	1129	0.50 ± 0.07	0.50 ± 0.05	0.50 ± 0.04	0.40 ± 0.05	0.30 ± 0.02
11	isopulegol	7786-67-6	MS, RI	1140	1145	-	-	-	-	0.60 ± 0.15
12	citronellal	106-23-0	MS, RI, STD	1153	1153	0.20 ± 0.01	1.00 ± 0.11	1.40 ± 0.13	0.90 ± 0.12	5.70 ± 0.43
13	limonene oxide	1195-92-2	MS, RI	1175	1182	-	-	0.90 ± 0.01	-	0.90 ± 0.24
14	terpineol	98-55-5	MS, RI, STD	1190	1191	-	-	-	-	0.20 ± 0.05
15	decanal	112-31-2	MS, RI, STD	1202	1204	8.30 ± 0.17	1.80 ± 0.18	2.20 ± 0.02	1.60 ± 0.14	0.70 ± 0.24
16	citronellol	106-22-9	MS, RI, STD	1225	1227	1.90 ± 0.02	6.20 ± 0.55	6.60 ± 0.14	6.30 ± 0.63	28.24 ± 0.10
17	geraniol	106-24-1	MS, RI, STD	1250	1253	1.20 ± 0.03	5.50 ± 0.59	6.20 ± 0.22	6.30 ± 0.51	2.33 ± 0.74
18	decyl alcohol	112-30-1	MS, RI, STD	1262	1263	4.50 ± 0.18	-	-	-	-
19	geranial	141-27-5	MS, RI, STD	1268	1270	-	5.30 ± 0.53	7.60 ± 0.12	5.30 ± 0.31	12.50 ± 0.47
20	undecan-2-one	112-12-9	MS, RI, STD	1288	1293	0.20 ±0.03	-	-	-	-
21	undecanal	112-44-7	MS, RI, STD	1305	1306	0.20 ± 0.02	-	-	-	-
22	methyl nerate	1862-61-9	MS, RI, STD	1319	1323	6.70 ± 0.23	5.7 ± 0.6	4.40 ± 0.15	5.30 ± 0.13	6.10 ± 0.50
23	δ-elemene	20307-84-0	MS, RI	1334	1339	-	-	-	0.10 ± 0.02	-
24	α-cubebene	17699-14-8	MS, RI, STD	1349	1351	-	0.20 ± 0.01	0.10 ± 0.00	0.10 ± 0.01	-
25	α-copaene	3856-25-5	MS, RI, STD	1376	1378	0.60 ± 0.01	0.20 ± 0.02	0.20 ± 0.01	0.20 ± 0.02	-
26	ß-elemene	515-13-9	MS, RI	1388	1394	1.20 ± 0.04	2.10 ± 0.21	1.50 ± 0.01	1.80 ± 0.07	0.20 ± 0.1
27	β-caryophyllene	87-44-5	MS, RI, STD	1419	1423	5.80 ± 0.18	4.90 ± 0.55	4.10 ± 0.03	3.50 ± 0.06	0.50 ± 0.23
28	cadina-4(14),5-diene	54324-03-7	MS, RI	1430	1433	0.80 ± 0.06	1.80 ± 0.20	1.60 ± 0.03	1.30 ± 0.07	-
29	γ-elemene	3242-08-8	MS, RI	1432	1435	0.40 ± 0.02	1.10 ± 0.11	0.90 ± 0.01	0.90 ± 0.05	-
30	α-humulene	6753-98-6	MS, RI, STD	1456	1457	3.70 ± 0.21	1.90 ± 0.21	1.60 ± 0.01	1.50 ± 0.07	0.20 ± 0.09
31	alloaromadendrene	25246-27-9	MS, RI	1457	1465	0.90 ± 0.04	0.30 ± 0.03	0.10 ± 0.01	0.20 ± 0.08	-
32	germacrene D	23986-74-5	MS, RI, STD	1482	1485	0.50 ± 0.04	0.80 ± 0.10	1.10 ± 0.12	1.40 ± 0.18	0.20 ± 0.1

Table 4. *Cont.*

N°	Compounds	Cas Number	Identification	Ria	Rib	Fruits				
						July	Aug	Sep	Oct	Nov
33	ß-selinene	17066-67-0	MS, RI	1483	1490	-	-	0.10 ± 0.02	0.10 ± 0.04	-
34	α-selinene	473-13-2	MS, RI	1488	1498	-	-	0.40 ± 0.01	0.40 ± 0.09	-
35	γ-cadinene	39029-4-9	MS, RI	1513	1517	0.90 ± 0.06	1.10 ± 0.08	0.70 ± 0.01	0.80 ± 0.05	0.20 ± 0.07
36	δ-cadinene	483-76-1	MS, RI	1524	1526	2.70 ± 0.14	3.30 ± 0.35	2.20 ± 0.02	2.70 ± 0.05	0.60 ± 0.15
37	α-calacorene	21391-99-1	MS, RI	1542	1547	0.50 ± 0.01	-	-	-	-
38	elemol	639-99-6	MS, RI, STD	1547	1552	0.10 ± 0.02	0.40 ± 0.04	0.20 ± 0.02	-	-
39	spathulenol	6750-60-3	MS, RI	1578	1582	5.20 ± 0.06	-	-	-	-
40	caryophyllene oxide	1139-30-6	MS, RI, STD	1583	1588	9.60 ± 0.29	-	-	-	-
41	ζ-cadinol	5937-11-1	MS, RI	1639	1645	1.20 ± 0.31	0.30 ± 0.06	0.10 ± 0.01	0.40 ± 0.04	-
42	α-cadinol	481-34-5	MS, RI	1650	1659	1.80 ± 0.24	0.50 ± 0.05	-	0.70 ± 0.08	-
43	cadalene	483-78-3	MS, RI	1678	1679	-	-	-	0.10 ± 0.02	-
	Monoterpene hydrocarbons (%)					28.80	51.50	51.50	54.17	34.90
	Oxygenated monoterpene (%)					9.80	20.50	26.10	21.30	54.67
	Sesquiterpene hydrocarbons (%)					18.10	17.70	14.60	15.10	1.90
	Oxygenated sesquiterpenes (%)					25.10	6.90	4.70	6.90	6.10
	Others (%)					17.50	1.80	2.20	1.60	0.70
	Identified compounds (%)					99.30	98.40	99.10	99.07	98.27
	Yield (%)					**1.42**	**1.51**	**1.22**	**1.13**	**1.14**

Identification methods: RIa, theoretical kovats indices (Pubchem and NIST); RIb, calculated kovats indices; MS, mass spectra comparison with PAL 600®libraries; STD, retention time and mass spectra comparison with commercially available standards; RI, retention index comparison with the literature; CAS number; -, Under perception threshold.

The essential oils hydrodistillated from *Z. leprieurii* fruits during different months mainly contained monoterpenes hydrocarbons, which is in agreement with the chemical composition of fruit essential oils of the same species studied in Cameroon. Indeed, two different studies conducted in two distinct Cameroon sites showed citronellol (29.90% [20]; 17.37% [17]) and (*E*)-β-ocimene (44% [40]; 90.30% [49]) as the major compounds. Nevertheless, the chemical compositions characterized here are different from those already described. As essential oils obtained here and in previous studies were not hydrodistillated from plants growing in the same place and during the same period, differences in chemical compositions can be explained by climatic and environmental factors depending on each country, while one also cannot exclude a genetic influence that would combine with the other factors of variability.

The chemical analysis of essential oils hydrodistillated from the different organs of *Z. leprieurii* from Côte d'Ivoire highlighted the presence of a wide range of compounds, most of them already known for their different interesting biological activities. The presence of those molecules can explain the various uses of *Z. leprieurii* in traditional medicine for the treatment of many different affections, as mentioned above. The main molecules found in Ivorian *Z. leprieurii* essential oils and their known biological properties are presented in Figure 2.

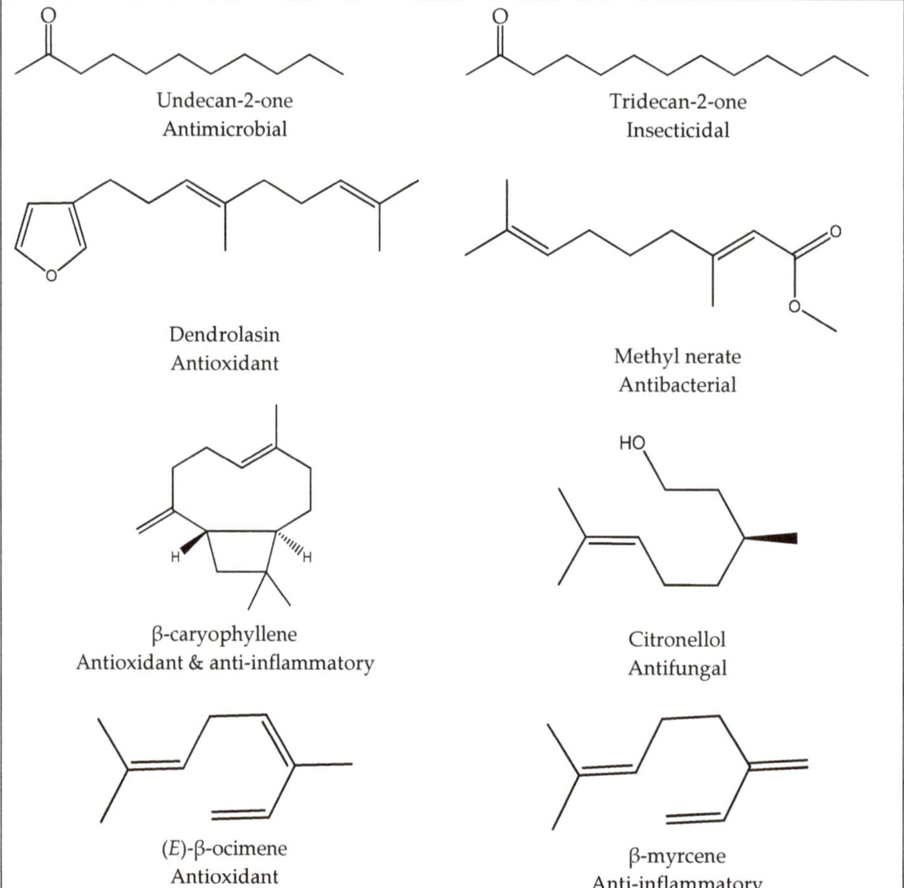

Figure 2. Some major seasonal compounds present in the leaf, trunk bark and fruit essential oils of *Z. leprieurii* from Côte d'Ivoire and their known biological activities [45,50–52].

3.2. Seasonal Effect on Essential Oil Composition

HCA and PCA analysis were performed to investigate the seasonal effect on essential oil compositions.

The HCA dendrogram (Figure 3), based on the Euclidean distance between collected samples, showed three distinct clusters, each one specific to one plant organ: (i) cluster I for fruits; (ii) cluster II for trunk bark; and (iii) cluster III for leaves. This shows that there is a significant difference in the composition of essential oils hydrodistillated from different *Z. leprieurii* organs. In addition, a seasonal effect was observed among each group, showing variation in the compositions of essential oils hydrodistillated from the same organ during the collection period. This seasonal effect was higher for the leaf essential oils, as the intra-class variance for those samples (2.50) was higher than for the fruit (1.77) and for the trunk bark (0.35) samples. It is possible that this higher variance for leaf essential oil samples is related to the fact that new leaves are produced all year long, while fruits are only produced at certain times of the year and the trunk bark develops very slowly over a period of several months or years. Leaves are then more susceptible to seasonal variations, such as levels of light exposure, state of maturity and water stress, than fruits and the trunk bark. Trunk bark essential oils have a lower chemical variability, probably due to the fact that the trunk bark formation is slow, and thus less impacted by environmental factors [53]. Results are supported by the fact that seasonal differences in the chemical composition of essential oils from fruits, trunk bark and leaves have already been highlighted for other *Zanthoxylum* species [21,23]. Indeed, while the chemical compositions of essential oils are genetically determined, it can be considerably modified by factors such as temperature, light, seasonality, water availability and nutrition. Biosynthesis of different compounds can be induced by environmental stimuli, which can change metabolic pathways [54,55].

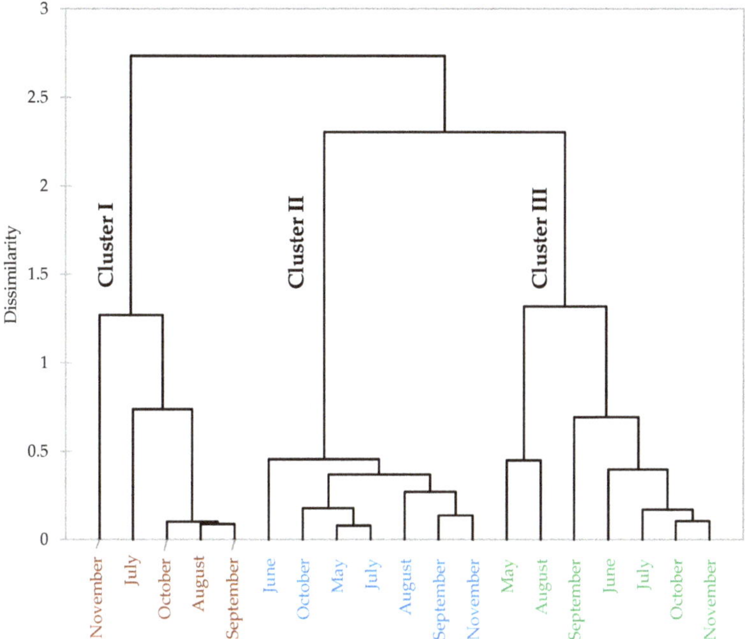

Figure 3. Dendrogram representing *Zanthoxylum leprieurii* essential oil samples. Cluster I: fruits; cluster II: trunk bark; and cluster III: leaves.

For the PCA analysis, the chemical composition data were projected through linear combinations of the 15 variables that were identified in all samples. Results showed that the first two axes (F1 and F2) explained 64.32% of the total variance (F1: 36.41% and F2: 27.92%). PCA results (Figure 4) showed three different specific clusters, each one being represented by one plant organ. Fruit essential oil samples in cluster I were mainly composed of β-myrcene (36.78 ± 14.67%), citronellol (9.98 ± 10.75%), geraniol (6.14 ± 4.42%), methyl nerate (5.64 ± 0.86%) and geraniol (4.36 ± 2.32%). Cluster II included trunk bark essential oil samples dominated by methylketones with tridecan-2-one (61.56 ± 12.65%) as the principal component. However, α- humulene (7.67 ± 2.71%), β-caryophyllene (6.82 ± 4.25%) and tridecan-2-ol (5.95 ± 2.58%) were also present in significant amounts. Cluster III included the leaf oil samples that were mainly composed of tridecan-2-one (24.11 ± 10.47%), β-caryophyllene (13.97 ± 4.94 %), dendrolasin (8.34 ± 4.72) and α-humulene (4.82 ± 1.29%). This group also showed high levels of (*E*)-β-ocimene (5.35 ± 8.56%), undecan-2-one (4.20 ± 3.34%), linalool (3.73 ± 2.89%), thymol (3.31 ± 4.91%), α-farnesene (3.16 ± 2.98%) and β-elemene (3.13 ± 1.83%).

Observations (axis F1 and F2 : 64.32 %)

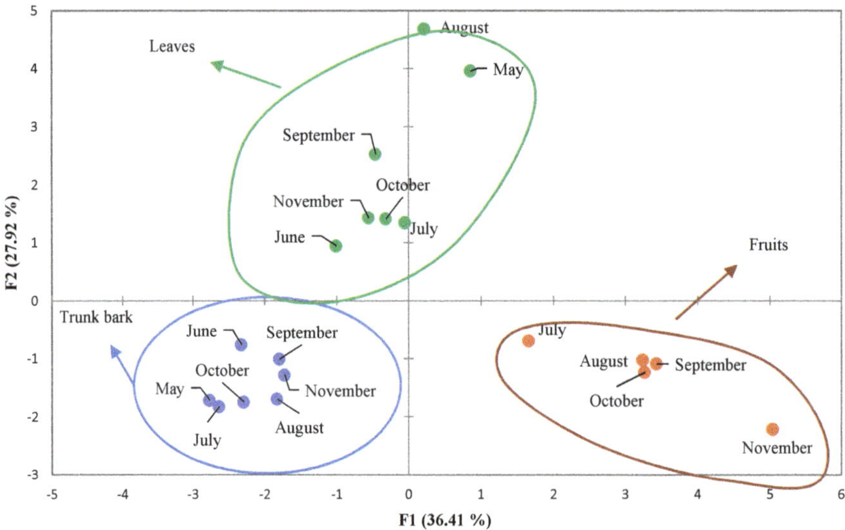

Figure 4. Principal component analysis of the chemical composition of essential oils hydrodistillated from *Zanthoxylum leprieurii* leaves, trunk bark and fruits from Côte d'Ivoire; described according to months and major compounds.

3.3. Essential oil Biological Activities

As mentioned previously, *Z. leprieurii* is widely used in traditional medicine for the treatment of different afflictions. Several authors have already supported those uses by reporting interesting biological properties of essential oils and solvent extracts obtained from this species growing in different places. However, it was shown here that *Z. leprieurii* essential oil chemical composition varies widely depending on the organ of the plant used and depending on the collection month. It is thus important to evaluate the biological activities of the essential oil hydrodistillated in this study, with regards to their compositions. Antioxidant, anti-inflammatory, insecticidal and anti-malarial properties of essential oils hydrodistillated from *Z. leprieurii* growing in Côte d'Ivoire were then evaluated. Essential oils used for the biological activity tests were selected based on their chemical composition. The August-selected leaf essential oil sample was characterized by high amounts of tridecan-2-one (30.20%), β-caryophyllene (13.70%) and thymol (13.30%). The major compounds of the chosen July trunk bark sample were tridecan-2-one (51.40%), (*E*,*E*)-farnesol (11.10%) and β-caryophyllene (8.50%). Finally, the July fruit sample was characterized by high proportions of β-myrcene (16.40%), caryophyllene oxide (9.60%), (*E*)-β-ocimene (8.30%) and decanal (8.30%).

3.3.1. Antioxidant Activity

DPPH Free Radical Scavenging Assay

According to Rice-Evans [56], the antioxidant activity of a compound corresponds to its ability to resist oxidation. The free radical scavenging ability of selected essential oils from *Z. leprieurii* leaves, trunk bark and fruits were determined using DPPH with Trolox as a positive control.

The results (Table 5) showed that all essential oil samples were able to reduce the stable DPPH radical to yellow diphenylpicrylhydrazine, with the scavenging effects increasing with higher essential oil concentrations (p-value < 0.05). Leaf essential oil had the highest antioxidant activity (IC_{50}: 33.12 ± 0.07 µg/mL), followed by trunk bark oil (IC_{50}: 65.68 ± 0.12 µg/mL) and fruit oil (IC_{50}: 103,55 ± 0.35 µg/mL). The comparison with the Trolox standard (29.13 ± 0.04 µg/mL) showed that the selected leaf essential oil sample has a high antioxidant activity. This activity could be due to the high contents in β-caryophyllene and thymol of this essential oil, as both of those molecules are already known for their antioxidant properties [57]. Those molecules were either present in lower quantities, or absent in the other tested essential oil samples (trunk bark and fruit).

High DPPH free radical scavenging activity was also described in leaf essential oils from other *Zanthoxylum* species; with an IC_{50} value of 27.00 ± 0.1 µg/mL for Indian samples [58]. However, our results strongly differed to those of Tchabong [40], who obtained IC_{50} values of 770 µg/mL and 1800 µg/mL for *Z. leprieurii* fruit and leaf oils from Cameroon, respectively. Those differences in antioxidant properties of essential oil samples from the same species and families collected at different sites and at different periods are probably due to differences in their chemical compositions. Those differences may come from the studied organ, as we showed that essential oil composition variability mainly comes from the chosen organ, but also from genetic factors and/or environmental factors.

Table 5. Biological properties of essential oils hydrodistillated from different *Z. leprieurii* organs. DPPH: 2,2-diphenyl-1-picrylhydrazyl, LOX: lipoxygenase, BSA: bovine serum albumin.

Organs and Standards	Biological Activities IC_{50} (µL/mL)			
	DPPH	**LOX Denaturation**	**BSA Denaturation**	**Anti-Plasmodial**
Leaves	33.12 ± 0.07	26.26 ± 0.04	26.08 ± 0.12	62.3 ± 3.4
Trunk Bark	65.68 ± 0.12	28.40 ± 0.02	35.07 ± 0.15	36.29 ± 4.2
Fruits	103.55 ± 0.35	32.42 ± 0.15	26.68 ± 0.09	>100
Trolox	28.13 ± 0.04			
Quercetin		21.57 ± 0.10		
Diclofenac			21.90 ± 0.08	
Artemisinin				0.004 ± 0.001

Ferric-Reducing Antioxidant Power

The ferric-reducing antioxidant power (FRAP) of essential oils extracted from leaves, trunk bark and fruits of *Z. leprieurii* was studied here for the first time. Results (Figure 5) showed that essential oils exhibited strong antioxidant activities, which were higher with increasing oil concentrations (p-value < 0.05) [59]. Fruit and leaf essential oils exhibited higher FRAP activity than trunk bark oils. All these organs were compared to the Trolox, which represented the standard.

Two different assays, DPPH and FRAP, were conducted in this study to evaluate the antioxidant potential of essential oils hydrodistillated from *Z. leprieurii* leaves, trunk bark and fruits. The two different tests resulted in dissimilar results, as leaf and fruit oils gave the highest and the lowest antioxidant activities with the DPPH free radical scavenging assay, respectively; while in the ferric-reducing antioxidant power assay, the highest antioxidant activities were obtained with leaf and fruit oils. Variations in the antioxidant activities of essential oils evaluated by DPPH and FRAP methods are probably due to the differences in reagents used by each method [60]. Indeed, the DPPH assay evaluates the ability of essential oils to scavenge free radicals, while the FRAP method assesses essential oils' reducing power. The results obtained here showed that essential oils hydrodistillated from different *Z. leprieurii* organs have interesting antioxidant properties, which originate from two different modes of action: free radical scavenging and reducing abilities. The various compounds in essential oils hydrodistillated from *Z. leprieurii* organs are probably the origin of those different antioxidant activities [61,62]. For example, quantities of (E)-β-ocimene, perillene and caryophyllene oxide, which are known for their antioxidant properties [63,64], were found in the fruit oil sample, which were present in much lower proportions or completely absent in other essential oils.

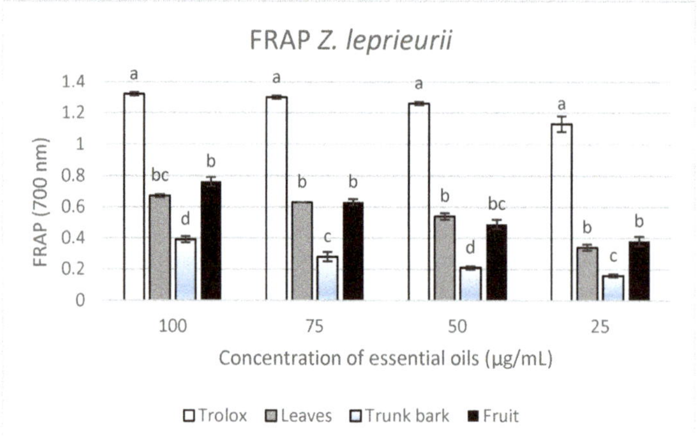

Figure 5. The ferric-reducing power of essential oils hydrodistillated from leaves, trunk bark and fruits of *Z. leprieurii*. Mean values and standard deviation values were presented (*n* = 3). For a same concentration, data with the same letter were not significantly different from each other according to Duncan's test (*p*-value < 0.05).

3.3.2. Anti-Inflammatory Activity

In order to assess the anti-inflammatory potential of *Z. leprieurii* essential oils, their lipoxygenase inhibitory activity was evaluated, and the anti-denaturation method of bovine albumin serum (BSA) was also used.

Lipoxygenase Denaturation Inhibition Activity

The tested essential oils showed high to moderate lipoxygenase inhibitory activity (IC$_{50}$: 26.26 ± 0.04 µg/mL, 28.40 ± 0.02 µg/mL and 32.42 ± 0.15 µg/mL for leaf, trunk bark and fruit oils, respectively) when compared to standard Quercetin (21.57 ± 0.10 µg/mL) (Table 5). These results show that *Z. leprieurii* essential oils have anti-inflammatory properties, as has also been previously described with *Z. leprieurii* growing in different places [65,66].

Inhibition of Albumin Denaturation

In vitro anti-inflammatory properties of *Z. leprieurii* trunk bark, leaf and fruit essential oils were evaluated by the anti-denaturation method of bovine albumin serum (BSA) for the first time, in comparison with the control Diclofenac (IC50: 21.90 ± 0.08 µg/mL). The results (Table 5) showed that *Z. leprieurii* leaf, fruit and trunk bark essential oils have high-to-moderate anti-inflammatory activities, with IC_{50} values of 26.08 ± 0.12 µg/mL, 26.68 ± 0.09 µg/mL and 35.07 ± 0.15 µg/mL, respectively. Moreover, the percentage of BSA protection was dependent on essential oil concentrations (*p*-value < 0.05). The origin of these high lipoxygenase inhibitory activities could be the difference in organ content of monoterpenes, methylketones and sesquiterpenes, which are known for their anti-inflammatory activities [67–69].

3.3.3. Insecticidal Activity

Losses due to insect infestation during grain storage are a serious problem around the world, and more acutely in developing countries. Consumption of grains is not the only loss caused by insects, as a high level of pest detritus also leads to grains being unfit for human consumption in terms of quality. It could be estimated that one third of the world's food production is destroyed by insects every year, which represents more than $100 billion. The highest losses occur in developing countries (43%), such as Côte d'Ivoire [70]. In the tropical zone, average losses range from 20% to 30%, while in the temperate zones, losses are from 5% to 10% [71]. Moreover, the trend to use natural insecticides to avoid chemical residues in food is growing.

The insecticidal activities of *Z. leprieurii* trunk bark, leaf and fruit essential oils were evaluated against *Sitophilus granarius*, one of the most damaging pests of stored cereals in the world. This insect is a primary pest, as it is able to drill holes in grains, laying its eggs inside them and allowing secondary pests to develop [33].

Results showed that all essential oils were efficient to kill insects in 24 h, with trunk bark oil showing the highest insecticidal activity (LC_{50} = 8.87 µL/mL) in comparison with leaf and fruit essential oils (LC_{50} = 15.77 µL/mL and 11.26 µL/mL, respectively); those activities were slightly lower than those of the chemical insecticide Talisma UL (LC_{50} = 3.44 µL/mL). Moreover, in comparison with cinnamon (*Cinnamomum zeylanicum*) and clove (*Syzygium aromaticum*) essential oils, generally described as exhibiting high insecticidal activities [72], LC_{50} of *Z. leprieurii* essential oils is lower, showing better insecticidal activities of the latter and thus promising prospects for application in the protection of stored foodstuffs. Concerning LC_{90} and LC_{95}, results showed that the chemical insecticide (LC_{90} = 27.83 µL/mL, LC_{95} = 56.66 µL/mL) was less effective than leaf essential oil (LC_{90} = 26.27 µL/mL, LC_{95} = 31.26 µL/mL) and trunk bark essential oil (LC_{90} = 23.69 µL/mL, LC_{95} = 33.10 µL/mL), but more effective than fruit essential oils (LC_{90} = 93.20 µL/mL, LC_{95} = 191.20 µL/mL). These data indicate that the insecticidal effect of essential oils varies depending on the chemical composition and synergistic effects occurring between the compounds [73]. In this study, essential oils hydrodistillated from *Z. leprieurii* organs had an interesting effect on *Sitophilus granarius* adults, as insecticidal activities were better than those of *Z. fagara* and *Z. monoplyllum* (LC_{50} of 153.9 µL/mL and 140.1 µL/mL, respectively) [74]; and the LC_{50} was better than those reported on larvicidal activity [75] with *Z. leprieurii* extracts and *Z. avicennae* essential oil [76]. It should be noted that chemical composition of essential oils is different among these *Zanthoxylum* species and, according to the author [77], mortality evolution showed that toxicity depends on aspects such as the chemical composition and the target insect sensitivity.

The repulsive effect of *Z. leprieurii* trunk bark, leaf and fruit essential oils and chemical insecticides were also evaluated by the McDonald method. The results (Table 6) showed a high repulsive effect for the trunk bark essential oil (88.83%), followed by leaf essential oil (76.66%) and fruit essential oil (61.00%), in comparison with the low repulsive effect of the chemical insecticide Talisma UL (24.78%). Furthermore, repellent properties were dose–response correlated and high when compared to other essential oils considered to be highly repulsive [78].

Table 6. Repulsion percentage of *Sitophilus granarius* after 2 h of treatment with essential oils and Talisma UL. Effect of substance tested [35].

Tested Substances	Average Repulsion (%)	Class	Effect of Substance Tested
Leaf essential oil	76.66	IV	Repulsive
Trunk bark essential oil	88.83	V	Highly repulsive
Fruit essential oil	61.00	III	Mildly repulsive
Talisma UL	24.78	II	Weakly repulsive

3.3.4. Anti-Plasmodial Activity

The anti-plasmodial activity of *Z. leprieurii* trunk bark, leaf and fruit essential oils was evaluated here for the first time. The results (Table 5) showed that trunk bark essential oil has a moderate anti-plasmodial activity (IC_{50}: 37.49 ± 4.2 µg/mL), and leaf essential oil has a low activity (IC_{50}: 59.30 ± 3.4 µg/mL), in comparison with the artemisinin standard (IC_{50}: 0.004 ± 0.001 µg/mL). No significant anti-plasmodial activity was highlighted for the fruit essential oil ($IC_{50} > 100$). The moderate trunk bark anti-plasmodial activity may be due to methylketones, as tridecan-2-one is the dominant compound in this oil. However, no studies have yet shown the anti-plasmodial activity of this molecule. Moreover, it is possible that the highlighted activity comes from the presence of minor compounds, as well as from the synergy between different molecules. Nevertheless, studies were carried out on *Z. leprieurii* and other species of *Z. chalybeum* and *Z. zanthoxyloides* plant extracts, showing high anti-plasmodial activities [79–82], all of which supports the effective use of *Zanthoxylum* species in traditional medicine for the treatment of malaria.

4. Conclusions

In this study, the variability in the chemical composition of leaf, trunk bark and fruit essential oils hydrodistillated from Ivorian *Z. leprieurii* was studied for the first time over seven months for leaves and trunk bark, and five months for fruits. Results showed that essential oils were mainly dominated by sesquiterpenes (β-caryophyllene, dendrolasin and thymol), methylketones (tridecan-2-one and undecane-2 one) and monoterpenes (β-myrcene, (E)-β-ocinene and perillene) in leaf, trunk bark and fruit samples, respectively. Statistical PCA and HCA analysis showed that the variability in essential oil compositions mainly comes from the organ, as all samples were clustered in three groups, each one corresponding to one organ. However, differences in essential oil compositions inside each cluster were highlighted, showing the probable impact of the seasonal effect on essential oil compositions. Those differences in essential oil compositions may be due to different seasonal parameters, as it was shown here that the temperature, precipitations and humidity were not constant during the plant collecting period. However, it is also known that biotic factors, such as pest attacks, widely impact essential oil chemical compositions. Those were not recorded during this study, but may also be at the origin of essential oil variability. As a perspective, it would be interesting to study their impact on *Z. leprieurii* essential oil variability. Moreover, the study was conducted with plants growing on the same site. The comparison of the present results with those of the existing literature considering *Z. leprieurii* plants growing in other countries showed totally different essential oil compositions, showing that genetic differences might also induce dissimilar essential oil compositions, resulting in distinct essential oil chemotypes.

Z. leprieurii is widely used in traditional medicine for the treatment of different diseases, such as rheumatism, tuberculosis, urinary infections and generalized body pains. In order to explain those uses, in-vitro biological activities of hydrodistillated essential oils were studied here. Results obtained in this study showed strong antioxidant, anti-inflammatory and moderate anti-plasmodial activities. Moreover, expected results also showed high differences in the biological activities of essential oils linked with their differences in chemical composition, which should be taken into account in future research on *Z. leprieurii* essential oil biological activities, but also to find the proper plant harvesting moment for a use in traditional medicine. However, while those results are promising and confirm the relevance of the traditional uses of these plants, in-vitro experiments should be supported by in-vivo tests, as differences can be observed between in-vitro and in-vivo test results.

Grain storage is particularly problematic, as pests cause large losses. Ivoirian essential oils from *Z. leprieurii* demonstrated an interesting repellent effect and contact toxicity properties against *Sitophilus granarius* with the essential oils extracted from the three organs tested (trunk bark, leaves and fruits). All these essential oils are promising candidates for developing new plant insecticides to protect stored products. Moreover, according to De Lucas and colleagues [83], parts of the plant could also be used in silos directly without the extraction step of the essential oils to control pest losses. Indeed, this practice is widely used in Africa because plant material is readily available and usable without any transformation. Moreover, it is easy to separate the plant material added to the silo from the grain for use as food or feed. It would then be interesting to study the insecticidal properties of *Z. leprieurii* organs in that way.

In conclusion, *Z. leprieurii* from Côte d'Ivoire as a medicinal and aromatic plant provides interesting sources of biologically active compounds, such as antioxidants, anti-inflammatory agents and natural insecticides. The results obtained here support the current uses of this plant in traditional medicine, but also highlight the importance of the location and the season on chemical composition and thus biological properties.

Author Contributions: Conceptualization, E.A.T., M.-L.F. and Z.F.T.; methodology, E.A.T.; software, E.A.T., G.B.B., E.L.W. and H.M.; validation, M.-L.F. and Z.F.T.; formal analysis, E.A.T.; investigation, E.A.T. and F.N.; resources, E.A.T. and M.-L.F.; data curation, E.A.T., M.G., H.M., G.B.B., E.L.W. and A.L.; writing—original draft preparation, E.A.T., M.G. and H.M.; writing—review and editing, M.G., M.-L.F. and Z.F.T.; visualization, E.A.T. and M.G.; supervision, M.-L.F., M.F. and Z.F.T.; project administration, M.-L.F. and Z.F.T.; funding acquisition, M.-L.F. and Z.F.T. All authors have read and agreed to the published version of the manuscript.

Funding: This research was funded by the Education, Audiovisual and Culture Executive Agency (EACEA) trough EOHUB project 600873-EPP-1-2018-1ES-EPPKA2-KA.

Acknowledgments: We would like to thank the members of the Laboratory of Biological Organic Chemistry of the University Felix Houphouet Boigny of Cocody (Côte d'Ivoire) and all the laboratory staff of Chemistry of Natural Molecules, University of Liege (Gembloux Agro-Bio Tech) for their scientific contribution particularly Danny Trisman, Saskia Sergeant, Thomas Bertrand, Franck Michels, Marie Davin, Laurie Josselin, Pierre-Yves Werrie and Clement Burgeon. We are also grateful to Felicien for his support.

Conflicts of Interest: The authors declare no conflict of interest.

References

1. Gonçalves, M.A.M. Micromorphology and in vitro Antibacterial Evaluation of *Zanthoxylum* and *Hymenocardia* Species. Ph.D. Thesis, Universidade de Lisboa, Lisboa, Portugal, 2018.

2. Dongmo, P.M.J.; Tchoumbougnang, F.; Sonwa, E.T.; Kenfack, S.M.; Zollo, P.H.A.; Menut, C. Antioxidant and anti-inflammatory potential of essential oils of some *Zanthoxylum* (Rutaceae) of Cameroon. *Int. J. Essent. Oil* **2008**, *2*, 82–88.

3. Sriwichai, T.; Sookwong, P.; Siddiqui, M.W.; Sommano, S.R. Aromatic profiling of *Zanthoxylum myriacanthum* (makwhaen) essential oils from dried fruits using different initial drying techniques. *Ind. Crops Prod.* **2019**, *133*, 284–291. [CrossRef]

4. Bunalema, L.; Obakiro, S.; Tabuti, J.R.; Waako, P. Knowledge on plants used traditionally in the treatment of tuberculosis in Uganda. *J. Ethnopharmacol.* **2014**, *151*, 999–1004. [CrossRef] [PubMed]

5. Burkil, H.M. Royal Botanic Gardens; Kew Year. In *The Useful Plants of West Tropical Africa*; Royal Botanic Gardens: London, UK, 1994; Volume 2.

6. Tine, Y.; Renucci, F.; Costa, J.; Wélé, A.; Paolini, J. A Method for LC-MS/MS profiling of coumarins in *Zanthoxylum xanthoxyloides* (Lam.) B. Zepernich and Timler extracts and essential oils. *Molecules* **2017**, *22*, 174. [CrossRef] [PubMed]

7. Mester, I. *Chemistry and Chemical Taxonomy of the Rutales*; Waterman, P.G., Grundon, M.F., Eds.; Academic Press: London, UK, 1983; pp. 31–96.

8. Kpomah, E.D.; Uwakwe, A.A.; Abbey, B.W. Aphrodisiac studies of diherbal mixture of *Zanthoxylum leprieurii* Guill. & Perr. And *Piper guineense* Schumach. & Thonn. on male wistar rats. *GJRMI* **2012**, *1*, 381.

9. Agyare, C.; Kisseih, E.; Kyere, I.Y.; Ossei, P.P.S. Medicinal plants used in wound care: Assessment of wound healing and antimicrobial properties of *Zanthoxylum Leprieurii*. *IBSPR* **2014**, *2*, 81–89.

10. Bunalema, L.; Fotso, G.W.; Waako, P.; Tabuti, J.; Yeboah, S.O. Potential of *Zanthoxylum leprieurii* as a source of active compounds against drug resistant *Mycobacterium tuberculosis*. *BMC Complement. Altern. Med.* **2017**, *17*, 89. [CrossRef]

11. Guetchueng, S.T.; Nahar, L.; Ritchie, K.J.; Ismail, F.; Wansi, J.D.; Evans, A.R.; Sarker, S.D. Kaurane diterpenes from the fruits of *Zanthoxylum leprieurii* (Rutaceae). *Rec. Nat. Prod.* **2017**, *11*, 304–309.

12. Zondegoumba, E.N.T.; Dibahteu, W.L.; de Araujo, A.; Vidari, G.; Liu, Y.; Luo, S.; Li, S.; Junior, F.J.B.M.; Scotti, L.; Tullius, M. Cytotoxic and Schistosomidal Activities of Extract, Fractions and Isolated Compounds from *Zanthoxylum Leprieurii* (Rutaceae). *IJSBAR* **2019**, *44*, 209–222.

13. Ngoumfo, R.M.; Jouda, J.B.; Mouafo, F.T.; Komguem, J.; Mbazoa, C.D.; Shiao, T.C.; Choudhary, M.I.; Laatsch, H.; Legault, J.; Pichette, A.; et al. In vitro cytotoxic activity of isolated acridones alkaloids from *Zanthoxylum leprieurii* Guill. et Perr. *Bioorg. Med. Chem.* **2010**, *18*, 3601–3605. [CrossRef]

14. Adesina, S.K. The Nigerian *Zanthoxylum*: Chemical and biological values. *AJTCAM* **2005**, *2*, 282–301. [CrossRef]

15. Misra, L.N.; Wouatsa, N.V.; Kumar, S.; Kumar, R.V.; Tchoumbougnang, F. Antibacterial, cytotoxic activities and chemical composition of fruits of two Cameroonian *Zanthoxylum* species. *J. Ethnopharmacol.* **2013**, *148*, 74–80. [CrossRef] [PubMed]

16. Tatsadjieu, L.N.; Ngang, J.E.; Ngassoum, M.B.; Etoa, F.X. Antibacterial and antifungal activity of *Xylopia aethiopica*, *Monodora myristica*, *Zanthoxylum xanthoxyloides* and *Zanthoxylum leprieurii* from Cameroon. *Fitoterapia* **2003**, *74*, 469–472. [CrossRef]

17. Gardini, F.; Belletti, N.; Ndagijimana, M.; Guerzoni, M.E.; Tchoumbougnang, F.; Zollo, P.H.A.; Micci, C.; Lanciotti, R.; Kamdem, S.L.S. Composition of four essential oils obtained from plants from Cameroon, and their bactericidal and bacteriostatic activity against *Listeria monocytogenes*, *Salmonella enteritidis* and *Staphylococcus aureus*. *Afr. J. Microbiol. Res.* **2009**, *3*, 264–271.

18. Tanoh, E.A.; Nea, F.; Yapi, T.A.; Boué, G.B.; Jean-Brice, B.; Tomi, F.; Tonzibo, Z.F. Essential Oil of *Zanthoxylum lepreurii* Guill. & Perr. Rich in Undecan-2-One and Tridecan-2-One. *J. Essent. Oil-Bear. Plants* **2018**, *21*, 1397–1402.

19. Oyedeji, A.O.; Lawal, O.A.; Adeniyi, B.A.; Alaka, S.A.; Tetede, E. Essential oil composition of three *Zanthoxylum* species. *J. Essent. Oil Res.* **2008**, *20*, 69–71. [CrossRef]

20. Fogang, H.P.; Tapondjou, L.A.; Womeni, H.M.; Quassinti, L.; Bramucci, M.; Vitali, L.A.; Petrelli, D.; Lupidi, G.; Maggi, F.; Papa, F.; et al. Characterization and biological activity of essential oils from fruits of *Zanthoxylum xanthoxyloides* Lam. and *Z. leprieurii* Guill. & Perr., two culinary plants from Cameroon. *Flavour Fragr. J.* **2012**, *27*, 171–179.

21. Eiter, L.C.; Fadamiro, H.; Setzer, W.N. Seasonal variation in the leaf essential oil composition of *Zanthoxylum clava-herculis* growing in Huntsville, Alabama. *Nat. Prod. Commun.* **2010**, *5*, 457–460. [CrossRef]

22. Kim, J.H. Seasonal variations in the content and composition of essential oil from *Zanthoxylum Piperitum*. *J. Ecol. Field Biol.* **2012**, *35*, 195–201. [CrossRef]

23. Bhatt, V.; Sharma, S.; Kumar, N.; Sharma, U.; Singh, B. Chemical Composition of Essential Oil among Seven Populations of *Zanthoxylum armatum* from Himachal Pradesh: Chemotypic and Seasonal Variation. *Nat. Prod. Commun.* **2017**, *12*, 1643–1646.

24. Benini, C.; Ringuet, M.; Wathelet, J.P.; Lognay, G.; du Jardin, P.; Fauconnier, M.L. Variations in the essential oils from ylang-ylang (*Cananga odorata* [Lam.] Hook f. & Thomson forma genuina) in the Western Indian Ocean islands. *Flavour Fragr. J.* **2012**, *27*, 356–366.

25. Bettaieb Rebey, I.; Bourgou, S.; Aidi Wannes, W.; Hamrouni Selami, I.; Saidani Tounsi, M.; Marzouk, B.; Fauconnier, M.L.; Ksouri, R. Comparative assessment of phytochemical profiles and antioxidant properties of Tunisian and Egyptian anise (*Pimpinella anisum* L.) seeds. *Plant Biosyst.* **2018**, *152*, 971–978. [CrossRef]

26. Lins, L.; Dal Maso, S.; Foncoux, B.; Kamili, A.; Laurin, Y.; Genva, M.; Jijakli, M.H.; De Clerck, C.; Fauconnier, M.-L.; Deleu, M. Insights into the relationships between herbicide activities, molecular structure and membrane interaction of cinnamon and citronella essential oils components. *Int. J. Mol. Sci.* **2019**, *20*, 4007. [CrossRef] [PubMed]

27. Tanoh, E.A.; Nea, F.; Kenne Kemene, T.; Genva, M.; Saive, M.; Tonzibo, F.Z.; Fauconnier, M.L. Antioxidant and lipoxygenase inhibitory activities of essential oils from endemic plants of Côte d'Ivoire: *Zanthoxylum*

mezoneurispinosum Ake Assi and *Zanthoxylum psammophilum* Ake Assi. *Molecules* **2019**, *24*, 2445. [CrossRef] [PubMed]

28. Wang, B.; Qu, J.; Feng, S.; Chen, T.; Yuan, M.; Huang, Y.; Ding, C. Seasonal Variations in the Chemical Composition of *Liangshan Olive* Leaves and Their Antioxidant and Anticancer Activities. *Foods* **2019**, *8*, 657. [CrossRef]

29. Lee, J.H.; Kang, B.S.; Hwang, K.H.; Kim, G.H. Evaluation for anti-inflammatory effects of *Siegesbeckia glabrescens* extract in vitro. *Food Agric. Immunol.* **2011**, *22*, 145–160. [CrossRef]

30. Nikhila, G.S.; Sangeetha, G. Anti-inflammatory properties of the root tubers of *Gloriosa superba* and its conservation through micropropagation. *J. Med. Plants Res.* **2015**, *9*, 1–7. [CrossRef]

31. Kar, B.; Kumar, R.S.; Karmakar, I.; Dola, N.; Bala, A.; Mazumder, U.K.; Hadar, P.K. Antioxidant and in vitro anti-inflammatory activities of *Mimusops elengi* leaves. *Asian Pac. J. Trop. Biomed.* **2012**, *2*, S976–S980. [CrossRef]

32. Nea, F.; Tanoh, E.A.; Wognin, E.L.; Kemene, T.K.; Genva, M.; Saive, M.; Tonzibo, Z.F.; Fauconnier, M.L. A new chemotype of *Lantana rhodesiensis* Moldenke essential oil from Côte d'Ivoire: Chemical composition and biological activities. *Ind. Crops Prod.* **2019**, *141*, 111766. [CrossRef]

33. Fleurat-Lessard, F. Gestion intégrée de la protection des stocks de céréales contre les insectes sans traitement insecticide rémanent. *Phytoma* **2018**, *716*, 32–40.

34. Darwish, Y.A.; Omar, Y.M.; Hassan, R.E.; Mahmoud, M.A. Repellent effects of certain plant essential oil, plant extracts and inorganic salts to granary weevil, *Sitophilus granarius* (L.). *Arch. Arch. Phytopathol. Pflanzenschutz* **2013**, *46*, 1949–1957. [CrossRef]

35. Mc Donald, L.L.; Guyr, H.; Speire, R.R. Preliminary evaluation of new candidate materials as toxicants, repellents, and attractants against stored-product insects. *Mark. Res. Rep.* **1970**, *882*, 189.

36. Ledoux, A.; St-Gelais, A.; Cieckiewicz, E.; Jansen, O.; Bordignon, A.; Illien, B.; Di Giovanni, N.; Marvilliers, A.; Hoareau, F.; Pendeville, H.; et al. Antimalarial activities of alkyl cyclohexenone derivatives isolated from the leaves of *Poupartia borbonica*. *J. Nat. Prod.* **2017**, *80*, 1750–1757. [CrossRef] [PubMed]

37. BEI Reagent Search. Available online: https://www.beiresources.org/Catalog/BEIParasiticProtozoa/MRA-102.aspx (accessed on 27 July 2017).

38. Bordignon, A.; Frédérich, M.; Ledoux, A.; Campos, P.E.; Clerc, P.; Hermann, T.; Quetin-Leclercq, J.; Cieckiewicz, E. In vitro antiplasmodial and cytotoxic activities of sesquiterpene lactones from *Vernonia fimbrillifera* Less. (Asteraceae). *Nat. Prod. Res.* **2018**, *32*, 1463–1466. [CrossRef] [PubMed]

39. Makler, M.T.; Ries, J.M.; Williams, J.A.; Bancroft, J.E.; Piper, R.C.; Gibbins, B.L.; Hinrichs, D.J. Parasite lactate dehydrogenase as an assay for *Plasmodium falciparum* drug sensitivity. *Am. J. Trop. Med. Hyg.* **1993**, *48*, 739–741. [CrossRef] [PubMed]

40. Tchabong, F.; Sameza, S.R.; Tchameni, M.L.; Mounbain, N.S.; Mouelle, F.; Jazet, S.A.; Menut, D.P.M.; Tchoumbougnang, C. Chemical composition, free radical scavenging and antifungal activity of *Zanthoxylum leprieurii* essential oils against *Epidermophyton floccosum* and *Microsporum gypseum*, two most prevalent cutaneous Mycosis. *J. Pharm.* **2018**, *8*, 13–19.

41. Shams, M.; Esfahan, S.Z.; Esfahan, E.Z.; Dashtaki, H.N.; Dursun, A.; Yildirim, E. Effects of climatic factors on the quantity of essential oil and dry matter yield of coriander (*Coriandrum sativum* L.). *INDJSRT* **2016**, *9*, 1–4. [CrossRef]

42. Azzaz Nabil, A.E.; El-Khateeb Ayman, Y.; Farag Ayman, A. Chemical composition and biological activity of the essential oil of *Cyperus articulatus*. *Int. J. Acad. Res.* **2014**, *6*, 265–269.

43. Choucry, M.A. Chemical composition and anticancer activity of *Achillea fragrantissima* (Forssk.) Sch. Bip. (Asteraceae) essential oil from Egypt. *J. Pharm. Phytother.* **2017**, *9*, 1–5.

44. Azizan, N.; Mohd Said, S.; Zainal Abidin, Z.; Jantan, I. Composition and antibacterial activity of the essential oils of *Orthosiphon stamineus* Benth and *Ficus deltoidea* Jack against *Pathogenic Oral Bacteria*. *Molecules* **2017**, *22*, 2135. [CrossRef]

45. Dahham, S.S.; Tabana, Y.M.; Iqbal, M.A.; Ahamed, M.B.; Ezzat, M.O.; Majid, A.S.; Majid, A.M. The anticancer, antioxidant and antimicrobial properties of the sesquiterpene β-caryophyllene from the essential oil of Aquilaria crassna. *Molecules* **2015**, *20*, 11808–11829. [CrossRef] [PubMed]

46. Park, J.; Rodríguez-Moyá, M.; Li, M.; Pichersky, E.; San, K.Y.; Gonzalez, R. Synthesis of methyl ketones by metabolically engineered *Escherichia coli*. *J. Ind. Microbiol. Biotechnol.* **2012**, *39*, 1703–1712. [CrossRef] [PubMed]

47. Pareja, M.; Qvarfordt, E.; Webster, B.; Mayon, P.; Pickett, J.; Birkett, M.; Glinwood, R. Herbivory by a phloem-feeding insect inhibits floral volatile production. *PLoS ONE* **2012**, *7*, e31971. [CrossRef] [PubMed]

48. Saini, M.; Wang, Z.W.; Chiang, C.J.; Chao, Y.P. Metabolic engineering of *Escherichia coli* for production of butyric acid. *J. Agric. Food Chem.* **2014**, *62*, 4342–4348. [CrossRef] [PubMed]

49. Lamaty, G.; Menut, C.; Bessiere, J.M.; Aknin, M. Aromatic plants of tropical Central Africa. II. A comparative study of the volatile constituents of *Zanthoxylum leprieurii* (Guill. et Perr.) Engl. and *Zanthoxylum tessmannii* Engl. leaves and fruit pericarps from the Cameroon. *Flavour Fragr. J.* **1989**, *4*, 203–205. [CrossRef]

50. Pereira, F.D.; Mendes, J.M.; Lima, I.O.; Mota, K.S.; Oliveira, W.A.; Lima, E.D. Antifungal activity of geraniol and citronellol, two monoterpenes alcohols, against *Trichophyton rubrum* involves inhibition of ergosterol biosynthesis. *Pharm. Biol.* **2015**, *53*, 228–234. [CrossRef]

51. Watson, A.D. Lipidomics: A global approach to lipid analysis in biological systems. *J. Lipid Res.* **2006**, *47*, 2101–2111. [CrossRef]

52. Mirghaed, A.T.; Yasari, M.; Mirzargar, S.S.; Hoseini, S.M. Rainbow trout (Oncorhynchus mykiss) anesthesia with myrcene: Efficacy and physiological responses in comparison with eugenol. *Fish Physiol. Biochem.* **2018**, *44*, 919–926. [CrossRef]

53. Geng, S.; Cui, Z.; Huang, X.; Chen, Y.; Xu, D.; Xiong, P. Variations in essential oil yield and composition during *Cinnamomum cassia* bark growth. *Ind. Crops Prod.* **2011**, *33*, 248–252. [CrossRef]

54. Gobbo-Neto, L.; Lopes, N.P. Plantas medicinais: Fatores de influência no conteúdo de metabólitos secundários. *Quim. Nova.* **2007**, *30*, 374–381. [CrossRef]

55. Lima, H.R.P.; Kaplan, M.A.C.; Cruz, A.V.D.M. Influência dos fatores abióticos na produção e variabilidade de terpenóides em plantas. *Flor. Am.* **2012**, *10*, 71–77.

56. Rice-evans, C.A.; Miller, N.J.; Bolwell, P.G.; Bramley, P.M.; Pridham, J.B. The relative antioxidant activities of plant-derived polyphenolic flavonoids. *Free Radic. Res.* **1995**, *22*, 375–383. [CrossRef] [PubMed]

57. Torres-Martínez, R.; García-Rodríguez, Y.M.; Ríos-Chávez, P.; Saavedra-Molina, A.; López-Meza, J.E.; Ochoa-Zarzosa, A.; Garciglia, R.S. Antioxidant activity of the essential oil and its major terpenes of *Satureja macrostema* (Moc. and Sessé ex Benth.) Briq. *Pharmacogn. Mag.* **2017**, *13*, S875–S880.

58. Negi, J.S.; Bisht, V.K.; Bhandari, A.K.; Bisht, R.; Kandari Negi, S. Major constituents, antioxidant and antibacterial activities of *Zanthoxylum armatum* DC. essential oil. *IJPT* **2012**, *11*, 68–72.

59. Gogoi, R.; Loying, R.; Sarma, N.; Munda, S.; Pandey, S.K.; Lal, M. A comparative study on antioxidant, anti-inflammatory, genotoxicity, anti-microbial activities and chemical composition of fruit and leaf essential oils of *Litsea cubeba* Pers from North-east India. *Ind. Crops Prod.* **2018**, *125*, 131–139. [CrossRef]

60. Ouedraogo, R.A.; Koala, M.; Dabire, C.; Hema, A.; Bazie, V.B.E.J.T.; Outtara, L.P.; Nebie, R.H. Teneur en phénols totaux et activité antioxydante des extraits des trois principales variétés d'oignons (*Allium cepa* L.) cultivées dans la région du Centre-Nord du Burkina Faso. *IJBCS* **2015**, *9*, 281–291. [CrossRef]

61. Huang, D.; Ou, B.; et Prior, R.L. The chemistry behind antioxidant capacity assays. *J. Agric. Food Chem.* **2005**, *53*, 1841–1856. [CrossRef]

62. Foti, M.C.; Daquino, C.; Geraci, C. Electron-transfer reaction of cinnamic acids and their methyl esters with the DPPH• radical in alcoholic solutions. *J. Org. Chem.* **2004**, *69*, 2309–2314. [CrossRef]

63. Ghasemi, P.A.; Izadi, A.; Malek, P.F.; Hamedi, B. Chemical composition, antioxidant and antibacterial activities of essential oils from *Ferulago angulata*. *Pharm. Biol.* **2016**, *54*, 2515–2520. [CrossRef]

64. Kunwar, G.; Prakash, O.; Chandra, M.; Pant, A.K. Chemical composition, antifungal and antioxidant activities of *Perilla frutescens* (L.) syn. *P. ocimoides L.* collected from different regions of Indian Himalaya. *AJTM* **2013**, *8*, 88–98.

65. Negi, J.S.; Bisht, V.K.; Bhandari, A.K.; Singh, P.; Sundriyal, R.C. Chemical constituents and biological activities of the genus *Zanthoxylum*: A review. *AJPAC* **2011**, *5*, 412–416.

66. Adebayo, S.A.; Dzoyem, J.P.; Shai, L.J.; Eloff, J.N. The anti-inflammatory and antioxidant activity of 25 plant species used traditionally to treat pain in southern African. *BMC Complement. Altern. Med.* **2015**, *15*, 159. [CrossRef] [PubMed]

67. Da Silva, K.A.B.S.; Klein-Junior, L.C.; Cruz, S.M.; Cáceres, A.; Quintão, N.L.M.; Delle Monache, F.; Cechinel-Filho, V. Anti-inflammatory and anti-hyperalgesic evaluation of the condiment laurel (*Litsea guatemalensis Mez.*) and its chemical composition. *Food Chem.* **2012**, *132*, 1980–1986. [CrossRef]

68. Chen, J.; Wang, W.; Shi, C.; Fang, J. A comparative study of sodium houttuyfonate and 2-undecanone for their in vitro and in vivo anti-inflammatory activities and stabilities. *Int. J. Mol. Sci.* **2014**, *15*, 22978–22994. [CrossRef] [PubMed]

69. Kim, D.S.; Lee, H.J.; Jeon, Y.D.; Han, Y.H.; Kee, J.Y.; Kim, H.J.; Shin, H.J.; Kang, J.; Lee, B.S.; Kim, S.H.; et al. Alpha-pinene exhibits anti-inflammatory activity through the suppression of MAPKs and the NF-κB pathway in mouse peritoneal macrophages. *Am. J. Chin. Med.* **2015**, *43*, 731–742. [CrossRef] [PubMed]

70. Johnson, F.; Oussou, K.R.; Kanko, C.; Tonzibo, Z.F.; Foua-Bi, K.; Tano, Y. Bioefficacité des Huiles Essentielles de Trois Espèces végétales (Ocimum gratissimum, Ocimum canum et Hyptis suaveolens), de la Famille des Labiées dans la lutte contre *Sitophilus zeamais*. *Eur. J. Sci. Res.* **2017**, *150*, 273–284.

71. Yallappa, R.; Nandagopal, B.; Thimmappa, S. Botanicals as grain protectants. *Psyche J. Entomol.* **2012**, *2012*, 646740.

72. Plata-Rueda, A.; Campos, J.M.; da Silva Rolim, G.; Martínez, L.C.; Dos Santos, M.H.; Fernandes, F.L.; Serrão, J.E.; Zanuncio, J.C. Terpenoid constituents of cinnamon and clove essential oils cause toxic effects and behavior repellency response on granary weevil, *Sitophilus granarius*. *Ecotoxicol. Environ. Saf.* **2018**, *156*, 263–270. [CrossRef]

73. Mansour, S.A.; El-Sharkawy, A.Z.; Abdel-Hamid, N.A. Toxicity of essential plant oils, in comparison with conventional insecticides, against the desert locust, *Schistocerca gregaria* (Forskål). *Ind. Crops Prod.* **2015**, *63*, 92–99. [CrossRef]

74. Prieto, J.A.; Patiño, O.J.; Delgado, W.A.; Moreno, J.P.; Cuca, L.E. Chemical composition, insecticidal, and antifungal activities of fruit essential oils of three Colombian *Zanthoxylum* species. *Chil. J. Agric. Res.* **2011**, *71*, 73–82. [CrossRef]

75. Talontsi, F.M.; Matasyoh, J.C.; Ngoumfo, R.M.; Chepkorir, R. Mosquito larvicidal activity of alkaloids from *Zanthoxylum lemairei* against the malaria vector *Anopheles gambiae*. *Pestic. Biochem. Physiol.* **2011**, *99*, 82–85. [CrossRef]

76. Liu, X.C.; Liu, Q.Y.; Zhou, L.; Liu, Q.R.; Liu, Z.L. Chemical composition of *Zanthoxylum avicennae* essential oil and its larvicidal activity on *Aedes albopictus* Skuse. *Trop. J. Pharm. Res.* **2014**, *13*, 399–404. [CrossRef]

77. Koba, K.W.P.; Poutouli, C.; Raynaud, P.; Yaka, P. Propriétés insecticides de l'huile essentielle d'*Aeollanthus pubescens* Benth sur les chenilles de deux Lépidoptères: *Selepa docilsi butler* (noctuidae) et *Scrobipalpa ergassima mayr.* (geleduidae). *J. Rech. Sci. Univ. Lomé.* **2007**, *9*, 19–25.

78. Toudert-Taleb, K.; Hedjal-Chebheb, M.; Hami, H.; Debras, J.F.; Kellouche, A. Composition of essential oils extracted from six aromatic plants of Kabylian origin (Algeria) and evaluation of their bioactivity on *Callosobruchus maculatus* (*Fabricius*, 1775) (Coleoptera: *Bruchidae*). *Afr. Entomol.* **2014**, *22*, 417–427. [CrossRef]

79. Tchinda, A.T.; Fuendjiep, V.; Sajjad, A.; Matchawe, C.; Wafo, P.; Khan, S.; Tane, P.; Choudhary, M.I. Bioactive compounds from the fruits of *Zanthoxylum leprieurii*. *PharmacolOnline* **2009**, *1*, 406–415.

80. Goodman, C.D.; Hoang, A.T.; Diallo, D.; Malterud, K.E.; McFadden, G.I.; Wangensteen, H. Anti-plasmodial Effects of *Zanthoxylum zanthoxyloides*. *Planta Med.* **2019**, *85*, 1073–1079. [CrossRef]

81. Muganga, R.; Angenot, L.; Tits, M.; Frederich, M. Antiplasmodial and cytotoxic activities of Rwandan medicinal plants used in the treatment of malaria. *J. Ethnopharmacol.* **2010**, *128*, 52–57. [CrossRef]

82. Adebayo, J.O.; Krettli, A.U. Potential antimalarials from Nigerian plants: A review. *J. Ethnopharmacol.* **2011**, *133*, 289–302. [CrossRef]

83. DE Luca, Y. Ingrédients naturels de preservation des grains stockes dans les pays en voie de developpement. *J. D'agriculture Tradit. et de Bot. Appliquée* **1979**, *26*, 20–52. [CrossRef]

 *foods*

Article

Corn-Starch-Based Materials Incorporated with Cinnamon Oil Emulsion: Physico-Chemical Characterization and Biological Activity

Edaena Pamela Díaz-Galindo [1], Aleksandra Nesic [2,3,*], Silvia Bautista-Baños [4], Octavio Dublan García [1] and Gustavo Cabrera-Barjas [2,*]

[1] Facultad de Química, Universidad Autónoma del Estado de México, Km 115 Carr. Toluca-Ixtlahuaca. El Cerrillo Piedras Blancas, Toluca 50295, Mexico; pam.dg12@hotmail.com (E.P.D.-G.); odublang@uaemex.mx (O.D.G.)

[2] Unidad de Desarrollo Tecnológico, UDT, Universidad de Concepción, Avda. Cordillera No. 2634, Parque Industrial Coronel, Coronel 4191996, Chile

[3] Vinca Institute for Nuclear Sciences, University of Belgrade, Mike Petrovica-Alasa 12–14, Belgrade 11000, Serbia

[4] Instituto Politécnico Nacional, Centro de Desarrollo de Productos Bióticos (CEPROBI), Carretera Yautepec-Jojutla, Km. 6, calle CEPROBI No. 8, Col. San Isidro, Yautepec, Morelos 62731, Mexico; silviabb2008@hotmail.com

* Correspondence: a.nesic@udt.cl (A.N.); g.cabrera@udt.cl (G.C.-B.)

Received: 18 March 2020; Accepted: 4 April 2020; Published: 10 April 2020

Abstract: Active packaging represents a large and diverse group of materials, with its main role being to prolong the shelf-life of food products. In this work, active biomaterials based on thermoplastic starch-containing cinnamon oil emulsions were prepared by the compression molding technique. The thermal, mechanical, and antifungal properties of obtained materials were evaluated. The results showed that the encapsulation of cinnamon oil emulsions did not influence the thermal stability of materials. Mechanical resistance to break was reduced by 27.4%, while elongation at break was increased by 44.0% by the addition of cinnamon oil emulsion. Moreover, the novel material provided a decrease in the growth rate of *Botrytis cinerea* by 66%, suggesting potential application in food packaging as an active biomaterial layer to hinder further contamination of fruits during the storage and transport period.

Keywords: starch films; active food packaging films; cinnamon oil emulsions; *Botrytis cinerea*

---

1. Introduction

In the last decade, the Chilean fruit industry has been consolidated as one of the main international leaders in the export of fresh fruits, particularly grapes, strawberries, and raspberries. Fruit exports accounted for 27% of the sector's total export value in 2016, with an export value of US$16 billion, which makes this sector the most important in the country, being surpassed only by the mining industry [1]. However, the appearance of gray mold caused by the fungal contamination of fruits poses a big problem, accounting for approximately 20% of fruit losses during storage and transport. *Botrytis cinerea* is the most widespread fungal disease on fruits and is mainly manifested in the post-harvest period.

Recently, active biodegradable packaging has gained more importance for fruit storage directly after harvesting, in order to minimize the appearance of gray mold and losses during transport. Namely, this type of packaging can be directly in contact with the surface of food products or with the headspace between the package and the food products. Moreover, active materials can appear in the form of sachets/capsules that contain an antifungal agent that is inserted into a package or in the form of inner coating of the packaging material. The role of the antifungal agent is to reduce, inhibit, or hinder the growth of fungi that may be present in the packed fruits [2–5]. Moreover, special attention is paid to the use of bioactive components such as plant-derived essential oils, due to their high antifungal/antimicrobial and antioxidant activity [6–10]. Incorporation of essential oils into polymer package presents a big technological challenge because of their evaporation during the melting processing of the polymer. Such a challenge can be solved by the inclusion of essential oil in the biopolymer matrix and the encapsulation of stable emulsion into a thermoplastic polymer that is processed at a lower temperature than a temperature at which essential oil evaporates/degrades. Moreover, this material could prevent easy penetration of volatiles into the food, protecting the items from coming in contact with substances that could affect their taste and odor.

Among all biopolymers, starch is a very promising candidate for the processing of biodegradable food packaging materials [11,12]. Starch can be found in plants such as corn, wheat, rice, and peas, so its use is expanded due to its low price and easy availability. Starch has thermoplastic behavior in the presence of plasticizers and when elevated temperature and shear are applied. In fact, the processing of thermoplastic starch (TPS) is possible by the use of conventional techniques for synthetic polymers, such as compression-molding, injection molding, and extrusion blow molding [13]. TPS-based materials have been commercialized over the last decade and are currently used in the food sector as single-use packages, for example, egg trays, plates, and cups. One of the greatest benefits of TPS is that it can be processed at significantly lower temperatures (90–140 °C) in comparison to other bioplastics/plastic materials (180–230 °C), allowing safe operation with volatile bioactive components (degradation around 180 °C), thus minimizing the loss during processing.

In this work, cinnamon oil was used as an antifungal component, because has already been proven to be an efficient bioactive agent toward *Botrytis cinerea* [14–16]. In order to minimize the losses during processing, stable water in oil emulsion was prepared in the presence of mucilage. Mucilage extracted from chia seeds shows high emulsifying activity and may also act as a stabilizer of emulsions [17–21]. Furthermore, stable emulsions were incorporated into TPS, in order to obtain antifungal biodegradable material that could be used as an inner layer of active packaging. Thermal, mechanical, and antifungal properties toward *Botrytis cinerea* were assessed.

2. Materials and Methods

The cinnamon essential oil was supplied by Cedrosa (Estado de México, Mexico). Chia seeds were purchased from the local market in Mexico. Starch from corn was purchased from Buffalo® 034,010 (CornProducts Chile Inducorn S.A., Santiago, Chile). Glycerol was obtained from OCN company (Qindao, China).

2.1. Chia Mucilage Extraction

The extraction of mucilage was performed according to the method proposed by Velázquez-Gutiérrez et al. [22]. Namely, 40 g of chia seeds were soaked in 800 mL of Mili-Q water. The pH of the mixture was adjusted to 8 by using 0.1 M NaOH solution. The mixture was stirred for 2 h at a constant temperature of 80 °C. Afterward, the mucilage was separated from the seed, and the filtrate was centrifuged for 8 min at 524× g. The supernatant was decanted and analyzed. The extracted mucilage was frozen, and afterward, the sample was dehydrated using a freeze-dryer for 48 h. The dehydrated products were stored in desiccators with P_2O_5 in order to prevent any moisture absorption until experiments that required usage of these products were performed.

2.2. Oil-in-Water (O/W) Emulsion Preparation

A certain amount of obtained mucilage was dissolved in water at room temperature for 12 h, with continuous stirring, in order to obtain different concentrations of aqueous solution (0.2–1.5 wt%). The emulsions were made by mixing the mucilage solutions as an aqueous phase and the cinnamon essential oil as a lipid phase with a laboratory T-25 digital Ultraturrax at 9600 rpm for 2 min. The concentration of cinnamon oil in water varied from 1 to 5 v/v (1/99; 2/98; 3/97; 4/96 and 5/95 v/v oil/water). The total volume of the aqueous/water phase was 50 mL. In order to check the stability of emulsions, the creaming index was monitored at specific storage time (0, 30, and 60 days). When creaming occurred during storage time at room temperature (25 °C), emulsions were homogenized to re-disperse the cream layer before the analysis. Three samples per each emulsion formulation were tested, and the deviation was less than 3%.

2.3. Characterization of Emulsions

2.3.1. Creaming Index

Each emulsion was evaluated to detect visible parameters such as color, creaming, coalescence, and/or separation of phases. After the emulsions were homogenized and centrifuged for 10 min at 524× g, the creaming index (CI, %) was checked and calculated according to the following Equation (1):

$$CI(\%) = \frac{H_t}{H_o} \times 100 \tag{1}$$

where H_o is the total height of the emulsion layer in vials and H_t is the height of the cream layer. Analyses were performed in triplicate.

2.3.2. Thermal Stability

The thermal stability (TS, %) of emulsions was evaluated by subsequent heating of the emulsion in a water bath at 80 °C for 30 min and subsequent cooling down to room temperature (20 ± 2 °C), followed by centrifugation for 10 min at 524× g. The heights of the emulsified layer and cream layer were measured, and the TS was calculated according to the following Equation (2):

$$S(\%) = \frac{H_o - H_t}{H_o} \times 100 \tag{2}$$

where H_o is the total height of the emulsion layer in vials and H_t is the height of the cream layer. Analyses were performed in triplicate.

2.3.3. In Vitro Antifungal Activity of Emulsions

Twenty milliliters of sterilized potato dextrose agar (PDA) was placed in Petri dishes (100 × 15 mm). A volume of 0.1 mL of the emulsion was uniformly dispersed in the culture medium PDA (Bioxon) on six Petri dishes per treatment and allowed to dry. A disc of 5 mm in diameter of *B. cinerea* was placed in the center of the Petri dishes and incubated at 25 ± 2 °C until control (with sterile water) reached its maximum development. The plates were sealed with Parafilm® to avoid vapor leakage.

Mycelial growth over time was measured daily using a Vernier caliper. The test ended when the mycelium completely covered the Petri dish in the control sample. Six repetitions per treatment were carried out. The mycelial growth inhibition index (IM) was calculated according to the following Equation (3):

$$IM(\%) = \frac{C_C - C_T}{C_C} \times 100 \tag{3}$$

where C_C is the control's growth, and C_T is the growth in the treatment group. Analyses were carried out in triplicate.

2.4. Preparation of Thermo-Plasticized Starch-Emulsion Plates

The first step in the preparation of materials was the thermo-plasticization of starch. The starch was homogenized with 150 g of glycerol and 50 g of water in a high-speed blade mixer (Cool Mixer, Labtech model LCM-24) at 45 °C and a speed of 2800 rpm. This sample was coded as TPS and used to prepare control the TPS plate by a compression molding technique. According to the preliminary results related to emulsions, the two best formulations were chosen to be incorporated into the starch matrix during the thermo-plasticization process. The same procedure was followed to obtain the thermo-plasticized starch loaded with emulsions, as for the control TPS sample.

Plates were made by the use of Labtech LP-20B hydraulic press. The 40 g of thermo-plasticized starch samples was placed between two stainless steel molds that were covered with a Teflon sheet. The samples were compressed with an applied pressure of 70 bar for 3 min at 140 °C. The resulting plates were cooled for 1 min before unmolding. The thickness of the obtained materials was approximately 0.5 mm.

2.5. Characterization of Plates

2.5.1. TGA

Thermogravimetric analysis (TGA) was performed by NETZSCH TG 209 F3 Tarsus®. The operating conditions were as follows: nitrogen flow of 10 mL/min, temperature heating range from 30 to 500 °C, and a heating rate of 10 C/min. All measurements were performed in triplicate, and obtained parameters were repeatable within ±3%.

2.5.2. Mechanical Analysis

Mechanical analysis was performed on a Universal test machine KARG Industrie Technik Smartens 005) according to the ASTMD638 (2010) standard test method. The test conditions were as follows: 23 ± 2 °C, 45 ± 5% RH, crosshead speed 2 mm/min. The measurements were carried out in sextuplicate. The standard deviation for the tested parameters was ± 10%.

2.5.3. In Vitro Antifungal Activity

The PDA culture medium and cultivation of *B. cinerea* was prepared as described in Section 2.3.3. The antifungal films were cut into 1 cm diameter pieces and attached to the inside cover of the Petri dishes. The Petri dishes were then sealed with Parafilm colony diameters (cm) in each Petri dish within the time they were monitored. As a control sample, starch films without antifungal compounds were used. Analyses were carried out in triplicate.

3. Results

3.1. Emulsions

The concentration of mucilage was shown to play a significant role in the stabilization of emulsions. In fact, emulsion was only obtained when the used mucilage content was above 0.75 wt%. Table 1 presents the values of the creaming index (CI) and thermal stability of emulsion formulations containing 1 wt% and 1.5 wt% of mucilage at zero-day, after 30 days and 60 days of storage at room temperature. The highest stability of emulsions was obtained when 1.5 wt% of mucilage was used, since creaming did not appear even after 60 days of storage, and the evaluated thermal stability was 99%. Other authors previously reported 120 days stability of w/o emulsion when chia mucilage was added at 0.75 and 1 wt%, respectively [18]. These results are in agreement with those obtained by Guiotto et al. [23] who prepared w/o emulsion and added chia mucilage (0.75 wt%) as a stabilizer. The mucilage addition contributed to the stabilization of the emulsion CI for 120 days. The authors correlated these results in a three-dimensional network, which showed reduced oil droplets mobility inside the emulsion. It has been previously reported that the emulsifying properties of chia mucilage could be associated with a certain protein content in its structure. Such proteins could contribute to the surface activity of chia mucilage dispersions [24].

Table 1. Stability of emulsions in different interval period.

Sample Code	Mucilage (wt %)	Cinnamon Oil (mL)	CI %			TS %		
			0.d	30.d	60.d	0.d	30.d	60.d
A1i	1	1	0	0	1	100	100	99
A2i	1	2	0	0	1	100	98	97
A3i	1	3	0	0	1	95	97	95
B1i	1.5	1	0	0	0	100	100	99
B2i	1.5	2	0	0	0	100	100	99
B3i	1.5	3	0	0	0	100	100	99

The effects of different emulsion formulations on the radial growth of *B. cinerea* are shown in Figure 1. The highest radial growth was obtained in the control sample (without emulsion application). All tested emulsions (see Table 1) completely inhibited the growth of *B. cinerea*. It is important to highlight that after two months of storage, all tested emulsions were again subjected to antifungal tests and again showed 100% growth inhibition. The high antifungal activity of cinnamon oil is already well known because of its chemical composition [25,26]. Namely, the main constituent of cinnamon oil is cinnamaldehyde, which contains an aldehyde group and a conjugated double bond outside the ring. These groups are responsible for the deactivation of enzymes in fungi [27]. Few studies have shown that cinnamon oil inhibits the biosynthesis of ergosterol, the major sterol constituent of the fungal plasma membrane, which leads to damage of the cell membrane structure, and consequently, the leakage of intracellular ions [28]. Hence, cinnamon oil stabilized by mucilage could be a good bioactive candidate for thermoplastic bio-packages to prevent or hinder the growth of *B. cinerea*. Since the best emulsion stability within the time showed B1i–B3i formulations, these formulations were chosen for further incorporation into thermoplastic starch (Table 2).

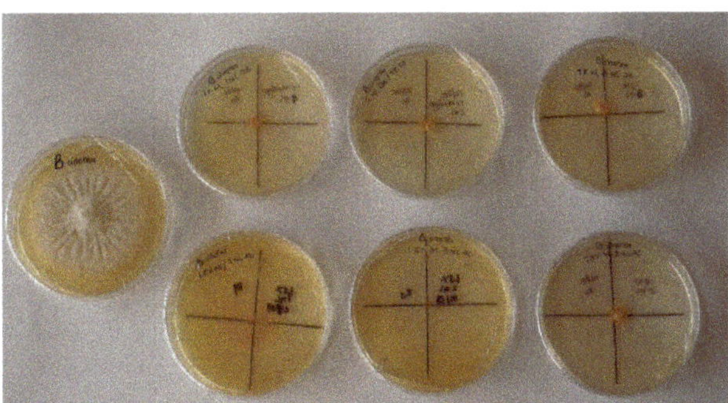

Figure 1. Antifungal activity of cinnamon oil emulsions.

Table 2. Compositions and codes of bioactive biodegradable plates.

Sample Code	Starch (kg)	Glycerol (g)	Water (g)	Emulsion (g)
Starch	0.5	150	50	0
Starch-B1i	0.5	150	0	50
Starch-B2i	0.5	150	0	50
Starch-B3i	0.5	150	0	50

3.2. Characterization of Starch/Emulsion Materials

3.2.1. Mechanical Analysis

The mechanical parameters, values of the tensile strength (TS), and percentage of elongation at break (e) of the materials are presented in Table 3. The neat thermoplasticized starch film exhibited an average tensile strength value of 2 MPa and an elongation at break value of 50.5%. The incorporation of emulsions into the starch matrix resulted in a decrease in tensile strength when compared with one of the neat starch films. The elongation at break value of films increased with the addition of essential oil emulsion in the starch matrix. As presented in Table 3, a reduction of approximately 24% in TS% value and an increase in plasticity by approximately 70% were obtained by encapsulation of B2i emulsion into starch. The majority of data from the literature provide evidence of a decrease in TS and an increase in elongation at break of films when essential oils are introduced in polysaccharide matrices, such as chitosan [29], starch [30], pectin [31], and alginate [32]. This trend was explained by the specific interactions between phenolic compounds from essential oils and functional groups from the biopolymer matrix that lead to more elastic matrices [33]. In fact, previous studies have reported that essential oils have a plasticizing effect on biopolymers and diminish the strong intermolecular chain–chain interactions in the polymer structure, thus imparting higher flexibility of films up to the break [34,35]. So far, data in the literature related to the incorporation of inclusion complexes into the thermoplastic biopolymer matrix are scarce. As a carrier of bioactive cinnamon oil and D-limonene, β-cyclodextrin was used and further incorporated into PLA [36] and PBS [37], respectively. However, there are no data related to the mechanical properties of these materials. Moreover, to the best of our knowledge, no data in the literature are found regarding the incorporation of emulsions into thermoplastic polymers.

Table 3. Mechanical parameters of starch-based plates.

Sample Code	TS (MPa)	e (%)
Starch	2.04	50.5
Starch-B1i	1.65	84.4
Starch-B2i	1.55	86.2
Starch-B3i	1.48	90.3

3.2.2. Thermal Analysis

The weight loss at 180 °C (W_{L180}), the temperature at which degradation starts (T_{onset}), the maximum weight loss temperature (T_{deg}), and char residue are reported in Table 4. Neat thermoplastic starch displayed two degradation steps (Figure 2). The first degradation step occurred up to 180 °C, where bonded and unbonded water was released, whereas the second step with T_{onset} at 280 °C and maximum degradation peak at 312 °C were related to starch chain decomposition. The addition of emulsion did not affect the thermal degradation profile of thermoplastic starch since there were no significant changes in T_{onset} and T_{deg} values. On the other side, a slight increase in weight loss up to 180 °C and char residue for starch-emulsion plates was observed. These results were expected because low concentrations of volatile components were introduced in thermoplastic starch. The unchanged thermal stability after the inclusion of essential oils/bioactive components were also observed in the literature for LDPE films incorporated with cinnamon and rosemary oil [38] for PLA films containing D-limonene [39] and for PLA films loaded with oregano oil [40].

Table 4. TGA parameters for starch-based plates.

Simple	W_{L180} (%)	T_{onset} (°C)	T_{deg} (°C)	Char Residue at 500 °C (%)
Starch	7.8	280	312	9.8
Starch-B1i	11	279	312	10.9
Starch-B2i	10.3	278	312	11.0
Starch-B3i	9.1	278	312	11.2

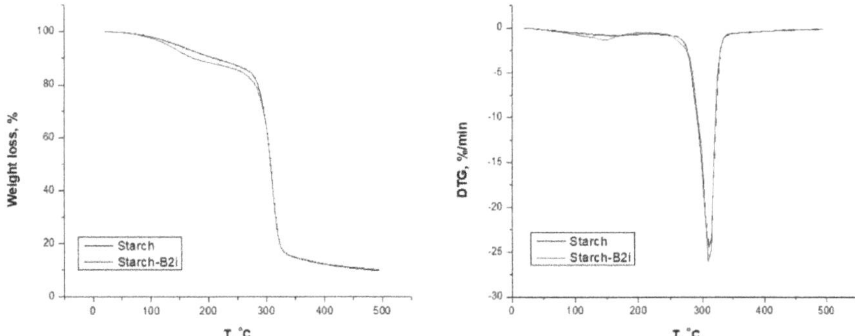

Figure 2. Thermal diagrams of starch and starch-emulsion materials.

3.2.3. Antifungal Activity

The main purpose of the antifungal tests was to evaluate the potential use of starch/emulsion plates as antifungal biodegradable layers/sachets in the food packaging industry, taking into account that *B. cinerea* is well known as a contaminant of fruits and vegetables. Figure 3 displays the fungal growth inhibition within the incubation time of *B. cinerea* at 25 °C. The neat starch plate did not show any antifungal activity, as was expected. Moreover, starch-emulsion plates did not show fungistatic activity but provided a lower rate of mycelium growth. However, it is important to underline that there was limited development of hyphae, and no spore germination was observed, which is important in the prevention of further acceleration of fungi contamination on fruits. The results revealed that mycelium growth inhibition (%) depended on the concentration of bioactive components included in the thermoplastic starch plates. With an increase in the cinnamon oil concentration in thermoplastic starch plates, a lower rate of *B. cinerea* growth was observed. In fact, the inhibition of mycelium growth was above 50% after 10 days of incubation only for samples Starch-B2i and Starch-B3i when compared with the control. This outcome could be explained by a low concentration of cinnamon oil in starch plates, ranging from 0.2 to 0.6 wt%. The inhibition of growth rate of Starch-B3i sample was improved by 66% in comparison with that of the control sample, which means that this material could be used in food packaging as a supporting layer inserted in the box, but only to hinder the further contamination of fruits during storage or transport, minimizing fruit loss and damage. In order to obtain biobased materials with higher antifungal efficiency, further optimization of the system is required. The main optimization of the plasticizer and emulsion ratio with respect to the starch matrix is necessary in order to avoid a further decrease in the mechanical stability of the final materials. In fact, introducing a higher amount of emulsion into the starch matrix would further increase the antifungal activity and elasticity of the material but would significantly reduce its tensile strength, which can be an undesirable effect from the industrial point of view. Moreover, a higher concentration of emulsion could cause olfactory and gustatory contamination of packed foods (off-flavor, off-odor) due to the migration of volatile compounds from package to food. Hence, besides mechanical and biological stability, further studies should include the evaluation of the side effects of materials containing cinnamon oil emulsion on the sensory properties of food (odor and flavor).

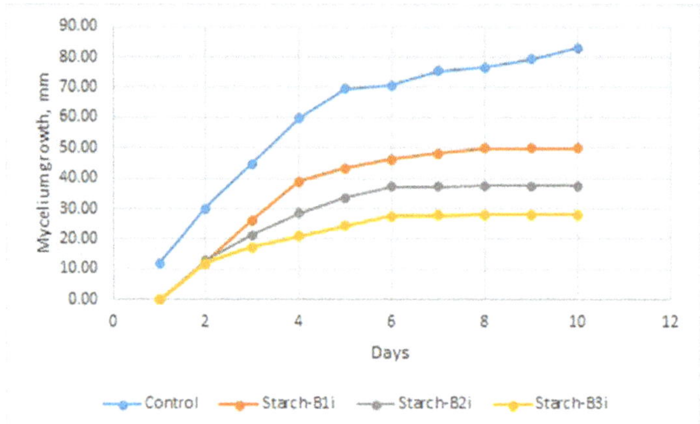

Figure 3. Mycelium growth of *B.cinerea* in the presence of starch and starch-emulsion materials.

4. Conclusions

The study investigated the potential use of thermoplastic starch incorporated with cinnamon oil emulsion as a bioactive antifungal material in the food packaging industry. The addition of cinnamon oil emulsion did not affect the thermal stability of thermoplastic starch. In contrast, the mechanical properties showed a clear enhancement in elongation of obtained bioactive material at the break point. Moreover, the highest loading of the emulsion into thermoplastic starch showed inhibition of the growth of *B. cinerea* in the "in vitro" antifungal test. These results demonstrate that thermoplastic starch loaded with cinnamon oil emulsion could be potentially used as a bioactive layer or emitter in the food packaging sector to hinder further infection of fruits.

Author Contributions: E.P.D.-G.: Investigation, Data curation, A.N.: Writing, reviewing and editing, supervision, G.C.-B.: Methodology-physical characterization of materials, Writing and reviewing, S.B.-B.: Methodology-microbiological analysis, O.D.G.: Supervision. All authors have read and agreed to the published version of the manuscript.

Funding: This research was funded by ANID CONICYT PIA/APOYO CCTE AFB170007 and ANID Fondecyt Regular 1191528.

Conflicts of Interest: The authors declare no conflict of interest.

References

1. Fruits and Vegetable Industry in Chile—Growth, Trends, and Forecast (2020–2025). Available online: https://www.mordorintelligence.com/industry-reports/colombia-fruit-vegetable-market (accessed on 18 March 2020).
2. Santagata, G.; Valerio, F.; Cimmino, A.; Dal Poggetto, G.; Masi, M.; Di Biase, M.; Malinconico, M.; Lavermicocca, P.; Evidente, A. Chemico-physical and antifungal properties of poly(butylene succinate)/cavoxin blend: Study of a novel bioactive polymeric based system. *Eur. Polym. J.* **2017**, *94*, 230–247. [CrossRef]
3. González, A.; Alvarez Igarzabal, C.I. Soy protein—Poly (lactic acid) bilayer films as biodegradable material for active food packaging. *Food Hydrocoll.* **2013**, *33*, 289–296. [CrossRef]
4. Nguyen Van Long, N.; Joly, C.; Dantigny, P. Active packaging with antifungal activities. *Int. J. Food Microbiol.* **2016**, *220*, 73–90. [CrossRef] [PubMed]
5. Manso, S.; Becerril, R.; Nerín, C.; Gómez-Lus, R. Influence of pH and temperature variations on vapor phase action of an antifungal food packaging against five mold strains. *Food Control* **2015**, *47*, 20–26. [CrossRef]
6. Requena, R.; Vargas, M.; Chiralt, A. Eugenol and carvacrol migration from PHBV films and antibacterial action in different food matrices. *Food Chem.* **2019**, *277*, 38–45. [CrossRef] [PubMed]
7. Das, S.; Gazdag, Z.; Szente, L.; Meggyes, M.; Horváth, G.; Lemli, B.; Kunsági-Máté, S.; Kuzma, M.; Kőszegi, T. Antioxidant and antimicrobial properties of randomly methylated β cyclodextrin—Captured essential oils. *Food Chem.* **2019**, *278*, 305–313. [CrossRef] [PubMed]

8. Hasheminejad, N.; Khodaiyan, F.; Safari, M. Improving the antifungal activity of clove essential oil encapsulated by chitosan nanoparticles. *Food Chem.* **2019**, *275*, 113–122. [CrossRef]

9. Vázquez-Sánchez, D.; Galvão, J.A.; Mazine, M.R.; Gloria, E.M.; Oetterer, M. Control of Staphylococcus aureus biofilms by the application of single and combined treatments based in plant essential oils. *Int. J. Food Microbiol.* **2018**, *286*, 128–138. [CrossRef]

10. Mari, M.; Bautista-Baños, S.; Sivakumar, D. Decay control in the postharvest system: Role of microbial and plant volatile organic compounds. *Postharvest Biol. Technol.* **2016**, *122*, 70–81. [CrossRef]

11. Lopez, O.; Garcia, M.A.; Villar, M.A.; Gentili, A.; Rodriguez, M.S.; Albertengo, L. Thermo-compression of biodegradable thermoplastic corn starch films containing chitin and chitosan. *LWT Food Sci. Technol.* **2014**, *57*, 106–115. [CrossRef]

12. Fabra, M.J.; López-Rubio, A.; Ambrosio-Martín, J.; Lagaron, J.M. Improving the barrier properties of thermoplastic corn starch-based films containing bacterial cellulose nanowhiskers by means of PHA electrospun coatings of interest in food packaging. *Food Hydrocoll.* **2016**, *61*, 261–268. [CrossRef]

13. Zhang, Y.; Rempel, C.; Liu, Q. Thermoplastic Starch Processing and Characteristics—A Review. *Crit. Rev. Food Sci. Nutr.* **2014**, *54*, 1353–1370. [CrossRef]

14. Combrinck, S.; Regnier, T.; Kamatou, G.P.P. In vitro activity of eighteen essential oils and some major components against common postharvest fungal pathogens of fruit. *Ind. Crops Prod.* **2011**, *33*, 344–349. [CrossRef]

15. Melgarejo-Flores, B.G.; Ortega-Ramírez, L.A.; Silva-Espinoza, B.A.; González-Aguilar, G.A.; Miranda, M.R.A.; Ayala-Zavala, J.F. Antifungal protection and antioxidant enhancement of table grapes treated with emulsions, vapors, and coatings of cinnamon leaf oil. *Postharvest Biol. Technol.* **2013**, *86*, 321–328. [CrossRef]

16. Tzortzakis, N.G. Impact of cinnamon oil-enrichment on microbial spoilage of fresh produce. *Innov. Food Sci. Emerg. Technol.* **2009**, *10*, 97–102. [CrossRef]

17. Quinzio, C.; Ayunta, C.; López de Mishima, B.; Iturriaga, L. Stability and rheology properties of oil-in-water emulsions prepared with mucilage extracted from *Opuntia ficus-indica* (L.). Miller. *Food Hydrocoll.* **2018**, *84*, 154–165. [CrossRef]

18. Capitani, M.I.; Nolasco, S.M.; Tomás, M.C. Stability of oil-in-water (O/W) emulsions with chia (*Salvia hispanica* L.) mucilage. *Food Hydrocoll.* **2016**, *61*, 537–546. [CrossRef]

19. Ma, F.; Zhang, Y.; Yao, Y.; Wen, Y.; Hu, W.; Zhang, J.; Liu, X.; Bell, A.E.; Tikkanen-Kaukanen, C. Chemical components and emulsification properties of mucilage from Dioscorea opposita Thunb. *Food Chem.* **2017**, *228*, 315–322. [CrossRef]

20. Campos, B.E.; Dias Ruivo, T.; da Silva Scapim, M.R.; Madrona, G.S.; de C. Bergamasco, R. Optimization of the mucilage extraction process from chia seeds and application in ice cream as a stabilizer and emulsifier. *LWT Food Sci. Technol.* **2016**, *65*, 874–883. [CrossRef]

21. Wu, Y.; Eskin, N.A.M.; Cui, W.; Pokharel, B. Emulsifying properties of water soluble yellow mustard mucilage: A comparative study with gum Arabic and citrus pectin. *Food Hydrocoll.* **2015**, *47*, 191–196. [CrossRef]

22. Velázquez-Gutiérrez, S.K.; Figueira, A.C.; Rodríguez-Huezo, M.E.; Román-Guerrero, A.; Carrillo-Navas, H.; Pérez-Alonso, C. Sorption isotherms, thermodynamic properties and glass transition temperature of mucilage extracted from chia seeds (*Salvia hispanica* L.). *Carbohydr. Polym.* **2015**, *121*, 411–419. [CrossRef]

23. Guiotto, E.N.; Capitani, M.I.; Nolasco, S.M.; Tomás, M.C. Stability of Oil-in-Water Emulsions with Sunflower (*Helianthus annuus* L.) and Chia (*Salvia hispanica* L.) By-Products. *J. Am. Oil Chem. Soc.* **2016**, *93*, 133–143. [CrossRef]

24. Timilsena, Y.P.; Adhikari, R.; Kasapis, S.; Adhikari, B. Rheological and microstructural properties of the chia seed polysaccharide. *Int. J. Biol. Macromol.* **2015**, *81*, 991–999. [CrossRef] [PubMed]

25. Sukatta, U.; Haruthaithanasan, V.; Chantarapanont, W.; Dilokkunanant, U.; Suppakul, P. Antifungal Activity of Clove and Cinnamon Oil and Their Synergistic Against Postharvest Decay Fungi of Grape in vitro. *Kasetsart J.-Nat. Sci.* **2008**, *42*, 169–174.

26. Mvuemba, H.N.; Green, S.E.; Tsopmo, A.; Avis, T.J. Antimicrobial efficacy of cinnamon, ginger, horseradish and nutmeg extracts against spoilage pathogens. *Phytoprotection* **2009**, *90*, 65. [CrossRef]

27. Bang, K.-H.; Lee, D.-W.; Park, H.-M.; Rhee, Y.-H. Inhibition of Fungal Cell Wall Synthesizing Enzymes by trans -Cinnamaldehyde. *Biosci. Biotechnol. Biochem.* **2000**, *64*, 1061–1063. [CrossRef]

28. Abdalla, W. Antibacterial and Antifungal Effect of Cinnamon. *Microbiol. Res. J. Int.* **2018**, *23*, 1–8. [CrossRef]

29. Souza, V.G.L.; Pires, J.R.A.; Rodrigues, P.F.; Lopes, A.A.S.; Fernandes, F.M.B.; Duarte, M.P.; Coelhoso, I.M.; Fernando, A.L. Bionanocomposites of chitosan/montmorillonite incorporated with Rosmarinus officinalis essential oil: Development and physical characterization. *Food Packag. Shelf Life* **2018**, *16*, 148–156. [CrossRef]

30. Li, J.; Ye, F.; Lei, L.; Zhao, G. Combined effects of octenylsuccination and oregano essential oil on sweet potato starch films with an emphasis on water resistance. *Int. J. Biol. Macromol.* **2018**, *115*, 547–553. [CrossRef]

31. Biddeci, G.; Cavallaro, G.; Di Blasi, F.; Lazzara, G.; Massaro, M.; Milioto, S.; Parisi, F.; Riela, S.; Spinelli, G. Halloysite nanotubes loaded with peppermint essential oil as filler for functional biopolymer film. *Carbohydr. Polym.* **2016**, *152*, 548–557. [CrossRef]

32. Benavides, S.; Villalobos-Carvajal, R.; Reyes, J.E. Physical, mechanical and antibacterial properties of alginate film: Effect of the crosslinking degree and oregano essential oil concentration. *J. Food Eng.* **2012**, *110*, 232–239. [CrossRef]

33. Bitencourt, C.M.; Fávaro-Trindade, C.S.; Sobral, P.J.A.; Carvalho, R.A. Gelatin-based films additivated with curcuma ethanol extract: Antioxidant activity and physical properties of films. *Food Hydrocoll.* **2014**, *40*, 145–152. [CrossRef]

34. Šuput, D.; Lazić, V.; Pezo, L.; Markov, S.; Vaštag, Ž.; Popović, L.; Radulović, A.; Ostojić, S.; Zlatanović, S.; Popović, S. Characterization of Starch Edible Films with Different Essential Oils Addition. *Polish J. Food Nutr. Sci.* **2016**, *66*, 277–286. [CrossRef]

35. Jiménez, A.; Fabra, M.J.; Talens, P.; Chiralt, A. Phase transitions in starch based films containing fatty acids. Effect on water sorption and mechanical behaviour. *Food Hydrocoll.* **2013**, *30*, 408–418. [CrossRef]

36. Wen, P.; Zhu, D.-H.; Feng, K.; Liu, F.-J.; Lou, W.-Y.; Li, N.; Zong, M.-H.; Wu, H. Fabrication of electrospun polylactic acid nanofilm incorporating cinnamon essential oil/β-cyclodextrin inclusion complex for antimicrobial packaging. *Food Chem.* **2016**, *196*, 996–1004. [CrossRef]

37. Mallardo, S.; De Vito, V.; Malinconico, M.; Volpe, M.G.; Santagata, G.; Di Lorenzo, M.L. Poly(butylene succinate)-based composites containing β-cyclodextrin/d-limonene inclusion complex. *Eur. Polym. J.* **2016**, *79*, 82–96. [CrossRef]

38. Dong, Z.; Xu, F.; Ahmed, I.; Li, Z.; Lin, H. Characterization and preservation performance of active polyethylene films containing rosemary and cinnamon essential oils for Pacific white shrimp packaging. *Food Control* **2018**, *92*, 37–46. [CrossRef]

39. Arrieta, M.P.; López, J.; Hernández, A.; Rayón, E. Ternary PLA–PHB–Limonene blends intended for biodegradable food packaging applications. *Eur. Polym. J.* **2014**, *50*, 255–270. [CrossRef]

40. Liu, D.; Li, H.; Jiang, L.; Chuan, Y.; Yuan, M.; Chen, H. Characterization of Active Packaging Films Made from Poly(Lactic Acid)/Poly(Trimethylene Carbonate) Incorporated with Oregano Essential Oil. *Molecules* **2016**, *21*, 695. [CrossRef]

Article

Toxicity and Synergistic Effect of *Elsholtzia ciliata* Essential Oil and Its Main Components against the Adult and Larval Stages of *Tribolium castaneum*

Jun-Yu Liang *, Jie Xu, Ying-Ying Yang, Ya-Zhou Shao, Feng Zhou and Jun-Long Wang

School of Life Science, Northwest Normal University, Lanzhou 730070, Gansu, China; XJ13893279431@163.com (J.X.); y17339832389@163.com (Y.-Y.Y.); 18893811951@163.com (Y.-Z.S.); fengzhou@nwnu.edu.cn (F.Z.); wangjunlong@nwnu.edu.cn (J.-L.W.)
* Correspondence: liangjunyu@nwnu.edu.cn

Received: 11 February 2020; Accepted: 10 March 2020; Published: 16 March 2020

Abstract: Investigations have indicated that storage pests pose a great threat to global food security by damaging food crops and other food products derived from plants. Essential oils are proven to have significant effects on a large number of stored grain insects. This study evaluated the contact toxicity and fumigant activity of the essential oil extract from the aerial parts of *Elsholtzia ciliata* and its two major biochemical components against adults and larvae of the food storage pest beetle *Tribolium castaneum*. Gas chromatography–mass spectrometry analysis revealed 16 different components derived from the essential oil of *E. ciliata*, which included carvone (31.63%), limonene (22.05%), and α-caryophyllene (15.47%). Contact toxicity assay showed that the essential oil extract exhibited a microgram-level of killing activity against *T. castaneum* adults (lethal dose 50 (LD_{50}) = 7.79 µg/adult) and larvae (LD_{50} = 24.87 µg/larva). Fumigant toxicity assay showed LD_{50} of 11.61 mg/L air for adults and 8.73 mg/L air for larvae. Carvone and limonene also exhibited various levels of bioactivity. A binary mixture (2:6) of carvone and limonene displayed obvious contact toxicity against *T. castaneum* adults (LD_{50} = 10.84 µg/adult) and larvae (LD_{50} = 30.62 µg/larva). Furthermore, carvone and limonene exhibited synergistic fumigant activity against *T. castaneum* larvae at a 1:7 ratio. Altogether, our results suggest that *E. ciliata* essential oil and its two monomers have a potential application value to eliminate *T. castaneum*.

Keywords: *Elsholtzia ciliata*; *Tribolium castaneum*; essential oil; carvone; limonene; insecticidal activity; synergistic effect

---

1. Introduction

Food security has always been a staple of discussion. Investigations have indicated that insects pose a great threat to global food security by damaging food crops and other food products derived from plants [1]. However, several pests show resistance, and the utilization of existing insecticides has more or less some side effects. For example, many of them can be lethal to nontarget organisms, and the residues of insecticides in crops also have negative impacts on human beings and the environment [2,3]. *Tribolium castaneum* is a species of beetle that is considered as a worldwide pest affecting mainly stored food products, such as grains, flour, and cereals, among others. These are dominant populations of insects found in stored traditional Chinese medicines [4]. *T. castaneum* can damage a great range of food and processed products, leading to agglomeration, discoloration, and spoilage, which result in serious economic losses [5]. The principal method to control these insects is the use of synthetic insecticides or fumigants. However, these methods may cause health hazards to warm-blooded animals, lead to environmental pollution, and potentially bring about insecticide-resistant insects, resulting in pest resurgence [6]. When dealing with food storage and preserving cultural relics and archives, it is essential to not only protect these materials from pests but to also reduce the extent of pesticide residues and avoid pollution. Therefore, an increasing number of researchers are searching and investigating different active natural products as botanical insecticides [7,8].

The essential oils extracted from various plants exhibit unique botanical and medicinal uses that, upon proper application, may not cause detrimental effects in humans and animal health as well as the environment. Essential oils are proven to have significant effects against a large number of stored grain insects, acting through ingestion [9] and contact toxicity [10,11]. The modes of action of plant essential oils on pests may include contact toxicity, fumigant, antifeedant, repellent, and growth-inhibiting activities [12,13]. Essential oils and their constituents from many plants have previously been confirmed to contain insecticidal or repellent activity, which inhibit the growth of insects that damage stored products [14–16]. Plant essential oils are often complex mixtures of terpenoids, and their bioactivity is likely to frequently be a result of synergy among constituents [17]. In addition, essential oils and their mono- and sesquiterpenoid constituents are fast-acting neurotoxins in insects, possibly interacting with multiple types of receptors [18]. Research has shown that, for rosemary (*Rosemarinus officinalis*) and lemongrass (*Cymbopogon citratus*) oils, synergy among major constituents results from increased penetration of toxicants through the insect's integument rather than through inhibition of detoxicative enzymes [19,20]. Moreover, these essential oils are volatile, and the products are also not risky for other organisms [21].

Elsholtzia ciliata (Thunb.) Hyland is a widely spread plant in China and is part of the herbal medicine collection with distinct special aroma [22–25]. The essential oils of *Elsholtzia* have certain poisonous activity on a variety of storage pests [26]. The *E. ciliata* essential oil was found to possess fumigant toxicity and contact toxicity against *Liposcelis bostrychophila*, with a lethal dose 50 (LC_{50}) value of 475.2 µg/L and 145.5 µg/cm^2, respectively [27]. The ether extract of *Elsholtzia stauntonii* had a strong fumigation effect on adult *Sitophilus zeamais* and *T. castaneum*. After four days of treatment, the adult mortality of *S. zeamais* reached over 95%, while it reached 100% for *T. castaneum* [28]. However, a literature survey showed no reports on insecticidal activity of the essential oil from the aerial parts of *E. ciliata* against *T. castaneum*. The present study was therefore undertaken to investigate the chemical components and insecticidal activities of the essential oil, including its active biochemical constituents against the food storage pest *T. castaneum*.

Carvone is a component of caraway (*Carum carvi Linnaeus*), dill (*Anethum graveolens Linnaeus*), and spearmint (*Mentha spicata Linnaeus*) seeds [29]. It is widely used in pesticides, food flavoring, feed flavoring, feed additive, personal care products, and veterinary medicine [30]. Limonene is listed in the Code of Federal Regulations as a generally recognized as safe (GRAS) substance for flavoring agents. It is commonly used in food items, such as fruit juices, soft drinks, baked goods, ice cream, and pudding [31], and it can be directly used in perfumes. It is also used in many flavor formulas with safety amount up to 30%, and the International Fragrance Association (IFRA) has no restrictions on it [32], although the potential occurrence of skin irritation necessitates regulation of this chemical as an ingredient in cosmetics. In conclusion, the use of limonene in cosmetics is safe under the current regulatory guidelines for cosmetics [33,34].

A literature survey showed some reports on insecticidal activity of carvone and limonene against insects. For instance, Fang et al. [35] stated that carvone and limonene had contact toxicity against *Sitophilus zeamais* with LD_{50} values of 2.79 µg/adult and 29.86 µg/adult, respectively. Carvone and limonene also possessed strong fumigant toxicity against *S. zeamais* (LC_{50} = 2.76 and 48.18 mg/L). Yang [36] found that, after 24 h exposure time, the mortalities of insects in carvone with three fumigant concentrations reached 100%. In addition, the limonene showed contact toxicity against *T. castaneum* adults with a LD_{50} value of 14.97 µg/adult [37].

2. Materials and Methods

2.1. Plant Materials and Extraction of Essential Oil

E. ciliata was gathered in Longxi County (35°1′ N latitude, 104°27′ E longitude, altitude 1880 m) in the Gansu province of China. To obtain the crude essential oil, the minced sample was connected to the distillation unit and condenser and maintained for 6 h. Anhydrous Na_2SO_4 was added to the crude essential oil to remove all water residue. The volume of the pure essential oil was recorded and the yield was calculated. The prepared essential oil was stored in the refrigerator at 4 °C until use.

2.2. Test Insects

T. castaneum adults were inoculated into a mixture of whole wheat flour and yeast flour at a mass ratio of 10:1 and cultured in a constant temperature incubator at 30 ± 1 °C with 75% ± 5% relative humidity for 24 h dark treatment. All adult beetles used in the experiment were considered as adult stage after an eclosion time of 1–2 weeks. On the other hand, the test larvae [38] were six instar larvae with an approximate length of 5–6 mm.

2.3. Gas Chromatography-Mass Spectrometry (GC-MS) Analysis

The GC-MS analysis was run on an Agilent 6890 N gas chromatograph connected to an Agilent 5973 N mass selective detector. They were equipped with a gas chromatography-flame ionization detector (GC-FID) and a HP-5MS (30 cm × 0.25 mm × 0.25 μm) capillary column. The essential oil sample was diluted in *n*-hexane to obtain a 1% solution. The injector temperature was maintained at 250 °C with the volume injected being 1 μL. The flow rate of carrier gas (helium) was 1.0 mL/min, with the mass spectra scanned from 50 to 550 *m/z*.

The retention indices (RI) were determined from gas chromatograms using a series of *n*-alkanes (C_5-C_{36}) under the same operating conditions. Based on RI, the chemical constituents were identified by comparing them with *n*-alkanes as a reference. The components of the essential oil were identified by matching their mass spectra with various computer libraries (Wiley 275 libraries, NIST 05, and RI from other literature) [39].

2.4. Contact Toxicity

The contact toxicity activities of *E. ciliata* essential oil and its main components were determined by the dot contact method [40]. The essential oil was diluted to five different concentration gradients (5%, 3.3%, 2.2%, 1.48%, 0.98%) with *n*-hexane. A 0.5 μL diluted solution was dropped on the torso of *T. castaneum* after being palsied by the freezing method. Then, the test insects were transferred to a glass bottle with a volume of 25 mL. *n*-Hexane and pyrethrin were used as negative and positive controls, respectively. Each concentration was repeated 5 times, and 10 test insects were used for each assay. After 24 h, the number of dead insects was recorded, and the mortality and corrected mortality were calculated. Insects that did not respond to a brush were considered dead. A similar experimental method was undertaken in testing the larval stage.

2.5. Fumigant Toxicity

Fumigant activities of *E. ciliata* essential oil and its main components against adults and larvae of *T. castaneum* were evaluated based on the method described by Wu et al. [41]. The essential oil was diluted with *n*-hexane to obtain five concentration gradients (10%, 6.6%, 4.4%, 2.9%, 1.77%). Diluted liquids of 10 μL were injected on the filter paper (2.0 cm^2) and placed on the inside of the bottle cap. The bottle cap was quickly screwed up and wrapped by the sealing film to form a closed space after 20 s. *n*-Hexane was used as a negative control, whereas methyl bromide and phoxim were used as positive controls for adults and larvae, respectively. Each concentration was repeated 5 times and tested in 10 test insects in each assay. After 24 h, the death of the test insects was observed and recorded, and the mortality and corrected mortality were calculated. The same experimental method was used to test the larval stage.

2.6. Two Main Components Compounding

We used the ten-point theory [42] that assumes that the half-lethal concentrations of A and B are determined by the virulence of a and b. Hence, the A + B mixture was evaluated by the co-toxic factor method. A total of 7 ratios were selected according to the corresponding concentration gradient order of 1:7, 2:6, 3:5, 4 4, 5:3, 6:2, and 7:1. The contact toxicity and fumigant toxicity methods were performed as described previously (Materials and Methods Sections 2.4 and 2.5). Three repetitions were done for each treatment, and a blank control was set.

2.7. Data Analysis

The LC_{50} (mg/L air) and the LD_{50} (μg/adult or larva) of the lethal activity were analyzed and calculated using SPSS 22.0 statistical software, and the corrected mortality was calculated by Abbott's formula. The determination of the synergistic effect was performed with combined toxicity evaluation using Sun Yunpei's co-toxicity method CTC (Co-toxicity index) [43]. The criteria were as follows: $80 \leq CTC \leq 120$ indicated an additive effect, $CTC > 120$ indicated a synergistic effect, and $CTC < 80$ indicated an antagonistic effect. The calculations were as follows:

①   Co-toxicity index (CTC) = ATI/TTI × 100%
②   Mixed virulence index (ATI) = standard drug LD_{50}/mixture (A+B) LD_{50} × 100%
③   Theoretical virulence index of (A+B) (TTI) = Va × Ma + Vb × Mb V_a = Virulence index of agent A, M_a = the mass fraction of agent A in the mixture V_b = Virulence index of agent B, M_b = the mass fraction of agent B in the mixture
④   Single dose virulence index (TI) = standard drug LD_{50}/LD_{50} for the test agent × 100%

2.8. Chemicals

Pyrethrins were purchased from Dr. Ehrenstorfer GmbH, Augsburg, Germany with a concentration of 27%. Phoxim were purchased from Dr. Ehrenstorfer GmbH, Augsburg, Germany with a purity of 98.0%; Carvone was purchased from Tishila (Shanghai) Chemical Industry Development Co., Ltd., China, with a purity of 99.0%. Limonene was purchased from Shanghai Aladdin Biochemical Technology Co., Ltd., China, with a purity of 95.0%.

3. Results

3.1. Chemical Compounds of E. ciliata Essential Oil

The essential oil extracted from the leaves of *E. ciliata* had a yield of 0.36% (*V*/m). The chemical compounds and relative contents of *E. ciliata* essential oil are shown in Table 1. In this study, we identified 16 compounds in *E. ciliata* essential oil, the main compounds were monoterpenoids and sesquiterpenes, with monoterpenoids accounted for 76.97%, sesquiterpenes accounted for 20.61%, and carvone was the highest monoterpenoid among all, while α-caryophyllene had the highest content of sesquiterpenes. What is more, we observed four major components of *E. ciliata* essential oil, namely, carvone (31.6%), limonene (22.05%), α-caryophyllene (15.47%), and dehydroelsholtzia ketone (14.86%). These components are distinct from previous works. For example, *E. ciliata* essential oil derived from Mao'er Mountain of northeastern China mainly constituted dehydroelsholtzia ketone (68.35%) and elsholtzia ketone (25.19%) [44]. More than 30 components were separated from the essential oil of *E. ciliata* in Changbai Mountains in northeastern China, and the main components were β-dehydrogeranione (51.77%) and elsholtzia ketone (33.33%) [45]. In addition, the elsholtzia ketone concentration in the essential oil from both Changbai Mountains and Mao'er Mountain in the Liu's research was higher than that in this experiment. The dehydroelsholtzia ketone in the essential oil from Mao'er Mountain (68.35%) in Liu's research was double that of this experiment. All the *E. ciliata* in the abovementioned works were gathered from Northeast China, while the *E. ciliata* studied in this paper was from Northwest China. The large climate difference between the two areas may be one of the reasons for the differences in essential oil composition. Moreover, the difference in harvesting time and growth years may also cause differences in essential oil components.

Table 1. Chemical composition of the essential oil from *E. ciliata*.

Number	Constituent	Retention Time/Min (Rt)	Ri *	Relative Content (%)
1	α-Pinene	3.394	932	0.55
2	β-Pinene	3.812	977	0.74
3	Myrcene	3.861	988	1.02
4	β-Phellandrene	3.966	1019	0.19
5	Limonene	4.464	1040	22.05
6	β-Ocimene	4.654	1061	4.08
7	Linalool	5.367	1090	0.83
8	Elsholtzia ketone	6.726	1199	1.02
9	Carvone	7.366	1216	31.63
10	Dehydroelsholtzia ketone	8.104	1277	14.86
11	Cubebene	9.180	1344	1.06
12	β-Bourbonene	9.635	1379	0.44
13	β-Caryophyllen	10.077	1414	2.92
14	α-Caryophyllene	10.397	1450	15.47
15	(-)-Humulene epoxide II	10.643	1454	0.25
16	α-Farnesene	11.965	1489	0.47
-	Total	-	-	97.58
	Others			2.42

* RI (retention index) as determined on a HP-5MS column using the homologous series of *n*-hydrocarbons.

3.2. Contact Activity

Table 2 shows the results of contact activities of *E. ciliata* essential oil and the two main components (carvone and limonene) against *T. castaneum* adults and larvae. The essential oil of *E. ciliata* showed obvious contact toxicity against *T. castaneum* adult and larval stages with LD_{50} of 7.79 μg/adult and 24.87 μg/larva, respectively. Among the two main components, carvone had stronger contact activity against adults (LD_{50} = 5.08 μg/adult), which was 7.59-fold higher than the effect of limonene (LD_{50} = 38.57 μg/adult). This result implies that carvone might have been a key component of *E. ciliata* essential oil involved in contact toxicity against *T. castaneum*. Although the contact activities of essential oil and carvone against *T. castaneum* adults was weaker than that of the positive control pyrethrin (LD_{50} = 0.09 μg/adult), the *E. ciliata* essential oil showed stronger contact effect than previously reported plants. For example, Wu et al. [41] found that the LD_{50} of *Platycladus orientalis* essential oil against *T. castaneum* was 48.59 μg/adult. The essential oils of *Murraya exotica* aerial parts showed contact toxicity against *T. castaneum* adults with LD_{50} values of 20.94 μg/adult [46]. Therefore, *E. ciliata* essential oil and its two main components (carvone and limonene) have strong contact toxicity against *T. castaneum*.

Table 2. Contact toxicity of *E. ciliata* essential oil and its main constituents against *T. castaneum*.

T. castaneum	Treatment	Ld_{50} (mg/Adult)	95% Fl (mg/Adult)	Slope ± Se	*p*-Value	Chi Square χ^2
Adult	Essential oil	7.79	6.96–8.65	4.14 ± 0.46	0.85	16.17
	Carvone	5.08	4.19–6.20	4.30 ± 0.46	0.01	44.15
	Limonene	38.57	34.48–43.09	3.84 ± 0.42	0.55	21.54
	Pyrethrin	0.09	0.08–0.11	2.48 ± 0.31	0.92	14.27
Larva	Essential oil	24.87	19.55–30.69	1.69 ± 0.22	0.64	24.72
	Carvone	33.03	26.55–41.26	1.86 ± 0.23	0.75	18.12
	Limonene	49.68	34.10–84.04	0.95 ± 0.15	0.54	26.70
	Pyrethrin	1.31	0.75–2.17	0.80 ± 0.10	0.82	16.72

3.3. Fumigation Activity

Fumigation activity of *E. ciliata* essential oil and its two components are shown in Table 3. Both *E. ciliata* essential oil and the two major components had obvious fumigant toxicity against *T. castaneum* adults and larvae, although *E. ciliata* essential oil had a stronger fumigating effect on *T. castaneum* larvae (LC_{50} = 8.73 mg/L air). The fumigant toxicity of carvone against adults (LC_{50} = 4.34 mg/L air) was significantly higher than that against larvae (LC_{50} = 28.71 mg/L air). Limonene also had obvious fumigation activity against adults, with a LC_{50} of 5.52 mg/L air. The fumigation effect of carvone and limonene was 2.68 and 2.1 times greater, respectively, than the effect of the essential oil against adults. When the two components were applied together, the fumigation activity increased significantly. A previous study has also reported that carvone and limonene have strong fumigation activity against *T. castaneum* [36]. Therefore, it can be inferred that carvone and limonene are two of the active ingredients containing fumigant toxicity against *T. castaneum*.

For the fumigation effect against larvae, *E. ciliata* essential oil had the best fumigation activity, which was 3.29 times higher than the effect of carvone and 2.36 times higher than that of limonene. The fumigation activity of *E. ciliata* essential oil and the two components appeared weak. The fumigation activity of essential oil was 6-fold weaker than the positive control, and the fumigation activities of carvone and limonene against *T. castaneum* adults was weaker than methyl bromide. However, compared with the fumigation effect of other essential oils, *E. ciliata* essential oil and the two monomers had relatively stronger activity. For instance, Han et al. [47] reported eugenol had contact toxicity against *T. castaneum* larvae and adults with LC_{50} values of 219.00 µL/mL and 363.08 µL/mL, respectively. In addition, Lv et al. [48] used Soxhlet extraction and ether as a solvent to extract essential oils from garlic, chili powder, citrus peel, and toon bark, which showed fumigation activity against *T. castaneum* larvae but not against adults. Given the characteristic of *E. ciliata* essential oil, it is most likely to develop a fumigant insecticide effect against the larvae of *T. castaneum*.

In summary, the contact toxicity of *E. ciliata* essential oil and its components against adult *T. castaneum* was significantly stronger than that against larvae. A pertinent point in this case is the completion of *T. castaneum* metamorphosis. The adults and larvae of *T. castaneum* are very different [49], especially in terms of self-protection mechanisms and body substances, such as the numerous enzymes that contribute to different degrees of tolerance to external stimuli. In addition, Liang et al. [50] also proved that these two forms differ greatly in their responses to various substances. As described in the literature, the main constituents of the defensive secretions of *T. castaneum* are methyl quinone, 1-pentadecene, 1,6-heptadecadiene, and paeonol. These compounds are repellent to adults whilst being attractive to larvae. Moreover, older adults are more sensitive to these compounds than young adults. Therefore, the whole process of metamorphosis diversifies the response to specific substances, which in turn leads to *E. ciliata* essential oil or its components having dramatically different contact activity against *T. castaneum* adults and larvae. In addition, according to the literature, monoterpenoids and sesquiterpenoid constituents are fast-acting neurotoxins in insects [18]. Both carvone and limonene are monoterpenoids, so it is speculated that carvone and limonene act as fast-acting neurotoxins on pests. In future research, the fumigating mechanism of carvone and limonene will be further explored. In addition, we shoule consider bioactive confrontation of high elsholtzia ketone or dehydroelsholtzia ketone *Elsholtzia* oils with those containing mostly carvone. We also need to consider chiral GC of oil and completion of R- and S-carvone together with R- and S-limonene to use in insect assays.

Table 3. Fumigant toxicity of *E. ciliata* essential oil and its main constituents against *T. castaneum*.

T. castaneum	Treatment	LC$_{50}$ (mg/L Air)	95% FL (mg/L Air)	Slope $\pm$ SE	*p*-Value	Chi Square X^2
Adult	Essential oil	11.61	9.21–14.01	4.39 $\pm$ 0.47	0.00	87.62
	Carvone	4.34	3.90–4.84	6.27 $\pm$ 0.83	0.98	7.89
	Limonene	5.52	2.75–9.22	1.69 $\pm$ 0.47	0.83	5.85
	Methyl bromide [a]	1.83	1.43–2.23	4.90 $\pm$ 0.51	0.89	8.67
Larva	Essential oil	8.73	6.62–11.25	1.42 $\pm$ 0.17	0.99	11.19
	Carvone	28.71	23.07–36.05	1.63 $\pm$ 0.15	0.36	35.41
	Limonene	20.64	16.96–25.56	1.71 $\pm$ 0.16	0.86	24.46
	Phoxim	1.05	1.23–2.08	1.65 $\pm$ 0.45	0.89	3.25

[a] The data for methyl bromide was derived from the literature with a consistent experimental method [51].

3.4. Carvone Mixed with Limonene and Its Contact Toxicity against T. castaneum Adult

After mixing carvone and limonene in seven different ratios, as shown in Table 4, we found that when the volume ratio of carvone to limonene was 2:6, the CTC value was 134.33, suggesting a synergistic effect ($\geq$120). On the other hand, when the volume ratio was 1:7, the CTC showed an additive effect (between 80 and 120). In other ratios, the respective CTCs were less than 80, suggesting an antagonistic effect. As shown in the results, the effect of the limonene mixture appeared unsatisfactory. One of the possible reasons may be that carvone and limonene work in a similar manner; as a result, the addition of limonene inhibits the contact toxicity effect of carvone. The proportion of carvone in *E. ciliata* essential oil was 1.67 times higher than that of limonene, which was equivalent to a compounding agent having a volume ratio of 5:3; the CTC was 67.43 (less than 80), indicating an antagonistic effect. This indicates that the contact toxicity of *E. ciliata* essential oil against *T. castaneum* adults may not be as strong as the contact activity of carvone.

Table 4. Contact toxicity and CTC () of carvone and limonene mixture against *T. castaneum* adults.

Volume Ratio	LD$_{50}$ (µg/Adult)	Slope $\pm$ SE	*p*-Value	ATI	TTI	CTC
1:7	24.60	2.935 $\pm$ 0.59	0.64	20.65	24.02	85.97
2:6	10.84	2.972 $\pm$ 0.51	0.54	46.85	34.88	134.33
3:5	34.43	1.856 $\pm$ 0.45	0.99	14.75	45.73	32.26
4:4	39.60	1.970 $\pm$ 0.47	0.99	12.83	113.17	11.33
5:3	140.30	1.605 $\pm$ 0.79	0.60	3.62	67.43	5.37
6:2	79.34	1.666 $\pm$ 0.95	0.95	6.40	78.29	8.18
7:1	434.82	1.495 $\pm$ 0.85	0.80	1.17	115.93	1.01

The contact toxicity of carvone against *T. castaneum* larvae displayed enhanced activity by combining in different ratios with limonene. As shown in Table 5, three of the seven ratios had CTC greater than 120 (synergism); these were 1:7, 2:6, and 7:1. In particular, carvone in a 1:7 ratio combination with limonene showed a significant increase in its activity over a single compound with a CTC value of 155. This combination provided strong contact toxicity with the corresponding LD$_{50}$ of 30.04 µg/larva after 24 h of incubation. Besides, when carvone and limonene were mixed in volume ratios of 2:6 and 7:1, the CTC values were 144.08 and 130.19, respectively. The CTCs of these effective combinations were all more than 120, suggesting a synergistic effect. However, when carvone and limonene were mixed in a ratio of 5:3, the CTCs were less than 80, with an antagonistic effect. The 5:3

ratio is similar to the carvone and limonene content ratio in essential oils. Essential oils have stronger contact toxicity against larvae than carvone and limonene, which appears to be a result of synergy among various constituents.

Table 5. Contact toxicity and CTC of carvone and limonene mixture against larvae of *T. castaneum*.

Volume Ratio	LD$_{50}$ (μg/Larva)	Slope ± SE	*p*-Value	ATI	TTI	CTC
1:7	30.04	2.323 ± 0.59	0.34	109.94	70.68	155.55
2:6	30.62	3.829 ± 0.73	0.71	107.87	74.87	144.08
3:5	84.30	1.145 ± 0.14	0.88	39.18	79.06	49.56
4:4	405.96	1.390 ± 0.12	0.46	8.14	83.24	9.77
5:3	66.30	3.074 ± 0.16	0.95	49.82	87.43	56.98
6:2	112.98	2.175 ± 0.94	0.90	29.24	91.62	31.91
7:1	26.48	5.321 ± 0.11	0.96	124.74	95.81	130.19

Table 6 shows the fumigation activity of carvone and limonene mixed in different ratios against the adult stage of *T. castaneum*. Out of these seven different ratios, the CTC of two particular ratios were greater than 120 (CTCs of 212.71 and 159.03), suggesting different degrees of synergism. Carvone + limonene at 1:7 ratio combination was found to be most effective in terms of fumigant toxicity against *T. castaneum* adults. This ratio provided strong fumigation activity with corresponding LC$_{50}$ of 2.51 mg/L air after 24 h of incubation. The CTC values of the other ratios of carvone and limonene were less than 80, showing an obvious antagonistic effect.

Table 6. Fumigant toxicity and CTC of carvone and limonene mixture against adult of *T. castaneum*.

Volume Ratio	LC$_{50}$ (mg/L Air)	Slope ± SE	*p*-Value	ATI	TTI	CTC
1:7	2.51	2.921 ± 0.48	0.00	172.91	81.29	212.71
2:6	3.25	4.793 ± 0.63	0.05	133.54	83.97	159.03
3:5	7.43	2.845 ± 0.93	0.80	58.41	86.64	67.42
4:4	6.55	2.656 ± 0.82	0.72	66.26	89.31	74.19
5:3	9.45	2.567 ± 0.76	0.88	45.93	91.98	49.93
6:2	11.39	2.814 ± 0.44	0.85	38.10	94.66	40.25
7:1	26.30	1.889 ± 0.40	0.72	16.50	97.33	16.95

After the carvone and limonene were mixed in different ratios, the fumigation activity and CTC of the larvae of *T. castaneum* were determined (Table 7). The mixtures of carvone and limonene at 5:3 ratio showed fumigant activity against adult *T. castuneum* (LC$_{50}$ = 20.58 mg/L air). Its CTC value was 89.65, and it appeared to show an additive effect. The values of CTC under other ratios were all less than 80 and thus suggested an antagonistic effect.

Table 7. Fumigant toxicity and CTC of carvone and limonene mixture against larvae of *T. castaneum*.

Volume Ratio	LC$_{50}$ (mg/L Air)	Slope ± SE	*p*-Value	ATI	TTI	CTC
1:7	38.56	1.846 ± 0.70	0.87	74.46	134.21	55.48
2:6	34.64	2.314 ± 0.77	0.80	82.88	129.32	64.09
3:5	31.08	2.201 ± 0.71	0.88	92.37	124.44	74.23
4:4	32.99	2.286 ± 0.72	0.94	87.03	119.55	72.79
5:3	27.93	3.041 ± 0.79	0.54	102.79	114.66	89.65
6:2	221.59	1.215 ± 0.85	0.81	12.96	109.76	11.80
7:1	52.91	3.189 ± 0.46	0.88	54.26	104.89	51.73

Figure 1 shows a general synergistic effect and antagonistic effect (to some degree) with different mixture ratios in terms of contact toxicity against the adult and larval stages of *T. castaneum*. The figure also indicates a deviation in the CTC value trends between adult and larval stages. We observed that when the mixture ratio was 2:6, the CTC values for both stages were greater than 120, which suggested synergism, particularly in larvae. The CTC value reached the maximum when carvone and limonene were mixed at a ratio of 1:7. This result implies that synergy for larvae is the best at a 1:7 ratio. However, at this ratio, the CTC value of adults was 85.97, indicating an additive effect. In addition, when the mixture ratios were 3:5 and 4:4, the CTC value of the contact killing effect in adult and larval stages decreased significantly. The CTC of larvae showed an upward trend after 4:4, reaching 130.19 at a ratio of 1:7, indicating a synergistic effect. On the contrary, the effect on adult *T. castaneum* declined after the mixture ratio of 4:4 and reached 1.01 at the ratio of 7:1, indicating a marked antagonistic effect. In conclusion, except at the ratio of 4:4, the CTC values of the *T. castaneum* larvae were slightly higher than those of adults in the same ratios. It can be deduced that the contact toxicity effect of carvone and limonene on the larvae of *T. castaneum* is generally better than that of adults at the same ratio.

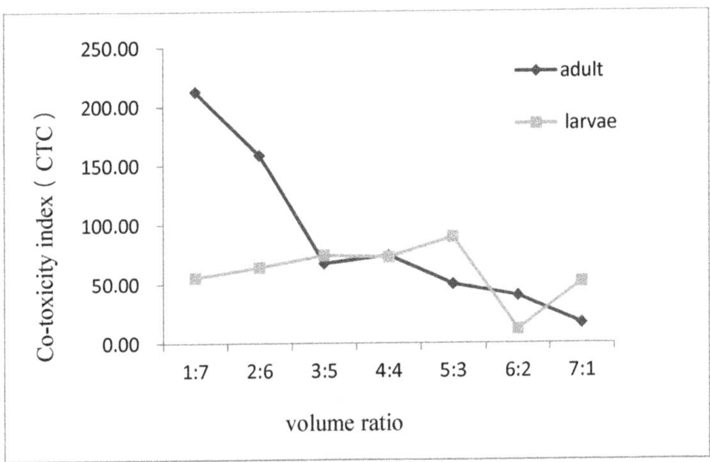

Figure 1. The CTC of contact activity of carvone and limonene at different ratios against adults and larvae of *T. castaneum*.

When carvone and limonene were mixed in different ratios, we observed obvious differences in fumigation activity against the adult and larval stages of *T. castaneum* (Figure 2). The CTC of adults reached the maximum value with the best synergistic effect at a ratio of 1:7. Moreover, the mixture showed a synergistic effect when the ratio was 2:6. After that, the value of CTC was less than 80, which indicated an antagonistic effect. However, the co-toxic effect against the larvae was generally weak or appeared antagonistic, except when the ratio was 5:3, which showed an additive effect. Generally, the CTC values of *T. castaneum* adults were slightly higher than the larvae using the same mixture ratio. Therefore, when carvone and limonene were mixed in the same ratio, its fumigation activity is better in adults than in larvae.

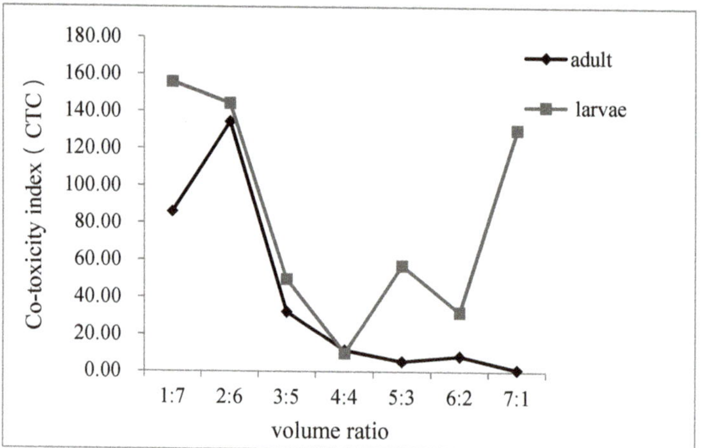

Figure 2. The CTC of fumigant activity of carvone and limonene at different ratios against adults and larvae of *T. castaneum*.

Through the mixture of the two major components, we identified the optimal mixing method that can effectively target *T. castaneum*. Changing the mixing ratio also changed the insecticide effects of the two compounds, but the effect of getting twice the result with half the effort was achieved for both plant essential oil mixed with compounds as well as essential oils mixed with essential oils. For example, the Commonwealth Scientific and Industrial Research Organisation (CSIRO) Institute of Entomology compared several natural plant extracts known to have insecticidal activity with ethyl formate and found that some plant products have a good synergistic effect [52]. The essential oils from plants have the advantages of having broad-spectrum insecticidal efficacy and being generally safe in humans, animals, and the environment. Carvone and limonene are derived from plant essential oil with the synergistic effect produced at a volume ratio of 2:6. The difference in the mode of action of the two substances against *T. castaneum* are important factors that influence its compounding effect. Exploring ways to make better use of mixed medicines will not only help overcome the high cost of plant essential oils but will also provide a theoretical basis for the practical application of the two medicines.

4. Conclusions

In this study, nine different components were identified from *E. ciliata* essential oil extract. The two main components, carvone and limonene, showed strong contact and fumigation activities against adults and larvae of *T. castaneum*. Meanwhile, *E. ciliata* essential oil also showed intense toxicity against the test insects. We also found that carvone might play a key role in the contact toxicity of *E. ciliata* essential oil against *T. castaneum*. Carvone and limonene exhibited synergistic effects at a volume ratio of 2:6. Altogether, our results suggest that *E. ciliata* essential oil extract and its two major components have a potential for downstream development as natural insecticides.

Author Contributions: Conceptualization, J.-Y.L., J.-L.W., and J.X.; Funding acquisition, J.-Y.L. and J.-L.W.; Investigation, J.-Y.L., J.X., Y.-Y.Y., Y.-Z.S. and F.Z.; Validation, J.-Y.L., Y.-Y.Y., Y.-Z.S., and F.Z.; Writing-original draft preparation, L.-J.Y. and J.X.; Writing-review & editing, J.-Y.L., J.X., Y.-Y.Y., and Y.-Z.S.; Supervision, J.-L.W. and F.Z.; Project administration, J.-Y.L.; All authors have read and agreed to the published version of the manuscript.

Funding: This work was supported by the National Natural Science Foundation of China (NO. 81660632) and the Natural Science Foundation of Gansu Province, China (NO. 18JR3RA092).

Conflicts of Interest: The authors declare no conflict of interest.

References

1. Athanassiou, C.G.; Arthur, F.H.; Throne, J.E. Efficacy of grain protectants against four psocid species on maize, rice and wheat. *Pest Manag.* **2009**, *65*, 1140–1146. [CrossRef] [PubMed]
2. Tapondjou, A.L.; Alder, C.; Bouda, H.; Fontem, D.A. Efficacy of powder and essential oil from *Chenopodium ambrosioides* leaves as post-harvest grain protectants against six-stored product beetles. *J. Stored Prod. Res.* **2002**, *38*, 395–402. [CrossRef]
3. Isman, M.B. Botanical insecticides, deterrents, and repellents in modern agriculture and an increasingly regulated world. *Annu. Rev. Entomol.* **2006**, *51*, 45–66. [CrossRef] [PubMed]
4. Wang, C.F.; Yang, K.; You, C.X.; Zhang, W.J.; Guo, S.S.; Geng, Z.F.; Du, S.S.; Wang, Y.Y. Chemical composition and insecticidal activity of essential oils from *Zanthoxylum dissitum* leaves and roots against three species of storage pests. *Molecules* **2015**, *20*, 7990–7999. [CrossRef] [PubMed]
5. Zhang, S.F.; Chen, H.J.; Xue, G.H. *Atlas of Beetles Associated with Stored Products*; China Agricultural Science and Technology Press: Beijing, China, 2008; p. 161. ISBN 978-78-0233-607-0.
6. Ebadollahi, A.; Safaralizadeh, M.H.; Pourmirza, A.A.; Gheibi, S.A. Toxicity of essential oil of *Agastache foeniculum* (Pursh) kuntze to *Oryzaephilus surinamensis* L. and *Lasioderma serricorne* F. *Plant Prot. Res.* **2010**, *50*, 215–219. [CrossRef]
7. Isman, M.B.; Singh, B.P. Botanical insecticides, deterrents, repellents and oils. *Ind. Crops Uses* **2010**, *20*, 433–445. [CrossRef]
8. Zettler, J.; Arthur, F.H. Chemical control of stored product insects with fumigants and residual treatments. *Crops Prot.* **2000**, *19*, 577–582. [CrossRef]
9. Fabres, A.; Janaina, C.M.D.S.; Fernandes, K.V.S.; Xavier-Filho, J.; Rezende, G.L.; Oliveira, A.E.A. Comparative performance of the red flour beetle *Tribolium castaneum* (Coleoptera: Tenebrionidae) on different plant diets. *J. Pestic. Sci.* **2014**, *87*, 495–506. [CrossRef]
10. Bossou, A.D.; Ahoussi, E.; Ruysbergh, E.; Adams, A.; Smagghe, G.D.; Kimpe, N.; Mangelinckx, S. Characterization of volatile compounds from three *Cymbopogon* species and *Eucalyptus citriodora* from benin and their insecticidal activities against *Tribolium castaneum*. *Ind. Crops Prod.* **2015**, *76*, 306–317. [CrossRef]
11. Lee, H.K.; Lee, H.S. Toxicities of active constituent isolated from *Thymus vulgaris* flowers and its structural derivatives against *Tribolium castaneum* (Herbst). *Appl. Biol. Chem.* **2016**, *59*, 821–826. [CrossRef]
12. Regnaltroger, C.; Vincent, C.; Arnason, J.T. Essential oils in insect control: Low–risk products in a High-Stakes World. *Annu. Rev. Entomol.* **2012**, *57*, 405–424. [CrossRef]
13. Chu, S.S.; Jiang, G.H.; Liu, W.L.; Liu, Z.L. Insecticidal activity of the root bark essential oil of *Periploca sepium* Bunge and its main component. *Nat. Prod. Res.* **2012**, *26*, 926–932. [CrossRef] [PubMed]
14. Omar, S.; Lalonde, M.; Marcotte, A. Insect growth-reducing and antifeedant activity in Eastern North America hardwood species and bioassay-guided isolation of active principles from *Prunus serotina*. *Agric. For. Entomol.* **2000**, *2*, 253–257. [CrossRef]
15. Ahmadi, M.; Abd-Alla, A.M.; Moharramipour, S. Combination of gamma radiation and essential oils from medicinal plants in managing *Tribolium castaneum* contamination of stored products. *Appl. Radiat. Isot.* **2013**, *78*, 16–20. [CrossRef] [PubMed]
16. Zhao, N.N.; Zhou, L.; Liu, Z.L.; Du, S.S.; Deng, Z.W. Evaluation of the toxicity of the essential oils of some common Chinese spices agaist *Liposcelis bostrychophila*. *Food Control* **2012**, *26*, 486–490. [CrossRef]
17. Tak, J.H.; Isman, M.B. Enhanced cuticular penetration as the mechanism for synergy of insecticidal constituents of rosemary essential oil in Trichoplusia ni. *Sci. Rep.* **2015**, *5*, 12690. [CrossRef]
18. Isman, M.B. Commercial development of plant essential oils and their constituents as active ingredients in bioinsecticides. *Phytochem. Rev.* **2019**, 1–7. [CrossRef]
19. Tak, J.H.; Isman, M.B. Metabolism of citral, the major constituent of lemongrass oil, in the cabbage looper, Trichoplusia ni, and effects of enzyme inhibitors on toxicity and metabolism. *Pestic. Biochem. Phys.* **2016**, *133*, 20–25. [CrossRef]
20. Tak, J.H.; Jovel, E.; Isman, M.B. Effects of rosemary, thyme and lemongrass oils and their major constituents on detoxifying enzyme activity and insecticidal activity in Trichoplusia ni. *Pestic. Biochem. Phys.* **2017**, *140*, 9–16. [CrossRef]

21. Aref, S.P.; Valizadegan, O.; Farashiani, M.E. The Insecticidal Effect of Essential Oil of *Eucalyptus floribundi* Against Two Major Stored Product Insect Pests; *Rhyzopertha dominica* (F.) and *Oryzaephilus surinamensis* (L.). *J. Plant Prot. Res.* **2015**, *55*, 35–41. [CrossRef]

22. Du, X.W.; Tan, J.C.; Cao, Y.F.; Zhang, X.H.; Chen, J.S. Research progress in plant essential oils. *Hunan Agric. Sci.* **2009**, *8*, 86–89. [CrossRef]

23. Neffati, A.; Bouhlel, I.; Sghaier, M.B.; Boubaker, J.; Limen, I.; Kilani, S.; Skandrani, I.; Bhouri, W.; Dauphin, J.L.; Barillier, D.; et al. Antigenotoxic and antioxidant activities of *Pituranthos chloranthus* essential oils. *Environ. Toxicol. Pharmacl.* **2009**, *27*, 187–194. [CrossRef] [PubMed]

24. Wu, K.G.; Luo, M.T.; Wei, H. Study on the antibacterial effect of 8 kinds of plant essential oils on common intestinal microorganisms in vitro. *Mod. Food Sci. Technol.* **2017**, *6*, 133–141.

25. Oke, F.; Aslim, B.; Ozturk, S.; Altundag, S. Essential oil composition, antimicrobial and antioxidant activities of *Satureja cuneifolia* Ten. *Food Chem.* **2009**, *112*, 874–879. [CrossRef]

26. Li, H.X.; Wei, M.S.; Yi, P.Y.; Ke, Z.G.; Man, Y.S. The control effect of 25 kinds of plant essential oils on the *Callosobruchus maculatus*. *Grain Storage* **2001**, *30*, 7–9.

27. Zhao, M.P.; Liu, X.C.; Lai, D.W.; Zhao, L.; Liu, Z.L. Analysis of the Essential Oil of *Elsholtzia ciliata* Aerial Parts and Its Insecticidal Activities against *Liposcelis bostrychophila*. *Helv. Chim. Acta* **2016**, *99*, 90–94. [CrossRef]

28. Lv, J.H.; Wu, S.H.; Yuan, L.Y.; Li, Y.F. Study on the effect of *Elsholtzia stauntonii benth* extract on the adults of *Sitophilus zeamais* and *Tribolium rubra*. *J. Henan Univ. Technol. Nat. Sci. E* **2008**, *29*, 31–34.

29. De Carvalho, C.C.; Da Fonseca, M.M.R. Carvone: Why and how should one bother to produce this terpene. *Food Chem.* **2006**, *95*, 413–422. [CrossRef]

30. Ma, C.; Liu, X.L.; Zhao, Z.G.; Zhang, L. Improvement on the synthesis of carvone perfume. *J. Southwest Univ. Nat. Sci. Ed.* **2003**, *308–309*, 340. [CrossRef]

31. Committee, E.S. Scientific Opinion on the safety assessment of carvone, considering all sources of exposure. *EFSA J.* **2014**, *12*, 1–74. [CrossRef]

32. Wang, W.J. Recent advances on limonene, a natural and active monoterpene. *China Food Addit.* **2005**, *1*, 33–37. [CrossRef]

33. Sun, J.D. D-limonene: Safety and clinical applications. *Altern. Med. Rev.* **2007**, *12*, 259–264. [CrossRef] [PubMed]

34. Kim, Y.W.; Kim, M.J.; Chung, B.Y. Safety Evaluation And Risk Assessment Of d-Limonene. *J. Toxicol. Environ. Health* **2013**, *16*, 17–38. [CrossRef] [PubMed]

35. Fang, R.; Jiang, C.H.; Wang, X.Y. Insecticidal Activity of Essential Oil of Carum Carvi Fruits from China and Its Main Components against Two Grain Storage Insects. *Molecules* **2010**, *15*, 9391–9402. [CrossRef]

36. Yang, B. *Study on Fumigation Activity of Plant Essential Oil. D*; Huazhong Agriculture University: Wuhan, China, 2008. [CrossRef]

37. Zhu, X.K.; Guo, S.S.; Zhang, Z. Insecticidal activities of the essential oil from Alpinia zerumbet leaves against Triboliun castaneum in storag. *Plant Prot.* **2017**, *6*, 151–155+162. [CrossRef]

38. Zhao, X.N.; Wang, J.Q. Study on the fumigant virulence of the oriental green pepper oil to the larvae of the *Tribolium castaneum*. *Packag. Eng.* **2012**, *33*, 9–13.

39. Adams, R.P. Identification of essential oil components by gas chromatography/mass spectrometry. *J. Am. Soc. Mass Spectr.* **2005**, *16*, 1902–1903. [CrossRef]

40. Yang, K.; Wang, C.F.; You, C.X. Bioactivity of essential oil of *Litsea cubeba* from China and its main compounds against two stored product insects. *J. Asia Pac. Entomol.* **2014**, *17*, 459–466. [CrossRef]

41. Wu, Y.; You, C.X.; Tian, Z.F. Insecticidal Activity of essential Oil from *Vitex negundo L. var. cannabifolia* Leaves on *Lasioderma serricorne*. *Plant Prot.* **2016**, *42*, 97–102. [CrossRef]

42. Chen, I.; Xu, H.H.; Li, Y.Y.; He, L.F.; Zhao, S.H. Discussion on the method of screening the best effective formula of pesticide compound. *Acta Phytophys. Sin.* **2000**, *27*, 349–354. [CrossRef]

43. Wang, S.X.; Yuan, H.B.; Ma, L.B.; Yang, C.H.; Reng, B.Z. Fumigation effect of plant essential oil mixed with ethyl formate on adults of *Sitophilus zeamais*. *J. Northeast Norm. Univ. Nat. Sci.* **2012**, *44*, 107–111.

44. Liu, X.P.; Jing, X.M. Study on chemical constituents and biological activities of *E.ciliata* essential oil. *J. Heilongjiang Bayi Agric. Univ.* **2018**, *30*, 1002–2090.

45. Jin, X.L.; Li, D.H. Component analysis of essential oil from wild *E. ciliata* fragrans in Changbai Mountain. *J. Yanbian Univ. Nat. Sci. Ed.* **1996**, *1*, 32–34.

46. Li, W.Q.; Jiang, C.H.; Chu, S.S.; Zuo, M.X.; Liu, Z.L. Chemical Composition and Toxicity against *Sitophilus zeamais* and *Tribolium castaneum* of the Essential Oil of Murraya exotica Aerial Parts. *Molecules* **2010**, *15*, 5831–5839. [CrossRef] [PubMed]

47. Han, Q.X.; Huang, S.S. The biological activity of eugenol on the *Tribolium castaneum*. *J. Chongqing Norm. Univ. Nat. Sci. Ed.* **2009**, *26*, 16–19. [CrossRef]

48. Lv, J.H.; Wang, X.M.; Bai, X.G.; Lu, Y.J. Control of four plants essential oils against the *Tribolium castaneum* research. *Henan Agric. Sci.* **2006**, *9*, 68–71. [CrossRef]

49. Huang, Y.P.; Chen, Q.; Chen, M.D. Exploration Experiments on the Effects of Feeding Conditions on the Development and Reproductive Capacity of *Tribolium castaneum*. *Biol. Teach.* **2012**, *37*, 58–59. [CrossRef]

50. Liang, Y.S. Behavioral responses of the main components of defensive secretions of adult and larvae of the genus *Tribolium castaneum*. *J. Chin. Cereals Oils Assoc.* **1995**, *4*, 18–22.

51. Chen, H.P.; Yang, K.; You, C.X.; Du, S.S.; Cai, Q.; He, Q.; Geng, Z.F.; Deng, Z.W. Chemical constituents and biological activities against *Tribolium castaneum (Herbst)* of the essential oil from *Citrus wilsonii* leaves. *J. Serbian Chem. Soc.* **2014**, *79*, 1213–1222. [CrossRef]

52. Guo, D.L.; Pu, W.; Yan, X.P.; Tao, C. Research progress of modified atmosphere and fumigation of foreign countries—A review of the 8th Int'l Conference on Controlled Atmosphere and Fumigation Conference (CAF 2004). *Grain Storage* **2004**, *32*, 44–48. [CrossRef]

Article

Volatile Transference and Antimicrobial Activity of Cheeses Made with Ewes' Milk Fortified with Essential Oils

Carmen C. Licon [1], **Armando Moro** [2], **Celia M. Librán** [3], **Ana M. Molina** [4], **Amaya Zalacain** [5], **M. Isabel Berruga** [4] **and Manuel Carmona** [6,*]

1   Department of Food Science and Nutrition, California State University, Fresno, 5300 N Campus Drive M/S FF17, Fresno, CA 93740, USA; cliconcano@mail.fresnostate.edu
2   Facultad de Ingeniería Agronómica, Universidad Técnica de Manabí, Avda. José María Urbina y Che Guevara, 130105 Portoviejo, Manabí, Ecuador; amoro@utmnaog.edu.ec
3   Food Product Quality Department, Consum S. Coop, Av. Alginet s/n, 46460 Silla, Valencia, Spain; cmlibran@consum.es
4   Food Quality Research Group, Institute for Regional Development (IDR), Universidad de Castilla-La Mancha, Campus Universitario, 02071 Albacete, Spain; ana.molina@uclm.es (A.M.M.); mariaisabel.berruga@uclm.es (M.I.B.)
5   Cátedra de Química Agrícola, E.T.S.I.A., Universidad de Castilla-La Mancha, Campus Universitario, 02071 Albacete, Spain; Amaya.zalacain@uclm.es
6   School of Architecture, Engineering and Design, Food Technology Lab, Universidad Europea de Madrid, C/Tajo s/n, Villaviciosa de Odón, 28670 Madrid, Spain
*   Correspondence: manuel.carmona@universidadeuropea.es

Received: 26 November 2019; Accepted: 26 December 2019; Published: 1 January 2020

Abstract: During the last decades, essential oils (EOs) have been proven to be a natural alternative to additives or pasteurization for the prevention of microbial spoilage in several food matrices. In this work, we tested the antimicrobial activity of EOs from *Melissa officinalis*, *Ocimum basilicum*, and *Thymus vulgaris* against three different microorganisms: *Escherichia coli*, *Clostridium tyrobutyricum*, and *Penicillium verrucosum*. Pressed ewes' cheese made from milk fortified with EOs (250 mg/kg) was used as a model. The carryover effect of each oil was studied by analyzing the volatile fraction of dairy samples along the cheese-making process using headspace stir bar sorptive extraction coupled to gas chromatography/mass spectrometry. Results showed that the EOs contained in *T. vulgaris* effectively reduced the counts of *C. tyrobutyricum* and inhibited completely the growth of *P. verrucosum* without affecting the natural flora present in the cheese. By contrast, the inhibitory effect of *M. officinalis* against lactic acid bacteria starter cultures rendered this oil unsuitable for this matrix.

Keywords: cheese; essential oils; *Escherichia coli*; *Clostridium tyrobutyricum*; *Penicillium verrucosum*; antimicrobial

---

1. Introduction

The cheese microbiota has an important role in the development of cheese flavor and texture. By contrast, exogenous microorganisms can have a negative impact on the organoleptic properties of cheese, with the potential for great economic loss. For example, the occurrence of coliforms (*Escherichia coli*, *Klebsiella aerogenes*) and sporulating butyric bacteria (*Clostridium tyrobutyricum*, *C. butyricum*, and *C. sporogenes*) is known to be responsible for early and late cheese blowing, respectively [1,2]. Also, some filamentous molds (*Penicillium comune*, *P. verrucosum*, and *P. nalgiovense*) of the dairy factory environment [3,4], which are usually found in cheese rind or interior, have been associated with the presence of mycotoxins, with a consequent human health risk [5,6]. Late cheese blowing is quite frequent in semi-hard and hard cheeses, including Grana Padano, Cheddar, and Manchego [7–10], and is characterized by the presence of numerous and irregular internal holes produced by CO_2 released from lactate metabolism [7,11]. In this context, *C. tyrobutyricum* is considered as a main spoiler agent markedly affecting the volatile profiles of cheese [12].

Several approaches are available to reduce the occurrence of late blowing cheese spoilage, such as pasteurization or the use of additives, including nitrates and lysozyme; however, none of these approaches is ideal. In the case of pasteurization, bacterial endospores can survive the pasteurization process and germinate as vegetative cells in cheese during ripening. Also, the addition of nitrates has been associated with the presence of nitrosamine in cheese, although the European Food Safety Authority has recently re-assessed the acceptable safe daily intake of nitrites and nitrates [13]. Lastly, lysozyme has antimicrobial effects on lactic acid bacteria during cheese ripening [14]. Given these constraints, the use of essential oils (EOs) as natural food preservatives has steadily gained recognition as an alternative to the aforementioned treatments, as they are designated as "Generally Recognized as Safe" by the Food and Drug Administration [15,16], and they have proven antibacterial [17] and antifungal [18,19] activity. That being said, the antimicrobial activity of EOs has been assayed mostly under in vitro conditions and against pathogenic microorganisms [20,21], and there is a paucity of studies focusing on food products, especially in cheese [22,23]. In this context, Hyldgaard et al. [15] have emphasized the importance of understanding the behavior of EOs in a food matrix—as differences have been reported between plant and animal food products [24,25]. Moreover, there is conflicting evidence between studies, even when using the same product type, likely because of compositional differences, for example, cheeses with different fat or moisture content [5,23,26]. The utility of EOs or their compounds in cheese production has been examined in several studies [22,23], including their use as surface covers [5], or added directly to a finished product [19,21,26] or microencapsulated [27]. Yet, very little is known about the impact of adding EOs directly to milk before cheesemaking. Hamedi et al. [28] showed that the efficacy of EOs against *Salmonella* spp. in cheese diminished significantly when the results were compared with those obtained using a laboratory medium. It would be reasonable to expect that the EOs used to combat spoilers or pathogens should also be tested against lactic acid bacteria and different starter cultures required for semi-hard and hard cheese making.

Against this background, the present study was designed to determine the antimicrobial activity and the transfer of chemical compounds to fortified cheeses of different EOs. We used *Melissa officinalis* (lemon balm), *Ocimum basilicum* (sweet basil), and *Thymus vulgaris* (common thyme), and three typical cheese spoilers, *E. coli*, *C. tyrobutyricum*, and *P. verrucosum*.

2. Materials and Methods

2.1. Plant Material and EO Production

The aerial parts of *M. officinalis*, *O. basilicum*, and *T. vulgaris* were supplied by Nutraceutical SRL (Brazov, Romania). The raw material was packed in sealed plastic bags and stored in the dark at room temperature until analysis. EOs were obtained by solvent-free microwave extraction (SFME) with a NEOS® apparatus (Milestone, Sorisole, Italy) using methodology previously employed by Moro et al. [29]. In total, 150 g of the plant was placed in the NEOS reactor with 250 mL of Milli-Q water to wet the dry plant sample. As its name implies, the technique does not use a solvent, but the plant must contain the water that drags the essential oils when heated by microwaves, the principle with which this equipment works. Exhaustive extraction of EOs was then performed (35 min): the extraction power was set at 600 W (5 min) and then at 250 W (30 min), and the temperature was monitored with an infrared sensor for avoiding overheating (95 ± 5 °C). The oil was collected in the device with graduation marks available to the equipment itself for this purpose. For the antimicrobial activity test, the EOs were filtered using 0.2-μm PTFE syringe filters (Millipore, Madrid, Spain) to ensure the absence of microorganisms before use.

2.2. Milk Samples

"Manchega" breed ewes' milk was used for cheese fabrication. Bulk tank milk was collected from a commercial farm in Albacete (Spain). Milk had the following compositional values (g/100 g): dry matter, 17.81; fat content, 6.80; and protein content, 5.61. The mean pH was 6.66, somatic cell counts were 603×10^3 cells/mL and 158×10^3 CFU/mL microbial load.

2.3. Microbial Strains

The following assayed strains were purchased from the Spanish Type Culture Collection (CECT, Burjassot, Valencia, Spain): *E. coli* CECT 4201, *C. tyrobutyricum* CECT 4011, and *P. verrucosum* CECT 2906.

2.4. Elaboration of Cheese Samples Fortified with EOs

Before beginning cheesemaking, vats of 30 L of milk were fortified with EO samples at a final concentration of 0.250 g/kg. EOs were mixed 1:1 with a commercial food emulsifier (Tween-20® Food quality, Panreac, Spain) selected because it is considered safe [30]. The control vat contained the emulsifier at the same concentration used in the experimental (EOs) tanks. Milk was heated to 20 °C for 30 min to facilitate oil solubilization, and a pressed ewes' milk cheese procedure was performed at a pilot dairy plant from Castilla-La Mancha University, according to Licón et al. [31], with some modifications. Briefly, the starter culture (CHOOZIT MA4001; Danisco, Sassenage, France) was added for 30 min with stirring, and the temperature was increased to 30 °C. At this point, commercial rennet (0.023% *v/v*) was added to the vat with vigorous stirring, and the milk was allowed to coagulate. Thirty minutes later, the curd was cut into 8–10 mm cubes, heated (37 °C), and stirred for 45 min before whey separation. Curd was press-molded for 4 h until reaching pH 5.2. Lastly, cheeses were salt brined at 9 °C and stored in a ripening chamber over four months at 12 °C and 80% humidity prior to performing the assays. The cheese chemical composition was determined using a Foss FoodScan analyzer (FoodScan Lab, FOSS, Hillerød, Denmark).

2.5. Volatile Extractions and HS-SBSE/GC/MS Analyses

EOs were directly injected (0.2 µL) into a gas chromatograph following the methodology of Moro et al. [29]. Milk and cheese volatile extraction was performed by the headspace stir bar sorptive extraction (HS-SBSE) method. For the former, 10 mL liquid dairy samples (milk and whey) were pipetted separately into headspace glass vials, whereas cheese volatile extraction was performed following the methodology of Licón et al. [32]. For all dairy samples, headspace glass vials were affixed with inserts for headspace exposition and supplemented with a 1×10^{-3} g/kg aqueous solution of the internal standard ethyl octanoate (Aldrich Chemical Co., Milwaukee, WI, USA). A polydimethylsiloxane (PDMS)-coated stir bar (0.5 mm film thickness, 10 mm length in liquid samples, and 20 mm length in cheese samples; Twister, Gersterl GmbH, Mülheim an der Ruhr, Germany) was placed into the insert, and headspace vials were sealed with an aluminum crimp cap. Before analysis, the glass inserts and vials were thoroughly cleaned and heat conditioned at 110 °C to avoid any odorous contamination. The extraction of volatile compounds was performed following conditions proposed by Moro et al. [33], stirring at 1000 rpm for 120 min (milk and whey) or 240 min (cheese) at 45 °C. The PDMS stir bars were rinsed with distilled water, dried with cellulose tissue, and finally transferred into thermal desorption tubes for the GC/MS analysis.

The extracted volatiles from dairy samples were desorbed in an automated thermal desorption system (Turbo Matrix ATM, PerkinElmer, Norwalk, CT, USA) under the following conditions: oven temperature, 280 °C; desorption time, 5 min; cold trap temperature, −30 °C; helium inlet flow rate, 45 mL/min. The volatiles were transferred into a Varian CP-3800 gas chromatograph (GC) equipped with a Saturn 2200 ion trap mass spectrometer (MS) (Varian Inc., Palo Alto, CA, USA) and an Elite-Volatiles Specialty phase capillary column (30 m × 0.25 mm i.d., 1.4 µm film thickness; PerkinElmer, Shelton, CT, USA). The column temperature was set at 35 °C for 2 min and then raised at 5 °C/min to 240 °C and held for 5 min. The detector temperature was 250 °C, and the helium carrier gas flow rate was 1 mL/min. The electron ionization mode at 70 eV was used for the MS analysis. The mass range varied from 35 to 300 *m/z*.

To avoid matrix interferences between the EOs and dairy matrix volatiles, the MS identification of volatiles was performed in single-ion-monitoring mode using their characteristic m/z values and by comparison of their mass spectra with those of pure compounds or reported in the NIST/ADAMS library. The identities of the EO components were established from the GC retention time (relative to Kovats index). Quantification was carried out in scan mode and expressed as the relative area using the correction factor for the internal standard (ethyl octanoate) area. The results of each volatile compound that was transferred to the dairy matrix were expressed as relative concentration area (g/kg) using the internal standard correction factor. Then the transference ratio or recovery yield (%) from milk to cheese of each compound that was found was calculated by the following Formula (1):

$$\text{recovery yield (\%)} = [\text{Xi (g/kg)}/\text{X (g/kg)}] \times 100 \tag{1}$$

where Xi indicates the presence of each compound in cheese, and X indicates the presence of the same compound in milk. Dairy samples were analyzed in triplicate.

2.6. Cheese Microbial Content

To enumerate the microbial content on ripened cheeses, a 10-g sample of each cheese was aseptically homogenized with 90 mL of sterile 0.1% (*w/v*) peptone water in an IUL Stomacher (IUL SA, Barcelona, Spain) for 60 s. Serial decimal dilutions of the homogenates were prepared with buffered peptone water (BPW) (Scharlau, Barcelona, Spain) and plated onto the corresponding media in duplicate using an Eddy Jet spiral plater (Eddy Jet v1.23, IUL SA, Barcelona, Spain). Total aerobic bacterial counts were performed on plate count agar (PCA; Panreac Química S.L.U., Barcelona, Spain) after incubation at 32 °C for 48 h under aerobic conditions. Lactic streptococci were plated on M17 agar (Biokar Diagnostics, Barcelona, Spain) with incubation at 37 °C for 48 h, under aerobic conditions. Brilliant Green Bile Agar was used for coliform incubation (BGB; Pronadisa Conda, Madrid, Spain) at 37 °C for 24 h, under aerobic conditions. *Clostridium* spp. was plated on a reinforced clostridial agar (RCA; Oxoid, Basingstoke, UK) and incubated at 37 °C for 48 h, under anaerobic conditions. Molds and yeasts were seeded in potato dextrose agar (PDA; Merck, Darmstadt, Germany) and incubated at 25 °C, during 96 h, in aerobic conditions. Microbial growth estimations were done with an automatic plate counter (Countermat Flash 4.2, IUL Intruments S.A., Barcelona, Spain), and the results were expressed as log cfu/g.

2.7. Antimicrobial Activity Test

The experimental procedure for antimicrobial activity determination is depicted in Figure 1, and allows the investigation of microbial spoilage, in the case of an external contamination such as that occurring in ripening chambers with molds. Nine cheese cubes of 27 mm^3 were obtained from each cheese using a cheese blocker (BOSKA, Bodegraven, Holland). The cubes were divided into three subgroups, with three cubes in each. Cubes were introduced into a sterile container and distributed as follows: Group 1, internal inoculation with *C. tyrobutyricum* at 10^3 cfu/g, incubated at 37 °C under anaerobic conditions (AnaeroGenTM, Oxoid LTD., Basingstoke, UK); Group 2, internal inoculation with *E. coli* at 10^3 cfu/g, incubated at 37 °C under aerobic conditions; Group 3, surface inoculation with *P. verrucosum* at 10^3 cfu/cm^2, incubated at 25 °C under aerobic conditions. *P. verrucosum* was inoculated onto the surface, given its inability to grow in the interior of the cheese.

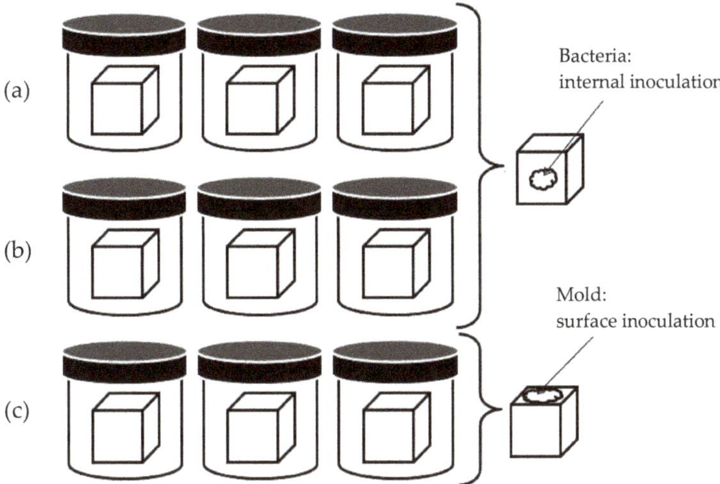

Figure 1. Antimicrobial activity assay performed by the inoculation of fortified cheeses with: (**a**) *Clostridium tyrobutyricum* (37 °C, anaerobic conditions), (**b**) *Escherichia coli* (37 °C, aerobic conditions) and (**c**) *Penicillium verrucosum* (25 °C, aerobic conditions).

2.8. Microorganism Inoculum Preparation

C. tyrobutyricum spore suspensions were obtained by prior prolonged incubation (1 week) on Reinforced Clostridial Medium (Oxoid LTD.). Subsequently, spores were harvested and cleaned following a procedure adapted from Yang et al. [34], which briefly consisted of double purification by centrifugation at 8000× *g* for 15 min at 4 °C. The final pellet was resuspended in sterilized distilled water, and the spore concentration of the suspension was determined by adapting the procedure of Anastasiou et al. [35] after 15 min heat treatment at 80 °C, by serial dilution in BPW. An *E. coli* suspension was obtained after 22 h of cultivation on Triptone Soy Medium (Oxoid LTD.); the colony-forming units were also established by serial dilution in BPW. In both cases, 1 mL aliquots of concentrated bacterial suspensions were stored at −20 °C in 15% of glycerol until needed for inoculation at a final concentration of 10^3 cfu/g.

P. verrucosum spore suspensions were sub-cultured weekly on Potato Dextrose Agar (Merck, Darmstadt, Germany) at 25 °C in the dark. Conidia were harvested according to Baratta et al. [36], and the spore suspension was adjusted to an optical density of 0.5 (λ = 530 nm), equivalent to 10^5 spores/mL. This suspension was employed for the immediate surface inoculation of cheese samples at a concentration of 10^3 cfu/g.

After 1 week of incubation, starters, total viable counts, and target microbial growth were determined in all cheese cubes. The experiment was performed in duplicate.

2.9. Statistical Analysis

Descriptive analysis and analysis of variance (ANOVA; $p < 0.001$) coupled to a Tukey' test ($p < 0.05$) were performed to determine group differences between the antimicrobial activity results using IBM Statistics SPSS software, v24 (SPSS Inc., Chicago, IL, USA).

3. Results and Discussion

3.1. Extraction and Composition Analysis of Essential Oils

EOs from aromatic plants are a complex mixture of volatile oils of low molecular weight that are obtained by steam distillation [37]. In the present study, EOs were obtained using a modern extraction technique based on solvent-free, microwave hydrodiffusion, also known as SFME or microwave hydrodiffusion and gravity [38]. The use of this technique offers several advantages over conventional hydrodistillation or solvent distillation, including the avoidance of artefacts during distillation, and also savings in energy and extraction time [38].

Chemical characterization of the EOs in terms of volatile composition was necessary before determining the transference ratio during cheesemaking. The total number of compounds identified in the EOs ranged from 14 in *O. basilicum* to 27 in *T. vulgaris* (Table 1), and they constituted over 87% of the total area composition.

According to chemical families of compounds, all EOs were represented mainly by monoterpenes—with 83.80% to 96.57% of the total peak area, respectively. Sesquiterpenes represented <3.2% of the total composition. In accordance with our previous study [33], the present results showed that all of the EOs were dominated by two or three major compounds (Table 1), representing up to 40% of the total area. These main compounds were commonly oxygenated monoterpenes, terpenes, which undergo biochemical modifications that add oxygen molecules and move or remove methyl groups [15]. In contrast to other studies [17,18], we found that the *O. basilicum* EO was described mainly by the aromatic compound 4-allyl-anisole also known as methyl chavicol (58.21%), rather than linalool (11.21%), which has been reported in larger amounts by other authors (20%–66%). In addition, we found a small amount (3.20%) of the sesquiterpene α-bergamotene (E)(Z).

Linalool is a linear monoterpene that is frequently found in volatile plant extracts. We found this in a range from 1.71% to 34.54% of the total area; the latter case was found for *T. vulgaris*, exceeding the concentration of thymol, which is usually the characteristic EO marker of this species [20,39]. The other family groups of compounds identified in this EO represented ~2% of the total composition.

Regarding the EOs of *M. officinalis*, nerol (35.85%) and neral (35.34%) were the major compounds identified, and the remaining compounds did not exceed 2.7% of the total area. These results differ from those of previous works [40,41], which suggested that citral—a mixture of neral and geranial—is the major compound [40,41]. Geranial and nerol are biosynthetically connected, as geranial is the aldehyde isomer of nerol.

The absence of or a smaller-than-expected amount of compounds has been reported by other authors, such as the absence of thymol in thyme oil, and the presence of other compounds, such as carvacrol, a phenolic monoterpene, or p-cymene, and γ-terpinene, precursors in its biogenetic pathway [42]. In this regard, some authors have highlighted the effect of plant chemotype on EO composition for the presence of thymol, thymol/linalool, and carvacrol chemotypes in different varieties of thyme [43]. Moreover, several studies have emphasized the importance of culture-growing conditions and harvesting, in addition to different varieties, when EOs are chemically characterized [17,42]. The extraction methodology is also known to affect the composition and quality of extracts, as the use of high temperatures can stimulate the hydrolysis and polymerization of some esters [44], whereas the use of solvents can leave residual substances that affect the biological properties of EOs [45]. Using the same extraction procedure as that used here, Okoh et al. [46] achieved better extraction yields and larger amounts of oxygenated monoterpenes than with EOs obtained by hydrodistillation, which may explain the compositional differences between studies.

Table 1. Volatile composition of essential oils (EOs) expressed as relative area (%).

Compounds	RT (min)	KI exp. *	m/z Pattern **	Melissa officinalis	Ocimum basilicum	Thymus vulgaris
Number of compounds				18	14	27
Monoterpenes family						
α-thujene	28.34	924	77/**93**/136	-	-	0.43
α-pinene	28.86	937	**93**/136	0.51	0.27	4.74
camphene	29.56	946	**93**/121/136	2.71	-	1.69
sabinene	30.51	970	**93**/77/41	0.02	0.23	1.27
β-pinene	30.77	975	41/**93**/107/121	2.22	0.52	8.13
myrcene	30.83	988	**41**/93/69	-	0.31	-
α-phelandrene	31.74	1002	**93**/77/136	-	-	0.8
α-terpinene	32.22	1014	**93**/121/136	-	-	5.12
β-ocimene (Z)	32.55	1017	79/**93**/136	0.43	-	-
p-cymene	32.64	1020	91/**119**	-	-	6.96
sylvestrene	32.66	1024	**41**/68/93/136	0.48	-	-
1,8 cineole	32.80	1026	43/108/139/**154**	-	5.85	-
β-phelandrene	32.90	1025	77/**93**/136	-	-	1.52
β-ocimene (E)	33.06	1032	79/**93**/136	0.31	0.84	0.11
γ-terpinene	33.74	1054	77/93/121/**136**	-	-	7.5
4-thujanol (Z)	34.16	1065	43/71/93/**139**/154	-	-	1.44
terpinolene	34.83	1086	43/**93**/121/136	0.3	-	2.47
linalool	35.09	1089	43/**71**/154	1.71	11.21	34.54
perillene	35.14	1093	**41**/69/81/150	0.12	-	-
4-thujanol (E)	35.37	1098	43/71/93/**139**/154	-	-	0.18
citronellal	37.04	1148	**41**/69/95/121/154	0.65	-	-
camphor	37.64	1141	**41**/95/152	-	0.54	0.3
borneol	38.01	1165	**95**/154	-	-	2.77
terpinen-4-ol	38.30	1174	43/**71**/154	0.28	0.33	10.54
α-terpineol	38.75	1186	43/**59**/93/136	0.56	-	2.14
4-allyl-anisole	38.84	1189	77/121/**148**	-	58.21	-
dihydro carvone (E)	39.09	1194	41/**67**/95/152	-	-	0.51
dihydro carvone (Z)	39.41	1200	41/**67**/95/152	-	-	0.56
linalyl acetate	40.25	1210	**43**/93/121	-	-	1.55
nerol	40.38	1227	**41**/69/154	35.85	-	-

Table 1. *Cont.*

Compounds	RT (min)	KI exp. *	*m/z* Pattern **	*Melissa officinalis*	*Ocimum basilicum*	*Thymus vulgaris*
carvone	40.747	1235	54/82/**93**/150	-	-	0.13
neral	41.31	1239	41/**69**/109/152	35.34	-	-
thymol	41.43	1281	**135**/150/65	-	-	0.66
bornyl acetate	41.91	1288	**41**/95/121	-	1.18	-
carvacrol	41.98	1298	41/**135**/150	-	-	0.51
eugenol	44.05	1356	43/131/149/**164**	-	3.03	-
neryl acetate	44.06	1359	41/**69**/93/154	7.54	-	-
methyl eugenol	45.03	1403	41/107/163/**178**	-	1.28	-
Sesquiterpenes						
α-bergamotene (E)(Z)	46.33	1432	41/**93**/119/204	-	3.2	-
β-caryophyllene (E)	46.48	1417	41/**93**/103/161	2.94	-	1.76
Others						
1-octen-3-ol	30.19	974	**43**/57	-	-	0.06
Total area of all identified compounds (%)				91.97	87	98.39
Total Monoterpenes (%)				89.03	83.8	96.57
Total Sesquiterpenes (%)				2.94	3.2	1.76
Total Others (%)				-	-	0.06

* Experimental Kovats index; ** in bold *m/z* used for quantification.

3.2. Volatile Composition of Dairy Samples

As previously reported by Tajkarimi et al. [47], the normal concentration range for spices and herbs used in food systems is between 0.05% and 0.1%. In the present study, an EO concentration of 0.25 g/kg was chosen to study the transference of volatile compounds during the cheese-making process, to prevent an excessive sensory impact and to provide antimicrobial activity. Indeed, the concentration of EOs is an important consideration, as it has been demonstrated that they may have an undesirable impact on cheese sensory properties by modifying the dynamics or activity of the microbial ecosystem during cheese making and ripening. This hypothesis derives from indirect observations in several trials of hard-cooked cheeses and experiments performed by Tornambé et al. [48], where EO concentration levels higher than 10 g/kg resulted in a high sensory impact and consequent rejection by consumers. Because specific surfactant actions are required to improve the affinity of the matrix for volatile compounds, particularly terpenes, we selected Tween®-20 as a polysorbate surfactant, whose stability and relative lack of toxicity allow it to be used as a detergent and emulsifier for culinary, scientific, and pharmacological purposes.

The methodology selected for the extraction and characterization of volatiles (HS-SBSE coupled with GC/MS) is a common technique in food volatile analysis, and it has been specifically optimized by Licón et al. [32] and Moro et al. [33] for pressed ewes' milk cheeses. This food matrix is quite complex, and several interactions can potentially take place between food components and EOs [15] due to the high fat and protein content of the cheese. For the present study, we only examined the volatiles present in the EOs, and the identification of other cheese compounds was dismissed. The results of the concentration of the main compounds identified in milk, cheese, and whey, together with the carryover percentages, are provided in Table 2.

The major compounds of the EOs (Table 1) corresponded to those identified in larger quantities in milk, cheese, and whey, whereas the minor compounds were below the method's limit of detection. The number of detected compounds in the different matrices ranged from 9 to 22, and between 82% and 95% of the compounds detected in the EOs were transferred to the dairy products. This transfer range was much broader than that described by Tornambé et al. [48] (43%) when a pasture plant EO was added to milk.

Regarding the different chemical families found in milk, monoterpenes were the most abundant in milk spiked with *M. officinalis* (47.76 mg/kg), *T. vulgaris* (249.81 mg/kg), and *O. basilicum* (82.71 mg/kg). For cheese and whey, different transference rates were obtained for each plant: for *M. officinalis*, monoterpene compounds (7.06%) in cheese and sesquiterpenes (30.61%) in whey showed the lowest and the highest carryover effects in this plant; for *T. vulgaris*, sesquiterpenes (16.67% and 39.58%) were the most abundant family of compounds in cheese and whey, respectively; whereas for *O. basilicum*, the best carryovers were observed for monoterpenes (28.44% and 23.15%) for cheese and whey, respectively. Transference of compounds in EOs to dairy matrices is challenging, as they are known to interact with fat, carbohydrate, and protein matrices in cheese [20,24]. Specifically, proteins and whey proteins can interact with compounds presenting with a hydroxyl group, restricting their ability to be transferred [20,23].

As individual compounds, the major content of *M. officinalis*-enriched dairy products (milk, cheese, whey) were nerol (17.56, 0.86, 3.57 mg/kg), neral (16.30, 0.86, 3.38 mg/kg), and camphene (8.60, 0.99, 1.36 mg/kg). Most of the compounds identified in *O. basilicum*-enriched milk were below 0.60 mg/kg, with the exception of 4-allyl-anisole (47.02 mg/kg), 1,8 cineole (15.49 mg/kg), and linalool (13.99 mg/kg). For *T. vulgaris*-enriched dairy products, a larger abundance of significant compounds was found, as eight compounds >10 mg/kg were detected in milk, reaching 6.5 mg/kg in cheese, and as high as 13 mg/kg in whey. The same was found for cheese and whey. However, these individual major compounds did not offer the best carryover ratios, and other minor compounds were better transferred: linalool (14.29%) in cheese, and β-caryophyllene (30.61%) in whey from *M. officinalis*, β-caryophyllene (16.67%) in cheese and 1,8 cineole (47.12%) in whey from *T. vulgaris*, and α-thujene (75.00%) in cheese and γ-terpinene (30.00%) in whey from *O. basilicum*. In the case of α-thujene, it has to be pointed out that it is a high transfer rate but for a very minority compound, which we do not even find in the essential oil of this plant. Maybe the enzymatic activity present in the milk could convert sabinene into α-thujene since they have great structural similarity. Indeed, it seems that the different functional groups of compounds also affected the transfer ratios, which were better for hydrocarbon monoterpenes than for oxygenated ones. Thus, better carryover ratios were reached by using EOs that are richer in hydrocarbons rather than oxygenated monoterpenes.

Table 2. Presence of compounds in fortified dairy products and transfers from milk to cheese and whey ($n = 3$).

| | *Melissa officinalis* | | | | | *Ocimum basilicum* | | | | | *Thymus vulgaris* | | | | |
| | Conc. (mg/kg) [†] | | | Transf. (%) [‡] | | Conc. (mg/kg) | | | Transf. (%) | | Conc. (mg/kg) | | | Transf. (%) | |
	M§	C	W	C	W	M	C	W	C	W	M	C	W	C	W
Number of compounds	11	9	10			19	18	19			22	22	18		
Monoterpene family															
α-thujene	-	-	-	-	-	0.04	0.03	0.01	75.00	25.00	2.19	0.28	0.22	12.79	10.05
α-pinene	1.53	0.18	0.23	11.76	15.03	0.47	0.33	0.09	70.21	19.15	21.10	2.58	2.19	12.23	10.38
camphene	8.60	0.99	1.36	11.51	15.81	0.13	0.08	0.03	61.54	23.08	7.79	0.92	0.77	11.81	9.88
sabinene	-	-	-	-	-	0.45	0.21	0.08	46.67	17.78	5.64	0.64	0.64	11.35	11.35
β-pinene	0.22	0.03	0.05	13.64	22.73	0.67	0.32	0.14	47.76	20.90	34.08	3.72	4.25	10.31	11.78
α-phelandrene	-	-	-	-	-	-	-	-	-	-	3.26	0.40	0.40	12.27	12.27
α-terpinene	-	-	-	-	-	-	-	-	-	-	22.52	2.50	2.80	11.10	12.43
p-cymene	-	-	-	-	-	0.27	0.12	0.05	44.44	18.52	13.04	1.48	1.65	11.35	12.65
sylvestrene	-	-	-	-	-	-	-	-	-	-	16.54	1.56	-	9.43	-
1.8 cineole	-	-	-	-	-	15.49	4.55	4.37	29.37	28.21	5.73	0.61	2.70	10.65	47.12
β-ocimene (E)	0.48	0.06	0.09	12.50	18.75	1.28	0.47	0.26	36.72	20.31	0.45	0.06	-	13.33	-
γ-terpinene	-	-	-	-	-	0.10	0.04	0.03	40.00	30.00	27.88	3.10	3.95	11.12	14.17
4-thujanol (Z)	-	-	-	-	-	-	-	-	-	-	0.52	0.03	-	5.77	-
terpinolene	0.45	0.04	0.06	8.89	13.33	0.26	0.34	0.06	130.77	23.08	7.19	0.88	1.17	12.24	16.27
linalool	1.19	0.17	0.36	14.29	30.25	13.99	3.85	3.62	27.52	25.88	61.83	6.57	12.98	10.63	20.99
4-thujanol (E)	-	-	-	-	-	-	-	-	-	-	1.99	0.13	-	6.53	-
camphor	-	-	-	-	-	0.61	0.18	0.15	29.51	24.59	0.85	0.09	0.17	10.59	20.00
borneol	-	-	-	-	-	0.15	0.04	0.03	26.67	20.00	2.14	0.30	0.49	14.02	22.90
terpinen-4-ol	0.15	-	0.02	-	13.33	0.20	0.05	0.04	25.00	20.00	13.03	1.45	3.31	11.13	25.40
α-terpineol	-	-	-	-	-	-	-	-	-	-	1.26	0.18	0.40	14.29	31.75
4-allyl-anisole	-	-	-	-	-	47.02	12.70	9.97	27.01	21.20	-	-	-	-	-
linalyl acetate	-	-	-	-	-	-	-	-	-	-	0.61	0.08	-	13.11	-
nerol	17.56	0.86	3.57	4.90	20.33	-	-	-	-	-	-	-	-	-	-
neral	16.30	0.86	3.38	5.28	20.74	-	-	-	-	-	-	-	-	-	-
eugenol	-	-	-	-	-	0.77	-	0.07	-	9.09	-	-	-	-	-
bornyl acetate	-	-	-	-	-	0.64	0.18	0.12	28.13	18.75	0.17	0.02	0.08	11.76	47.06

Table 2. *Cont.*

	Melissa officinalis					*Ocimum basilicum*					*Thymus vulgaris*				
	Conc. (mg/kg) [†]			Transf. (%) [‡]		Conc. (mg/kg)			Transf. (%)		Conc. (mg/kg)			Transf. (%)	
	M§	C	W	C	W	M	C	W	C	W	M	C	W	C	W
neryl acetate	1.28	0.18	0.34	14.06	26.56	-	-	-	-	-	-	-	-	-	-
methyl eugenol	-	-	-	-	-	0.17	0.03	0.03	17.65	17.65	-	-	-	-	-
Sesquiterpene family															
α-bergamotene (E)(Z)	-	-	-	-	-	0.68	0.18	0.14	26.47	20.59	-	-	-	-	-
β-caryophyllene (E)	0.49	0.06	0.15	12.24	30.61	-	-	-	-	-	0.48	0.08	0.19	16.67	39.58
Total identified compounds	48.25	3.33	9.61	7.11	19.92	83.39	23.70	19.29	28.42	23.13	250.29	27.66	38.36	11.05	15.33
Total monoterpenes	47.76	3.37	9.46	7.06	19.81	82.71	23.52	19.15	28.44	23.15	249.81	27.58	38.17	11.04	15.28
Total sesquiterpenes	0.49	0.06	0.15	12.24	30.61	0.68	0.18	0.14	26.47	20.59	0.48	0.08	0.19	16.67	39.58

[†] Concentration of the compound in the matrix expressed as mg/kg; [‡] Transfer of compounds from milk to cheese and whey; §M, C, W: Milk, Cheese, and Whey samples.

3.3. Antimicrobial Activity

The established concentration mean value of 10^3 was decided as a mid-point of known studies for the different species. In the case of *P. verrucosum*, the studies considered were those of Nielsen et al. [49] and Vazquez et al. [5]. The first ones inoculated Arzua-Ulloa cheeses with fungal species at the concentration of 1.5×10^3 spores/cm^2 and the second ones at 10^2 spores/cm^2. We decided to fit the inoculum at an intermediate level of 10^3 cfu/cm^2. For *E. coli*, several authors [21,50,51] used contamination levels in cheese or milk for cheese elaboration in the range from 10^1 cfu/g or mL to 10^5 cfu/g or mL. The average value of 10^3 seemed reasonable again, as it was also somewhat below the maximum contamination levels found for Clostridium in cheeses by several authors [9,52].

The antimicrobial effects of the plant EOs on the initial flora of fortified cheeses are shown in Figure 2. The antimicrobial effect of *M. officinalis* EOs was strong, whereas the effect of *T. vulgaris* EOs was milder, and the effect of *O. basilicum* EOs was intermediate. Additionally, *M. officinalis* and *O. basilicum* EOs showed the greatest inhibitory effect against clostridia microorganisms naturally occurring in the milk and cheese. Specifically, the EOs from *M. officinalis* and *O. basilicum* completely blocked the growth of Clostridium spp., whereas *T. vulgaris* tempered the growth of these bacteria by more than 1 log unit (2.25 and 3.47 log cfu/g in the *T. vulgaris*-fortified and control cheese, respectively). However, it was not possible to evaluate the inhibitory capacity on initial coliforms or molds as the milk was free of these two groups of microorganisms since none of them grew even in control cheeses.

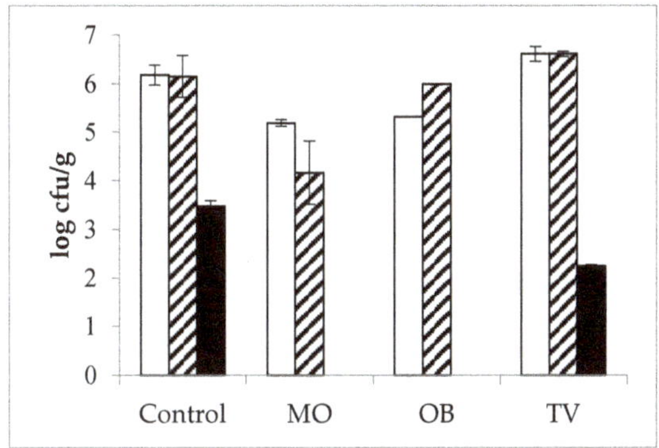

Figure 2. Microbial content (log cfu/g; mean ± SEM) in the control, *Melissa officinalis* (MO), *Ocimum basilicum* (OB), and *Thymus vulgaris* (TV) ripened cheeses. (Total Viable Counts: ☐; Lactic Acid Bacteria: ▨; *Clostridium* spp.: ■).

These findings indicate that late cheese blowing caused by clostridia development can be prevented by the tested EOs. Nevertheless, the robust antibacterial effect of *M. officinalis* EOs might negatively affect cheese ripening as it greatly influenced normal cheese flora development by reducing the starter bacteria content by nearly 2 log units (Figure 2). This imbalance in lactic streptococci might lead to flat flavors due to their lower activity in the ripening stages [53], paste defects deriving from slow acidification during cheese preparation [54], or even early cheese blowing as lactose consumption competition with coliforms would be lacking [55]. Indeed, when producing cheese, delays of more than 30 min were observed during the *M. officinalis* acidification process (data not shown). As mentioned, it was impossible to ascertain the effect of these EOs on coliforms, probably owing to the water activity of the four-month ripened cheeses preventing bacterial growth. Moreover, when compared against the control and *T. vulgaris*-fortified cheese, which had normal counts in a 150-day ripened cheese [56], the *O. basilicum* EOs had a mild effect on normal cheese flora (Figure 2).

The antimicrobial activity results of the fortified and control cheese samples after one week of incubation are shown in Figure 3. The effect of EOs on *Clostridium* spp. remained relevant (Figure 3a). In the Group 1 cubes (inoculated with *C. tyrobutyricum*), the addition of *O. basilicum* and *T. vulgaris* reduced the clostridial counts by more than 1 log unit as compared with the control samples (4.04 log cfu/g), whereas the *M. officinalis* cheeses had no clostridial counts. In our previous study on the anticlostridial activity of *M. officinalis* EOs in laboratory media, we found that the concentration of these EOs required to achieve total inhibition was ten times lower [57]. These results are in accordance with the fact that higher concentrations of EOs are needed in food matrices compared with those used in in vitro testing, highlighting the importance of performing simultaneous studies in vitro and in situ [58]. This inhibitory effect on clostridial growth reached in this assay was more robust than that described by other authors such as Deans and Ritchie [59], who tested pure oils in vitro, and were unable to demonstrate inhibition of C. sporogenes with any of the three tested EOs. By contrast, Baratta et al. [36] reported inhibitions with *O. basilicum* oil on another clostridial species, *C. perfringes*, which overall suggests varying resistance among strains.

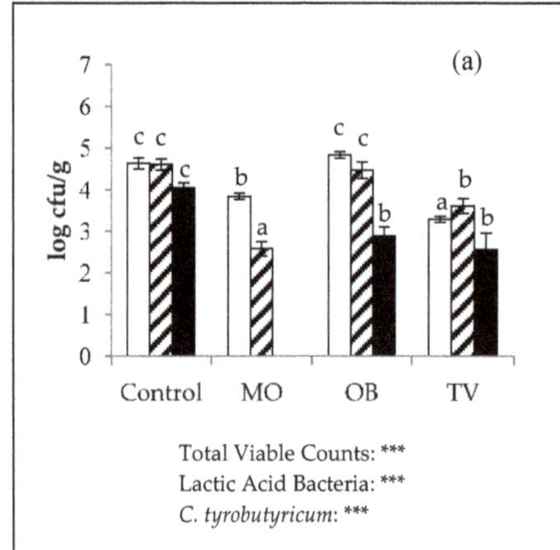
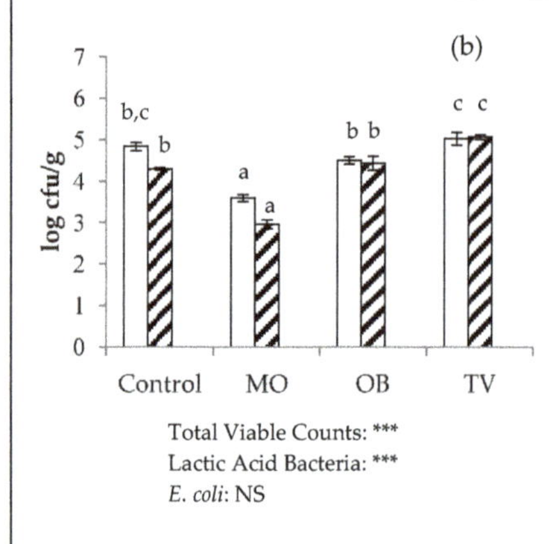
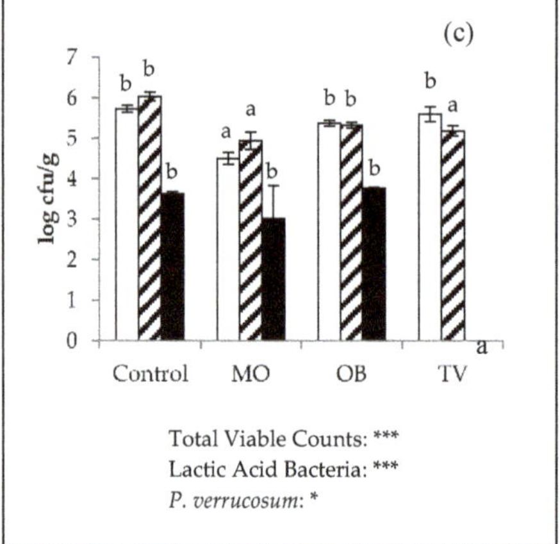

Figure 3. Microbial content (log cfu/g; mean ± SEM) in the control, *Melissa officinalis* (MO), *Ocimum basilicum* (OB), and *Thymus vulgaris* (TV) ripened cheeses inoculated and incubated for 1 week with (**a**) *Clostridium tyrobutyricum*, (**b**) *Escherichia coli*, and (**c**) *Penicillium verrucosum*. (Total Viable Counts: □; Lactic Acid Bacteria: ▨; Target microorganism: ■). ***, *, NS: Significance level $p < 0.001$, $p < 0.05$ and non-significant, respectively. a, b, c: Different values among the same microbial group are significantly different between essential oils applications ($p < 0.05$).

No growth was recorded for any of the cubes in Group 2 (inoculated with E. coli), which fits with the initial cheese enumeration of the coliforms (Figures 2 and 3b). It is commonly accepted that Gram-negative bacteria are more resistant than Gram-positive bacteria to EOs [23]. However, the results herein do not match with these observations, likely due to the harsh conditions of matured cheeses until coliform development; for instance, low pH, water activity, or lactose exhaustion [54].

Regarding the antifungal effect against *P. verrucosum*, we found a complete inhibition of growth in the *T. vulgaris*-fortified cheese, a slight reduction in the *M. officinalis* cheese (0.61 log unit) and no effect in the *O. basilicum* cheese (Figure 3c). These findings contrast with those obtained under in vitro conditions, where *O. basilicum* activity was the greatest, and *T. vulgaris* activity was the lowest [57]. Thus, the comparison of the effects of EOs on a cheese matrix and on laboratory media is important, as the activity may completely change.

Indeed, the activity of these EOs followed the same pattern in the cheese matrix as that observed in culture media against *C. tyrobutyricum*; thus, M. officinalis proved the most active, followed by *O. basilicum* and then *T. vulgaris* [60]. Cheese type can also have an effect on the antimicrobial potential of EOs, which was highlighted by Vázquez et al. [5], who found different effects of EO compounds when applied as cheese covers depending on cheese type. The authors of this study observed that it is possible to robustly inhibit *P. citrinum* in Arzúa-Olloa cheese with 200 µL/mL of eugenol, whereas no inhibition was observed for Cebreiro cheese, and the same was found when using thymol, the principal constituent of thyme oil [23,61]. These authors had to apply pure thyme oil to inhibit *Aspergillus parasiticus* growth in culture media.

Some other factors relating to the cheese matrix can completely alter the activity of EOs, which are in the main reduced as compared with laboratory media [24]. Several studies have demonstrated that food composition has a negative impact on EO efficacy, particularly carbohydrate, protein, and fat content [23,58]. In this line, low-fat cheeses are better for the action of EOs against Gram-positive bacteria but are worse for Gram-negative ones [26], and carbohydrates reduce the activity of EOs in other food matrices [24].

With the exception of the cheese samples incubated at 25 °C under aerobic conditions, the total viable counts and lactic streptococci generally decreased in relation to the initial cheese content (Figures 2 and 3). The decline in these bacterial counts ranged from 0.1 to 3.3 log units. Furthermore, these reductions seemed to be influenced by not only the addition of EOs but also by the incubation conditions (Figure 3). Indeed, the combined effect of an anaerobic environment and the addition of *M. officinalis* or *T. vulgaris* EOs led to the most marked reductions in microbial flora (Figure 3a). During a long ripening period, like that studied in this work, a reduction in starter microorganisms is due not only to their loss of viability but also to the release of intracellular enzymes [62]. These starter microorganisms, which are stored refrigerated for a long ripening period, generally acclimatize to low temperature. Hence, this selection for more cold-tolerant microorganisms can explain the lower inhibition noted in the cheese cubes incubated at lower temperatures. In addition, increasing the incubation temperature from 25 °C to 37 °C can trigger the evaporation of the volatile compounds transferred from EOs to cheese, thus increasing their content in the vapor phase and, consequently, inhibiting bacteria more efficiently, as formerly observed by other authors [63,64].

4. Conclusions

The present study demonstrates that most of the compounds present in the EOs from *M. officinalis*, *T. vulgaris*, and *O. basilicum* were transferred from milk to cheese and whey. The carryover results show hydrocarbon monoterpenes to be the best transferred compounds from milk to cheese (11%–53%) and whey (11%–20%), indicating that they are less affected by fat and casein matrices. Obtaining dairy products supplemented with aromatic compounds enhances their flavor, but also contributes to bioactive properties (antioxidant or antimicrobial) and are alternatives for the dairy industry. Therefore, further research is recommended to test these potential properties. This work also demonstrates the importance of conducting specific studies on the target food matrix in order to evaluate the antimicrobial activity of EOs. Occasionally their efficacy could be extrapolated, which was the case of the three EOs studied against *C. tyrobutyricum*, although lower concentrations are required when assaying in culture media. Yet with other microorganisms like *P. verrucosum*, extrapolation can lead to a misinterpretation of the potential of these EOs if only in vitro assays are performed to select the most appropriate ones because many matrix factors can impact the results.

The effect of these EOs on microorganisms that are crucial for proper cheese ripening must also be considered, given the risk of converting a good, natural solution for a technological problem into a new limitation. By considering these considerations, and the concentrations assayed, we conclude that the EOs of *M. officinalis* and *O. basilicum* display excellent activity that helps combat microorganisms that may cause late cheese blowing before and after inoculation, and they do not show post-inoculation inhibition against mold. However, the *M. officinalis* EOs are not recommended because they potently inhibit the starter cultures usually added during cheese manufacture. The most balanced EOs for combating the microbial cheese defects addressed in this work are those of *T. vulgaris*, which reduce the clostridia content, strongly inhibit mold growth, and do not damage lactic streptococci starters. Further studies are needed to better understand the precise effect of EOs from aromatic plants on cheese matrices to adjust the most adequate EOs concentration for consumer acceptability, as well as their effect on different cheese varieties or ripening stages.

Author Contributions: Conceptualization, A.M.M., A.Z., M.I.B. and M.C.; investigation, C.C.L., A.M., C.M.L., A.M.M. and A.Z.; writing—original draft preparation, A.M. and C.M.L.; writing—review and editing, C.C.L. and M.C.; supervision, M.I.B. and M.C.; project administration, M.C.; funding acquisition, M.I.B. All authors have read and agreed to the published version of the manuscript.

Funding: This research has been financially supported by the National Institute for Agricultural and Food Research and Technology (INIA, http://inia.es) by the project RTA2015-00018-C03-02.

Acknowledgments: M.C. thanks the support by the Spanish Ministry of Science, Innovation and Universities through the Ramón y Cajal Fellowship (RyC-2014-16307).

Conflicts of Interest: The authors declare no conflict of interest.

References

1. Mato Rodriguez, L.; Alatossava, T. Effects of copper on germination, growth and sporulation of Clostridium tyrobutyricum. *Food Microbiol.* **2010**, *27*, 434–437. [CrossRef] [PubMed]
2. Mullan, W.M.A. Causes and control of early gas production in cheddar cheese. *Int. J. Dairy Technol.* **2000**, *53*, 63–68. [CrossRef]
3. Kure, C.F.; Skaar, I.; Brendehaug, J. Mould contamination in production of semi-hard cheese. *Int. J. Food Microbiol.* **2004**, *93*, 41–49. [CrossRef] [PubMed]
4. Kure, C.F.; Wasteson, Y.; Brendehaug, J.; Skaar, I. Mould contaminants on Jarlsberg and Norvegia cheese blocks from four factories. *Int. J. Food Microbiol.* **2001**, *70*, 21–27. [CrossRef]
5. Vázquez, B.I.; Fente, C.; Franco, C.M.; Vázquez, M.J.; Cepeda, A. Inhibitory effects of eugenol and thymol on Penicillium citrinum strains in culture media and cheese. *Int. J. Food Microbiol.* **2001**, *67*, 157–163. [CrossRef]
6. Belitz, H.D.; Grosh, W.; Schieberle, P. Food Contamination. In *Food Chemistry*; Springer: Berlin, Germany, 2009.
7. Bassi, D.; Puglisi, E.; Cocconcelli, P.S. Understanding the bacterial communities of hard cheese with blowing defect. *Food Microbiol.* **2015**, *52*, 106–118. [CrossRef]
8. Klijn, N.; Nieuwenhof, F.F.; Hoolwerf, J.D.; van der Waals, C.B.; Weerkamp, A.H. Identification of Clostridium tyrobutyricum as the causative agent of late blowing in cheese by species-specific PCR amplification. *Appl. Environ. Microbiol.* **1995**, *61*, 2919–2924.
9. Garde, S.; Arias, R.; Gaya, P.; Nuñez, M. Occurrence of *Clostridium* spp. in ovine milk and Manchego cheese with late blowing defect: Identification and characterization of isolates. *Int. Dairy J.* **2011**, *21*, 272–278. [CrossRef]
10. Garde, S.; Ávila, M.; Arias, R.; Gaya, P.; Nuñez, M. Outgrowth inhibition of Clostridium beijerinckii spores by a bacteriocin-producing lactic culture in ovine milk cheese. *Int. J. Food Microbiol.* **2011**, *150*, 59–65. [CrossRef]

11. Garde, S.; Gaya, P.; Arias, R.; Nuñez, M. Enhanced PFGE protocol to study the genomic diversity of *Clostridium* spp. isolated from Manchego cheeses with late blowing defect. *Food Control* **2012**, *28*, 392–399. [CrossRef]

12. Gómez-Torres, N.; Garde, S.; Peirotén, Á.; Ávila, M. Impact of *Clostridium* spp. on cheese characteristics: Microbiology, color, formation of volatile compounds and off-flavors. *Food Control* **2015**, *56*, 186–194. [CrossRef]

13. EFSA. EFSA Confirms Safe Levels for Nitrites and Nitrates Added to Food. Available online: http://www.efsa.europa.eu/en/press/news/170615 (accessed on 13 July 2019).

14. D'Incecco, P.; Gatti, M.; Hogenboom, J.A.; Bottari, B.; Rosi, V.; Neviani, E.; Pellegrino, L. Lysozyme affects the microbial catabolism of free arginine in raw-milk hard cheeses. *Food Microbiol.* **2016**, *57*, 16–22. [CrossRef] [PubMed]

15. Hyldgaard, M.; Mygind, T.; Meyer, R.L. Essential oils in food preservation: Mode of action, synergies, and interactions with food matrix components. *Front. Microbiol.* **2012**, *3*, 12. [CrossRef] [PubMed]

16. Zantar, S.; Yedri, F.; Mrabet, R.; Laglaoui, A.; Bakkali, M.; Zerrouk, M.H. Effect of Thymus vulgaris and Origanum compactum essential oils on the shelf life of fresh goat cheese. *J. Essent. Oil Res.* **2014**, *26*, 76–84. [CrossRef]

17. Carović-Stanko, K.; Orlić, S.; Politeo, O.; Strikić, F.; Kolak, I.; Milos, M.; Satovic, Z. Composition and antibacterial activities of essential oils of seven Ocimum taxa. *Food Chem.* **2010**, *119*, 196–201. [CrossRef]

18. Hussain, A.I.; Anwar, F.; Hussain Sherazi, S.T.; Przybylski, R. Chemical composition, antioxidant and antimicrobial activities of basil (Ocimum basilicum) essential oils depends on seasonal variations. *Food Chem.* **2008**, *108*, 986–995. [CrossRef]

19. Caleja, C.; Barros, L.; Antonio, A.L.; Ciric, A.; Barreira, J.C.M.; Sokovic, M.; Oliveira, M.B.P.P.; Santos-Buelga, C.; Ferreira, I.C.F.R. Development of a functional dairy food: Exploring bioactive and preservation effects of chamomile (Matricaria recutita L.). *J. Funct. Foods* **2015**, *16*, 114–124. [CrossRef]

20. Burt, S. Essential oils: Their antibacterial properties and potential applications in foods—A review. *Int. J. Food Microbiol.* **2004**, *94*, 223–253. [CrossRef]

21. Govaris, A.; Botsoglou, E.; Sergelidis, D.; Chatzopoulou, P.S. Antibacterial activity of oregano and thyme essential oils against Listeria monocytogenes and Escherichia coli O157:H7 in feta cheese packaged under modified atmosphere. *LWT Food Sci. Technol.* **2011**, *44*, 1240–1244. [CrossRef]

22. Dupas, C.; Métoyer, B.; Hatmi, H.E.; Adt, I.; Mahgoub, S.A.; Dumas, E. Plants: A natural solution to enhance raw milk cheese preservation? *Food Res. Int.* **2019**, 108883. [CrossRef]

23. Khorshidian, N.; Yousefi, M.; Khanniri, E.; Mortazavian, A.M. Potential application of essential oils as antimicrobial preservatives in cheese. *Innov. Food Sci. Emerg. Tech.* **2018**, *45*, 62–72. [CrossRef]

24. Gutierrez, J.; Barry-Ryan, C.; Bourke, P. The antimicrobial efficacy of plant essential oil combinations and interactions with food ingredients. *Int. J. Food Microbiol.* **2008**, *124*, 91–97. [CrossRef]

25. Moreira, M.R.; Ponce, A.G.; Del Valle, C.E.; Roura, S.I. Effects of clove and tea tree oils on Escherichia coli O157:H7 in blanched spinach and minced cooked beef. *J. Food Process. Preserv.* **2007**, *31*, 379–391. [CrossRef]

26. Smith-Palmer, A.; Stewart, J.; Fyfe, L. The potential application of plant essential oils as natural food preservatives in soft cheese. *Food Microbiol.* **2001**, *18*, 463–470. [CrossRef]

27. Fernandes, R.V.D.B.; Guimarães, I.C.; Ferreira, C.L.R.; Botrel, D.A.; Borges, S.V.; de Souza, A.U. Microencapsulated Rosemary (Rosmarinus officinalis) Essential Oil as a Biopreservative in Minas Frescal Cheese. *J. Food Process. Preserv.* **2017**, *41*, e12759. [CrossRef]

28. Hamedi, H.; Razavi-Rohani, S.M.; Gandomi, H. Combination Effect of Essential Oils of Some Herbs With Monolaurin on Growth and Survival of Listeria Monocytogenes in Culture Media and Cheese. *J. Food Process. Preserv.* **2014**, *38*, 304–310. [CrossRef]

29. Moro, A.; Librán, C.M.; Berruga, M.I.; Zalacain, A.; Carmona, M. Mycotoxicogenic fungal inhibition by innovative cheese cover with aromatic plants. *J. Sci. Food Agric.* **2013**, *93*, 1112–1118. [CrossRef]

30. Regulation (EC) No 1333/2008 of the European Parliament and of the Council of 16 December 2008 on Food Additives. Last Consolidated Version 02008R1333-EN-28.10.2019-041.001. Available online: http://data.europa.eu/eli/reg/2008/1333/2019-10-28 (accessed on December 2019).

31. Licón, C.C.; Carmona, M.; Molina, A.; Berruga, M.I. Chemical, microbiological, textural, color, and sensory characteristics of pressed ewe milk cheeses with saffron (*Crocus sativus* L.) during ripening. *J. Dairy Sci.* **2012**, *95*, 4263–4274. [CrossRef]

32. Licón, C.C.; Hurtado de Mendoza, J.; Maggi, L.; Berruga, M.I.; Martín Aranda, R.M.; Carmona, M. Optimization of headspace sorptive extraction for the analysis of volatiles in pressed ewes' milk cheese. *Int. Dairy J.* **2012**, *23*, 53–61. [CrossRef]

33. Moro, A.; Librán, C.M.; Berruga, M.I.; Carmona, M.; Zalacain, A. Dairy matrix effect on the transference of rosemary (Rosmarinus officinalis) essential oil compounds during cheese making. *J. Sci. Food Agric.* **2015**, *95*, 1507–1513. [CrossRef]

34. Yang, W.-W.; Crow-Willard, E.N.; Ponce, A. Production and characterization of pure *Clostridium* spore suspensions. *J. Appl. Microbiol.* **2008**, *106*, 26–33. [CrossRef]

35. Anastasiou, R.; Aktypis, A.; Georgalaki, M.; Papadelli, M.; De Vuyst, L.; Tsakalidou, E. Inhibition of Clostridium tyrobutyricum by Streptococcus macedonicus ACA-DC 198 under conditions mimicking Kasseri cheese production and ripening. *Int. Dairy J.* **2009**, *19*, 330–335. [CrossRef]

36. Baratta, M.T.; Dorman, H.J.D.; Deans, S.G.; Figueiredo, A.C.; Barroso, J.G. Antimicrobial and antioxidant properties of some commercial essential oils. *Flavour Fragr. J.* **1998**, *13*, 235–244. [CrossRef]

37. Cassel, E.; Vargas, R.M.F.; Martinez, N.; Lorenzo, D.; Dellacassa, E. Steam distillation modeling for essential oil extraction process. *Ind. Crops Prod.* **2009**, *29*, 171–176. [CrossRef]

38. Singh Chouhan, K.B.; Tandey, R.; Sen, K.K.; Mehta, R.; Mandal, V. Critical analysis of microwave hydrodiffusion and gravity as a green tool for extraction of essential oils: Time to replace traditional distillation. *Trends Food Sci. Technol.* **2019**, *92*, 12–21. [CrossRef]

39. De Carvalho, R.J.; de Souza, G.T.; Honório, V.G.; de Sousa, J.P.; da Conceição, M.L.; Maganani, M.; de Souza, E.L. Comparative inhibitory effects of *Thymus vulgaris* L. essential oil against Staphylococcus aureus, Listeria monocytogenes and mesophilic starter co-culture in cheese-mimicking models. *Food Microbiol.* **2015**, *52*, 59–65. [CrossRef]

40. Turhan, H. 23-Lemon balm. In *Handbook of Herbs and Spices*; Peter, K.V., Ed.; Woodhead Publishing: Cambridge, UK, 2006; pp. 390–399. [CrossRef]

41. Carnat, A.P.; Carnat, A.; Fraisse, D.; Lamaison, J.L. The aromatic and polyphenolic composition of lemon balm (*Melissa officinalis* L. subsp. *officinalis*) tea. *Pharm. Acta Hel.* **1998**, *72*, 301–305. [CrossRef]

42. Stahl-Biskup, E.; Venskutonis, R.P. 19-Thyme. In *Handbook of Herbs and Spices*; Peter, K.V., Ed.; Woodhead Publishing: Cambridge, UK, 2004; pp. 297–321. [CrossRef]

43. Satyal, P.; Murray, B.L.; McFeetersand, R.L.; Setze, W.N. Essential Oil Characterization of Thymus vulgaris from Various Geographical Locations. *Foods* **2016**, *5*, 70. [CrossRef]

44. Handa, S.S.; Khanuja, S.; Longo, G.; Rakesh, D.D. *Extraction Technologies for Medicinal and Aromatic Plants*; International Centre for Science and High Technology: Trieste, Italy, 2008; pp. 21–25.

45. Gulluce, M.; Sahin, F.; Sokmen, M.; Ozer, H.; Daferera, D.; Sokmen, A.; Polissiou, M.; Adiguzel, A.; Ozkan, H. Antimicrobial and antioxidant properties of the essential oils and methanol extract from Mentha longifolia L. ssp. longifolia. *Food Chem.* **2007**, *103*, 1449–1456. [CrossRef]

46. Okoh, O.O.; Sadimenko, A.P.; Afolayan, A.J. Comparative evaluation of the antibacterial activities of the essential oils of Rosmarinus officinalis L. obtained by hydrodistillation and solvent free microwave extraction methods. *Food Chem.* **2010**, *120*, 308–312. [CrossRef]

47. Tajkarimi, M.M.; Ibrahim, S.A.; Cliver, D.O. Antimicrobial herb and spice compounds in food. *Food Control* **2010**, *21*, 1199–1218. [CrossRef]

48. Tornambé, G.; Cornu, A.; Verdier-Metz, I.; Pradel, P.; Kondjoyan, N.; Figueredo, G.; Hulin, S.; Martin, B. Addition of Pasture Plant Essential Oil in Milk: Influence on Chemical and Sensory Properties of Milk and Cheese. *J. Dairy Sci.* **2008**, *91*, 58–69. [CrossRef] [PubMed]

49. Nielsen, M.S.; Frisvad, J.C.; Nielsen, P.V. Protection by fungal starters against growth and secondary metabolite production of fungal spoilers of cheese. *Int. J. Food Microbiol.* **1998**, *42*, 91–99. [CrossRef]

50. Schlesser, J.E.; Gerdes, R.; Ravishankar, S.; Madsen, K.; Mowbray, J.; Teo, A.Y.-L. Survival of a Five-Strain Cocktail of Escherichia coli O157:H7 during the 60-Day Aging Period of Cheddar Cheese Made from Unpasteurized Milk. *J. Food Protec.* **2006**, *69*, 990–998. [CrossRef] [PubMed]

51. Ehsani, A.; Hashemi, M.; Naghibi, S.S.; Mohammadi, S.; Khalili Sadaghiani, S. Properties of *Bunium Persicum* Essential Oil and its Application in Iranian White Cheese Against *Listeria Monocytogenes* and *Escherichia Coli* O157:H7. *J. Food Safety* **2016**, *36*, 563–570. [CrossRef]

52. Le Bourhis, A.G.; Saunier, K.; Dore, J.; Carlier, J.P.; Chamba, J.F.; Popoff, M.R.; Tholozan, J.L. Development and validation of PCR primers to assess the diversity of *Clostridium* spp. in cheese by temporal temperature gradient gel electrophoresis. *App. Environ. Microbiol.* **2005**, *71*, 29–38. [CrossRef] [PubMed]

53. Fox, P.F.; Cogan, T.M. Factors that Affect the Quality of Cheese. In *Cheese: Chemistry, Physics and Microbiology*; Fox, P.F., McSweeney, P.L.H., Cogan, T.M., Guinee, T., Eds.; Elsevier Academic Press: London, UK, 2003; Volume 1, pp. 583–608.

54. Walstra, P.; Wouters, J.T.; Geurts, T.J. *Dairy Science and Technology*; Taylor & Francis Group: London, UK, 2006.

55. Sheehan, J. What causes the development of gas during ripening? In *Cheese Problems Solved: The Microbiology of Cheese Ripening*; PHL, M., Ed.; Woodhead Publishing Limited: Cambridge, UK, 2007; pp. 131–133.

56. Cogan, T.M. Cheese|Microbiology of cheese. In *Encyclopedia of Dairy Sciences*; John, W.F., Ed.; Academic Press: San Diego, CA, USA, 2011.

57. Librán, C.M.; Moro, A.; Molina, A.; Carmona, M.; Berruga, M.I. Empleo potencial de plantas aromáticas para combatir defectos en quesos de oveja. In *Parte I: Estudios In Vitro Frente a Penicillium Verrucosum. Proceedings of the XXXVII Congreso SEOC*; Ciudad Real, Spain, 2012; pp. 169–173.

58. Da Silva Dannenberg, G.; Funck, G.D.; Mattei, F.J.; da Silva, W.P.; Fiorentini, Â.M. Antimicrobial and antioxidant activity of essential oil from pink pepper tree (Schinus terebinthifolius Raddi) in vitro and in cheese experimentally contaminated with Listeria monocytogenes. *Innov. Food Sci. Emerg. Technol.* **2016**, *36*, 120–127. [CrossRef]

59. Deans, S.G.; Ritchie, G. Antibacterial properties of plant essential oils. *Int. J. Food Microbiol.* **1987**, *5*, 165–180. [CrossRef]

60. Librán, C.M.; Moro, A.; Carmona, M.; Molina, A.; Berruga, M.I. Empleo potencial de plantas aromáticas para combatir defectos en quesos de oveja. In *Parte II: Estudios In Vitro Frente a Microorganismos Causantes de Hinchazones. Proceedings of the XXXVII Congreso SEOC*; Ciudad Real, Spain, 2012; pp. 174–178.

61. Razzaghi-Abyaneh, M.; Shams-Ghahfarokhi, M.; Rezaee, M.-B.; Jaimand, K.; Alinezhad, S.; Saberi, R.; Yoshinari, T. Chemical composition and antiaflatoxigenic activity of *Carum carvi* L., Thymus vulgaris and Citrus aurantifolia essential oils. *Food Control* **2009**, *20*, 1018–1024. [CrossRef]

62. A, B.T.W. The Microbiology of cheese ripening. In *Cheese: Chemistry, Physics and Microbiology*; Fox, P.F., McSweeney, P.L.H., Cogan, T.M., Guinee, T.P., Eds.; Academic Press: Cambridge, UK, 2004.

63. Goñi, P.; López, P.; Sánchez, C.; Gómez-Lus, R.; Becerril, R.; Nerín, C. Antimicrobial activity in the vapour phase of a combination of cinnamon and clove essential oils. *Food Chem.* **2009**, *116*, 982–989. [CrossRef]

64. Reyes-Jurado, F.; Cervantes-Rincón, T.; Bach, H.; López-Malo, A.; Palou, E. Antimicrobial activity of Mexican oregano (Lippia berlandieri), thyme (Thymus vulgaris), and mustard (Brassica nigra) essential oils in gaseous phase. *Ind. Crops Prod.* **2019**, *131*, 90–95. [CrossRef]

Communication

Inhibition of *Escherichia coli* O157:H7 and *Salmonella enterica* Isolates on Spinach Leaf Surfaces Using Eugenol-Loaded Surfactant Micelles

Songsirin Ruengvisesh [1], Chris R. Kerth [2] and T. Matthew Taylor [2,*]

[1] Department of Nutrition and Food Science, Texas A&M University, College Station, TX 77843-2253, USA; songsirin@gmail.com
[2] Department of Animal Science, Texas A&M University, College Station, TX 77843-2471, USA; c-kerth@tamu.edu
* Correspondence: matt_taylor@tamu.edu; Tel.: +01-979-862-7678

Received: 23 October 2019; Accepted: 12 November 2019; Published: 15 November 2019

Abstract: Spinach and other leafy green vegetables have been linked to foodborne disease outbreaks of *Escherichia coli* O157:H7 and *Salmonella enterica* around the globe. In this study, the antimicrobial activities of surfactant micelles formed from the anionic surfactant sodium dodecyl sulfate (SDS), SDS micelle-loaded eugenol (1.0% eugenol), 1.0% free eugenol, 200 ppm free chlorine, and sterile water were tested against the human pathogens *E. coli* O157:H7 and *Salmonella* Saintpaul, and naturally occurring microorganisms, on spinach leaf surfaces during storage at 5 °C over 10 days. Spinach samples were immersed in antimicrobial treatment solution for 2.0 min at 25 °C, after which treatment solutions were drained off and samples were either subjected to analysis or prepared for refrigerated storage. Whereas empty SDS micelles produced moderate reductions in counts of both pathogens (2.1–3.2 $\log_{10}$ CFU/cm^2), free and micelle-entrapped eugenol treatments reduced pathogens by >5.0 $\log_{10}$ CFU/cm^2 to below the limit of detection (<0.5 $\log_{10}$ CFU/cm^2). Micelle-loaded eugenol produced the greatest numerical reductions in naturally contaminating aerobic bacteria, *Enterobacteriaceae*, and fungi, though these reductions did not differ statistically from reductions achieved by un-encapsulated eugenol and 200 ppm chlorine. Micelles-loaded eugenol could be used as a novel antimicrobial technology to decontaminate fresh spinach from microbial pathogens.

Keywords: micelles; plant-derived antimicrobial; Enteric pathogens; leafy greens

---

1. Introduction

The U.S. Centers for Disease Control and Prevention (CDC) has estimated that 47.8 million cases of foodborne illnesses occur annually in the U.S. due to known and unspecified foodborne disease agents [1]. Of these pathogens, *Escherichia coli* O157:H7 and non-typhoidal *Salmonella enterica* serotypes were deemed responsible for approximately 63,153 cases [2] and 1,027,561 cases of domestically acquired foodborne illnesses, respectively [3]. From 2006 to 2017 in the U.S., the numbers of foodborne disease cases associated with the shiga toxin-producing *E. coli* (STEC), and the various serovars of the non-typhoidal salmonellae, associated with fresh fruits and vegetables, has increased [4,5]. This increase could be partly due to improved surveillance for human pathogens [6], increased consumption of raw or minimally processed produce items, as well as other contributing factors (e.g., use of nontreated biological soil amendments or pathogen-contaminated irrigation water, and other practices which could increase pathogen transmission risks). Among many commodities, spinach and other leafy greens have been associated with multiple *E. coli* O157:H7 human disease outbreaks [7–9]. While less frequently associated with leafy greens in the U.S., multiple outbreaks of leafy green disease outbreaks involving multiple *Salmonella* spp. have been reported across many industrialized nations, summarized recently by Chaves et al. [10]. Foodborne disease outbreaks can cause substantial economic losses including medical expenses, lost wages, damage control costs for product recall and disposal of affected products, and production time loss [11].

Essential oils and their components (EOCs) are volatile, hydrophobic substances that can be extracted from various parts (e.g., flowers, leaves, rhizome, seeds, fruits, wood, and bark) of aromatic plants) [12]. Essential oils contain bioactive components that are derivatives of alcohols, ketones, aldehydes, esters, and phenols [12]. It has been reported that EOCs possess insecticidal, antioxidant, anti-inflammatory, anti-allergenic, anticancer, and antimicrobial properties, thereby potentially beneficial in medical, pharmaceutical, and food industries [13]. In foodstuffs, however, high concentrations of EOCs are often required to inactivate microorganisms due to the hydrophobic nature of some EOCs [14,15]. For example, eugenol is water-soluble up to only 4.93 g/L, though it is miscible in alcohols such as ethyl alcohol [16]. The requirement for use of elevated concentrations of EOCs can render EOCs impractical as food additives or sanitizers, as they may be excessively costly at usage concentrations and/or impart undesirable flavor and/or aroma to the food product [17,18]. Encapsulation has, therefore, been recommended for improving upon these negative characteristics of plant-derived antimicrobial agents, by increasing water-dispersibility, reduce the required dosage needed for foodborne pathogen inhibition, and provide protection to the antimicrobial agent from rapid volatilization [19–21]. Weiss et al. [14], in their review of nanoencapsulation strategies for food antimicrobials delivery to foods, recommended that encapsulating materials be inexpensively procured to offset the cost of additional processing needed to form the encapsulated structure. In this case, sodium dodecyl sulfate (SDS) can be purchased relatively inexpensively, and manufacture of micelles does not require highly costly equipment. In addition, consumer use of produce rinsing in the home prior to consumption would reduce the potential for undesirable flavor or mouthfeel consequences on micelle-treated produce surfaces. Thus, delivery methods for EOCs can be utilized to improve antimicrobial activities of EOCs in food systems so as to reduce the content of EOC required for antimicrobial functionality without significant compromise to sensory acceptability of treated commodities.

To enhance delivery of EOCs to microorganisms in foodstuffs, surfactants can be utilized to encapsulate EOCs [18,22,23]. Surfactants are surface-active, amphiphilic molecules that contain both hydrophilic and hydrophobic components; they can be classified as anionic, cationic, zwitterionic, or nonionic [24]. At low concentrations, surfactants adsorb to the aqueous phase of a lipid/water interface, lowering the surface tension [25]. When present at or above the critical micelle concentration (CMC), surfactant molecules will aggregate to form thermodynamically favored structures known as micelles. In micelle structures, hydrophobic molecules such as EOCs can be encapsulated inside the hydrophobic core, while hydrophilic headgroups of surfactants face outwardly contacting the aqueous phase [24,26].

In several studies, efficient pathogen inactivation using EOCs-encapsulated surfactant micelles/emulsion in foodstuffs has been reported [18,22,23,27]. Nonetheless, limited studies have been conducted to evaluate the antimicrobial activities of EOCs-containing micelles on the surfaces of fresh produce for the purpose of pathogen decontamination. Thus, the main objective of this study was to determine the efficacy of eugenol-loaded surfactant micelles, compared to other antimicrobial treatments, specifically non-encapsulated eugenol and 200 ppm free chlorine, to reduce numbers of inoculated *E. coli* O157:H7 and *S.* Saintpaul on surfaces of spinach leaves stored refrigerated. The second objective was to evaluate the efficacy of eugenol-containing micelles to reduce numbers of microbial hygiene indicator on leaf surfaces during refrigerated storage.

2. Materials and Methods

2.1. Preparation of Antimicrobial Micelles and Other Treatments

Eugenol-loaded micelles and other treatments (free eugenol, empty micelles, 200 ppm free chlorine, sterile distilled water) were prepared in identical manner to methods reported previously by our group [28]. Briefly, eugenol stock solution (70% *w/v*) was prepared by dissolution of eugenol (Sigma-Aldrich Co., St. Louis, MO, USA) in 95% ethyl alcohol (Koptec, King of Prussia, PA, USA), and stored at 5 °C until ready for use. Sodium dodecyl sulfate (SDS) micelles (1.0% *w/v*) were produced containing eugenol at 1.0% EOC according to previous methods [29]. After stirring until optical density at 632 nm stabilized, micelles were filter-sterilized by filtering through a 0.45 µm cellulose acetate filter. Micelles were then stored at 5 °C for no more than 36 h prior to use.

2.2. Revival of Bacterial Pathogens and Preliminary Assessment of Consistent Overnight Pathogen Growth for Pathogen Cocktail Preparation

Rifampicin-resistant (Rif^R; 100.0 µg/mL) *E. coli* O157:H7 (Strain K3999) from the pathogen isolate recovered from a 2006 U.S. spinach-borne disease outbreak and *S. enterica* serovar Saintpaul (Strain FDA/CFSAN 476398) from the 2008 U.S. peppers-transmitted disease outbreak were selected for spinach sample inoculation and decontamination. Pathogens were revived from cryo-storage (−80 °C) in the culture collection of the Food Microbiology Laboratory (Department of Animal Science, Texas A&M University, College Station, TX, USA) individually inoculating each isolate into a sterile 10.0 mL volume of Tryptic Soy Broth (TSB; Becton, Dickinson and Co., Franklin Lakes, NJ, USA) and incubating for 24 h at 35 °C without shaking. After incubation, a sterile loop was used to collect 10.0 µL of each culture; each was then aseptically passed into a new sterile 10.0 mL volume of TSB. These were subsequently incubated for 24 h at 35 °C. Following the second passage of cultures to complete revival and activation, equal volumes of microorganisms were blended into a cocktail for spinach surface inoculation, targeting an inoculation of approximately 6.0 log$_{10}$ CFU/cm^2. Preliminary tests were completed prior to experimental startup to verify researchers' ability to consistently produce predictable numbers of pathogen isolates following 24 h incubation in TSB at 35 °C, in order to reliably produce an inoculum. Following incubation of microorganisms, TSB volumes of each pathogen were serially diluted in 0.1% (*w/v*) peptone (Thermo-Fisher Scientific, Waltham, MA, USA) diluent and enumerated on Tryptic Soy Agar (TSA; Becton, Dickinson and Co.). Following 24 h incubation of inoculated TSA Petri plates at 35 °C, plates were counted and counts were log$_{10}$-transformed. The experiment was replicated in identical manner three times (*n* = 3) and numbers of each organism compared to one another to confirm that one pathogen would not contribute significantly more cells to the cocktail than the other. A cocktail of Rif^R *E. coli* O157:H7 and *S.* Saintpaul was subsequently prepared for spinach inoculation according to the method of Cálix-Lara et al. [30] without modification.

2.3. Antimicrobial Activity Testing for Antimicrobial Treatments on Pathogens-Inoculated and Noninoculated Spinach Leaf Samples Held under Refrigeration

Unwashed, freshly harvested spinach was purchased from a local fruit and vegetable distributor and transported immediately in insulated coolers containing cooling pouches to the Food Microbiology Laboratory (Department of Animal Science, Texas A&M University, College Station, TX, USA). For each sample, three pieces, each 10 cm^2, of spinach were aseptically excised using sterile scalpel and borer, placed in an empty sterile Petri dish, and spot-inoculated with approximately 7.0 log$_{10}$ CFU/mL cocktailed Rif^R *E. coli* O157:H7 and *S.* Saintpaul. Pathogen cocktail was spotted onto samples (ten spots at 10.0 µL), after which pathogen-inoculated samples were air-dried at ambient temperature (25 ± 1 °C) for 1.0 h to allow pathogen attachment to spinach leaf surfaces.

To test the sanitizing/growth inhibition efficacy of each treatment on pathogens or naturally occurring hygiene microorganisms, encapsulated eugenol (1.0% SDS + 1.0% eugenol-loaded micelles), free eugenol (1.0% eugenol), empty micelles (1.0% SDS), 200 ppm chlorine (adjusted to pH 7.0 with

0.1 N HCl), and sterile distilled water were individually applied to inoculated spinach samples in Petri dishes by immersing in 20 mL of treatment solution. Positive controls (pathogen inoculated without any treatment or non-inoculated spinach samples used for testing antimicrobial/sanitizing treatments against background microbiota) and negative controls (uninoculated sample without treatment) were included to determine pathogen attachment to spinach surfaces and confirm no naturally occurring 100.0 µg/mL Rif^R microbes, respectively. For day 0 samples, encapsulated eugenol, free eugenol, empty micelles, chlorine, and sterile distilled water were individually applied to Petri dishes via 2 min immersion with 20 mL of treatment solution, after which the solution was drained off and spinach samples immediately transferred to a sterile stomacher bag and mixed with 99 mL 0.1% (*w/v*) peptone diluent by pummeling in a stomacher (230 rpm) for 1 min.

For all non-day 0 samples, treatments were applied to spinach leaf samples in identical manner as for day 0-assigned samples, drained of treatment solution, and then transferred to new sterile Petri dishes, where they were stored at 5 ± 1 °C covered in saran film to afford oxygen transmission under dark conditions. Samples were withdrawn after 3, 5, 7, or 10 days of refrigerated storage for subsequent enumeration of inoculated pathogens or naturally occurring microbial organisms. As with day 0 samples, to enumerate pathogens, samples were placed in stomacher bags and pummeled with 99 mL of 0.1% peptone diluent for 1 min. Pummeled samples were serially diluted in 9 mL of 0.1% peptone diluent and dilutions were spread on surfaces of Lactose-Sulfite-Phenol Red-Rifampicin (LSPR) agar supplemented with 100.0 µg/mL rifampicin, in order to differentially enumerate *E. coli* O157:H7 colonies (cream-white with halo of fermented lactose) from *S.* Saintpaul colonies (black-centered colonies with no halo of lactose fermentation) [31]. Following 24 h incubation at 35 °C, colonies of Rif^R *E. coli* O157:H7 and *S.* Saintpaul were counted and recorded.

For enumeration of naturally occurring microbiota (aerobic bacteria, *Enterobacteriaceae*, and yeasts and molds) from non-inoculated, antimicrobial-treated spinach surface samples, resulting samples were serially diluted in 99 mL sterile 0.1% peptone diluent and 1.0 mL volumes were spread on 3M^{TM} Petrifilm^{TM} Aerobic Count Plates, 3M^{TM} Petrifilm^{TM} Enterobacteriaceae Count Plates, and 3M^{TM} Petrifilm^{TM} Yeast and Mold Count Plates. Aerobic Count Plate and Enterobacteriaceae Count Plate petrifilms were each incubated at 35 °C for 48 h, while Yeast and Mold Count Plate petrifilms were incubated at 25 °C for 5 days, all according to manufacturer instructions. Colonies were counted after incubation.

2.4. Statistical Analysis of Data

For preliminary data gathered for pathogen cocktail preparation (Section 2.2), mean counts of each pathogen ($n = 3$) were compared to one another by unpaired *t*-test (2-tailed, $p = 0.05$). All spinach decontamination experiments (Section 2.3) were replicated thrice identically; two independent samples were completed for each sample/treatment combination within a replicate ($n = 6$). The experiment was designed and completed as a full factorial, with $\alpha = 0.05$; spinach samples were randomly assigned to antimicrobial treatment and storage period conditions at experiment outset. All microbiological plate count data were $\log_{10}$-transformed prior to statistical analysis. The limit of detection for plating assays was 0.5 $\log_{10}$ CFU/cm^2. In cases where microbial numbers were below the limit of detection, the value of 0.4 $\log_{10}$ CFU/cm^2 was inserted for purposes of comparison of mean microbial counts by treatment and storage period. $\log_{10}$-transformed counts of each pathogen, or microbial hygiene indicator group, were compared for the main effects of antimicrobial treatment, storage period, and their interaction by a two-way analysis of variance (ANOVA). Statistically differing mean microorganism counts (pathogens, hygiene indicator grouping) were separated by Tukey's Honestly Significant Differences test at $p = 0.05$. Statistical analysis was completed on JMP Pro v.14 for Macintosh (SAS Institute, Inc., Cary, NC, USA).

3. Results

3.1. Consistency of Overnight Growth of Salmonella Saintpaul and E. coli O157:H7 Organisms for Cocktail Preparation

Mean populations of *E.coli* O157:H7 and *Salmonella* Saintpaul isolates following 24 h incubation at 35 °C during preliminary trials (Section 2.2) were 7.4 ± 0.2 and 7.6 ± 0.1 $\log_{10}$ CFU/mL, respectively. Mean plate counts of the pathogens following growth were not different from one another by *t*-test ($p = 0.156$), and were thus assessed to not provide non-differing counts of each pathogen to cocktail preparations for subsequent experiments on spinach leaves.

3.2. Inhibition of Salmonella Saintpaul on Spinach Surfaces by Antimicrobial Treatments over 10 Days of Refrigerated Storage

Table 1 presents the least-squares means of *Salmonella* Saintpaul populations on spinach leaf surfaces following treatment with SDS micelle-encapsulated eugenol, free eugenol, empty SDS micelles, 200 ppm chlorine, or sterile distilled water. For *Salmonella* reduction on spinach surfaces, overall, the trend of antimicrobial effects from greatest to least was Encap = Free-Eug $\geq$ 200 HOCl > SDS-Mic $\geq$ DW. Encapsulated eugenol, free eugenol, and chlorine exerted efficient residual effects in reducing pathogen populations to below or just over detectable levels after day 0 of storage. Only the free and micelle-encapsulated eugenol treatments reduced pathogens to below the limit of detection by plating ($0.5 \log_{10}$ CFU/cm^2). The population on the positive control (inoculated, nontreated) on day 0 of storage was $6.0 \log_{10}$ CFU/cm^2. On day 0, populations of *S.* Saintpaul after treatment with encapsulated eugenol, free eugenol, empty micelles, chlorine, and sterile water were varied, ranging from 1.8 to 5.6 $\log_{10}$ CFU/cm^2. Early in the experiment, free eugenol was equally effective as chlorine at reducing the pathogen on spinach, and produced a greater numerical reduction than did encapsulated eugenol in reducing *S.* Saintpaul (though counts of surviving pathogen between treatments did not differ). Conversely, neither empty SDS micelles nor sterile water reduced populations of *S.* Saintpaul ($p \geq 0.05$) on day 0 (Table 1). From days 3 until 10, all treatments resulted in *S.* Saintpaul declining in a treatment and time-specific manner, ultimately ranging at day 10 of storage from 0.4 to 4.7 $\log_{10}$ CFU/cm^2 (Table 1). Micelle-encapsulated eugenol, free eugenol, and 200 ppm chlorine were similarly effective in reducing *S.* Saintpaul populations and were more effective than empty SDS micelles and sterile water at days 3 through 10. Encapsulated eugenol and free eugenol initially reduced the pathogen compared to the control, and inhibited pathogen growth to undetectable numbers continuously from days 3 to 10. Compared to the control, water treatment increased the population of *S.* Saintpaul to 4.7 $\log_{10}$ CFU/cm^2 on day 10. Compared to the level of *S.* Saintpaul on day 0, the levels of *S.* Saintpaul on the positive control decreased from day 5 to 10 of storage ($p < 0.05$), likely the result of cold temperature storage in combination with potential for pathogen cells to be exposed to spinach-derived compounds with antimicrobial activity (e.g., organic acids, phytoaxelins, phenolic compounds).

Table 1. Least-squares means of surviving *Salmonella* Saintpaul ($\log_{10}$ CFU/cm^2) on spinach surfaces as a function of the interaction of antimicrobial treatment and days of aerobic storage at 5 °C.

Storage Period (Days)	Encap [1]	Free-Eug	SDS-Mic	200 HOCl	DW	Control
0	2.8GH [2]	1.8HI	5.4ABCD	2.0HI	5.6ABC	6.0A
3	0.4K	0.4K	4.7CDEF	0.7JK	5.2ABCD	5.8AB
5	0.4K	0.4K	4.5DEF	1.6IJ	4.8BCDEF	4.5CDEF
7	0.4K	0.5JK	4.0EF	0.9IJK	4.8BCDE	4.3DEF
10	0.4K	0.4K	3.6FG	0.5JK	4.7BCDEF	3.8EFG
$p \leq 0.0001$	Pooled Standard Error = 0.2					

[1] Antimicrobial treatments were: 1.0% sodium dodecyl sulfate (SDS) micelles loaded with 1.0% eugenol (Encap); 1.0% un-encapsulated eugenol (Free-Eug); 1.0% SDS micelles unloaded (SDS-Mic); 200 ppm pH 7.0 free chlorine (200 HOCl); sterile distilled water (DW); inoculated, nontreated (Control). [2] Values depict least-squares means calculated from three identically completed replicates, each containing duplicate identically processed independent samples ($n = 6$). Means read across columns and rows that do not share capitalized letters (A, B, C, …) differ by two-way analysis of variance and Tukey's Honestly Significant Differences Means Separation Test at $p = 0.05$.

3.3. Inhibition of E. coli O157:H7 on Spinach Surfaces by Antimicrobial Treatments over 10 Days of Refrigerated Storage

Similar trends were observed for *E. coli* O157:H7-inoculated spinach treated with antimicrobials (free, encapsulated) as those reported for *Salmonella*-inoculated spinach (Section 3.2). Table 2 depicts populations of *E. coli* O157:H7 on spinach samples after antimicrobial sanitizing treatment, over 10 days of refrigerated ($5 \pm 1°C$) storage. The initial population of *E. coli* O157:H7 on the positive control on day 0 was 6.0 $\log_{10}$ CFU/cm^2. On day 0, antimicrobial treatments, except sterile water, reduced populations of *E. coli* O157:H7 to numbers ranging from 2.3 to 5.0 $\log_{10}$ CFU/cm^2. As was the case with *Salmonella* Saintpaul testing, initially free eugenol treatment produced the greatest numerical reduction in pathogen counts. Moreover, similar to *Salmonella* testing, encapsulated eugenol-treated *E. coli* O157:H7 counts did not differ from those of the free eugenol-treated *E. coli* O157:H7 count, though numerical counts of *E. coli* O157:H7 were higher than like counts of *Salmonella* at day 0 for free and micelle-loaded eugenol treatments. From days 3 to 10, *E. coli* O157:H7 populations treated with either micelle-encapsulated or free eugenol bore non-detectable pathogen counts (0.4 $\log_{10}$ CFU/cm^2). Conversely, other treatments (sterile water, empty SDS micelles, and 2 00 ppm chlorine) produced smaller reductions in pathogen counts following their application. Encapsulated eugenol, free eugenol, and chlorine reduced pathogen counts to non-detection or near non-detection values within 7 days of refrigerated storage ($p \geq 0.05$); all were more effective than empty micelles or water ($p < 0.05$) on day 3. From days 5 to 10, all treatments but sterile water reduced populations of *E. coli* O157:H7 to lower levels than positive controls ($p < 0.05$). The levels of *E. coli* O157:H7 on untreated spinach samples decreased from 6.0 to 4.0 $\log_{10}$ CFU/cm^2 from day 0 to 10, a similar but less substantial decline as that observed for *S.* Saintpaul (Tables 1 and 2).

Table 2. Surviving *Escherichia coli* O157:H7 ($\log_{10}$ CFU/cm^2) on spinach surfaces as a function of the interaction of antimicrobial treatment and days of aerobic storage at 5 °C.

Storage Period (Days)	Encap [1]	Free-Eug	SDS-Mic	200 HOCl	DW	Control
0	3.1DEFG [2]	2.3GHI	5.0ABC	2.7FGH	5.3AB	6.0A
3	0.4K	0.4K	4.1CDE	0.7JK	4.7ABC	5.9A
5	0.4K	0.4K	3.8CDEF	1.5IJK	4.2BCD	4.6BC
7	0.4K	0.6JK	2.9EFGH	0.8JI	4.1CDE	4.4BC
10	0.4K	0.4K	1.7HIJ	0.6JK	3.9CDEF	4.0CDE
$p > 0.0001$	Pooled Standard Error = 0.3					

[1] Antimicrobial treatments were: 1.0% sodium dodecyl sulfate (SDS) micelles loaded with 1.0% eugenol (Encap); 1.0% unencapsulated eugenol (Free-Eug); 1.0% SDS micelles unloaded (SDS-Mic); 200 ppm pH 7.0 free chlorine (200 HOCl); sterile distilled water (DW); inoculated, nontreated (Control). [2] Values depict least-squares means calculated from three identically completed replicates, each containing duplicate identically processed independent samples ($n = 6$). Means read across columns and rows that do not share capitalized letters (A, B, C, …) differ by two-way analysis of variance and Tukey's Honestly Significant Differences Means Separation Test at $p = 0.05$.

3.4. Inhibition of Naturally Occurring Microbial Hygiene Indicator Groups on Treated Spinach over 10 Days of Refrigerated Storage

With respect to antimicrobial treatments and their impacts on naturally contaminating hygiene-indicating microorganisms, for aerobic bacteria and *Enterobacteriaceae*, treatments followed the trend from greatest to least antibacterial effects of Encap = Free-Eug $\geq$ 200 HOCl > DW > SDS-Mic (Figure 1). The antifungal effect of treatments on surfaces of spinach samples followed the trend of Encap = Free-Eug = 200 HOCl $\geq$ SDS-Mic > DW (Figure 1). In the case of spinach leaf samples that were utilized for determining the efficacy of antimicrobial treatments against naturally contaminating aerobic bacteria, *Enterobacteriaceae*, and fungi (yeasts/molds), microbial loads on spinach samples were significantly influenced by antimicrobial treatment for all groups of tested microorganisms. In all cases, encapsulated and free eugenol reduced organisms versus sterile water and the control, but surviving counts of aerobic bacteria, *Enterobacteriaceae* and fungi did not differ for micelle-loaded eugenol versus free eugenol (Figure 1). SDS micelles exerted some antimicrobial effect when compared with water or the control for all groups of microbes, though not to the extent observed for eugenol-including treatments or the 200 ppm free chlorine treatment. Indeed, for *Enterobacteriaceae*, SDS micelles appeared to produce a higher count of *Enterobacteriaceae* versus the control and water-treated samples, potentially resulting from de-clumping of cells by the surfactant, or higher initial loads on SDS micelles-treated spinach samples at the experiment initiation (Figure 1b). While no group of microorganisms was reduced to non-detectable levels, eugenol treatments resulted in the fewest numbers of hygiene indicator microbes on treated spinach, indicating potential for best outcomes related to protection of spinach keeping quality.

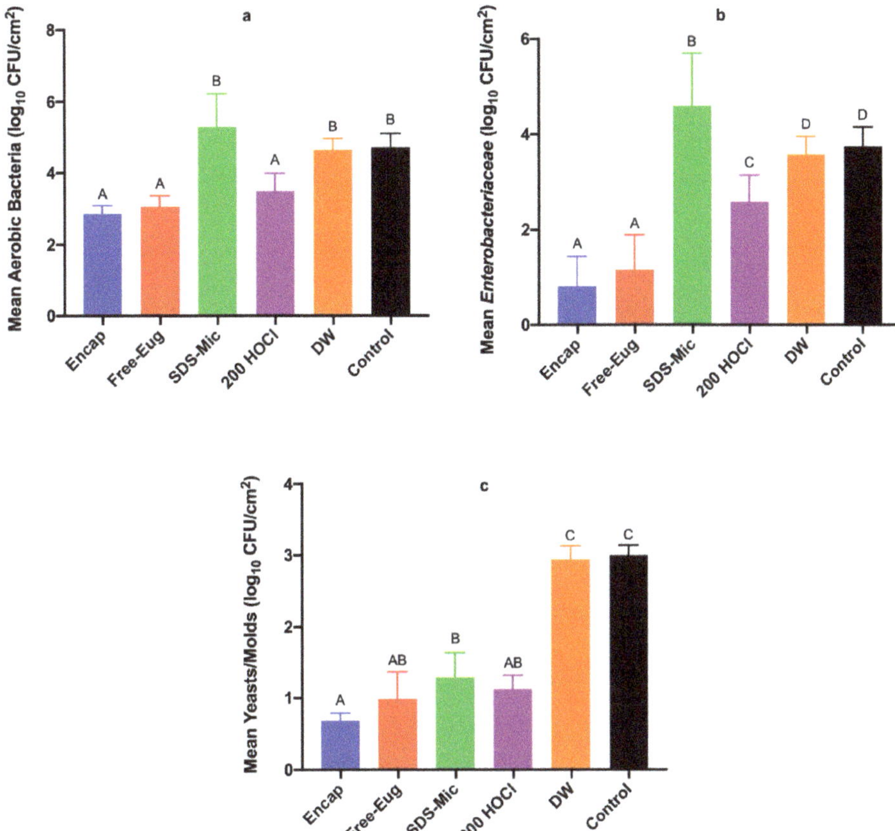

Figure 1. Means of naturally occurring microorganisms on spinach samples as function of antimicrobial treatment: (**a**) aerobic bacteria, (**b**) *Enterobacteriaceae*, and (**c**) yeasts and molds ($p < 0.0001$). Treatments were: 1.0% sodium dodecyl sulfate (SDS) micelles loaded with 1.0% eugenol (Encap); 1.0% unencapsulated eugenol (Free-Eug); 1.0% SDS micelles unloaded (SDS-Mic); 200 ppm pH 7.0 free chlorine (200 HOCl); sterile distilled water (DW); no treatment, non-inoculated (Control). Bars depict arithmetic means from three identical replications with duplicate independent samples per replicate ($n = 6$); error bars depict one sample standard deviation from the mean. Columns not sharing capitalized letters (A, B, C, D) differ at $p = 0.05$.

4. Discussion

Eugenol (4-allyl-2-methoxyphenol) is a naturally occurring phenolic EOC in clove oils and has been reported to exhibit effective antimicrobial activities against a wide range of microorganisms [32–34]. Reported mechanisms of action of EOCs against microorganisms have included cellular membrane disruption, alteration in membrane permeability, release of proteins and nucleic acids, and structural and morphological changes [32]. In this study, SDS was utilized to encapsulate 1% eugenol for inhibiting enteric bacterial pathogens and naturally occurring microorganisms on surfaces of spinach samples. SDS, an anionic surfactant, is a derivative of lauric acid and a mixture of sodium alkyl sulfates consisting of a 12-carbon tail attached to a sulfate head group, rendering it amphiphilic [35,36]. The possible functions of surfactant micelles in delivering an antimicrobial to pathogens may include: (1) enhanced dispersion of EOC in aqueous phase; (2) transport of EOCs to microbial membranes, and; (3) disruption of microbial membranes to enhance uptake of EOC [19,37–39]. Micelles themselves are covered by polar headgroups, making them amphiphilic structures [40]. However, the surfactant monomers of the micelles structures are amphiphilic and may thermodynamically bind to bacterial membrane components [40]. In this research, the antimicrobial activities of free and encapsulated eugenol did not significantly differ. Although eugenol is hydrophobic, it possesses slight water solubility (0.64 g/L) [41] and thus may have resulted in partial dissolution and dispersion of eugenol in wash water.

The rough surfaces of spinach [42], as well as cracks, pockets, crevices, and native openings (e.g., stomata), may favor microbial attachment and provide protection to microorganisms from antimicrobial intervention [43,44]. On leaf surfaces, there is a boundary layer, a thin layer of air influenced by the leaf surface [45]. The layer can vary in thickness and can influence the temperature, moisture, and speed of water vapor leaving the stomata through the motionless layer [45]. When spinach samples were treated with encapsulated or free eugenol, the antimicrobial EOC may have become trapped in a boundary layer and crevices. During storage, eugenol may have vaporized and exerted residual effect in inactivating microorganisms. The surface of spinach is covered with cuticle, a continuous extracellular membrane of polymerized lipids with associated waxes [46]. The hydrophobic nature of the waxy cuticle may have prevented chlorine, which is more hydrophilic, from inactivating microorganisms on spinach surfaces.

Hypochlorous acid (HOCl) is the principal form of available chlorine in an aqueous solution that exerts the greatest bactericidal activity against a wide range of microorganisms. To maintain available HOCl, the pH of the solution must be maintained in the range of 6.0 to 7.5 [47]. In this study, the pH of a chlorine solution was adjusted to 7.0 at the experiment's outset, prior to its application onto inoculated samples. Distilled water was used to prepare the chlorine solution, so the presence of organic matter was reduced. Thus, chlorine showed potent antibacterial effect in reducing pathogens and microbiota on fresh produce in the study. Indeed, chlorine treatment was as effective as eugenol-including treatments in the cases of aerobic bacteria and yeasts/molds but not for *Enterobacteriaceae*, wherein counts of microbes treated with 200 ppm chlorine did not statistically differ versus those treated either with micelle-loaded or free eugenol. Effects of chlorine on microbial inactivation in leafy greens have been reported throughout many refereed papers and expert reports. Zhang and Farber [48] reported the maximum log_{10} reduction of *L. monocytogenes* at 4 and 22° C to be 1.3 and 1.7 log_{10} CFU/g for lettuce and 0.9 and 1.2 log_{10} CFU/g for cabbage, respectively. In the current study, chlorine (200 ppm) produced greater reductions for inoculated pathogens versus naturally occurring *Enterobacteriaceae* (Tables 1 and 2; Figure 1), similar to results reported by other researchers testing 100–200 ppm HOCl on spinach [49,50], potentially resulting from differences in differing attachment strengths from naturally occurring versus inoculated pathogen cells, as well as potential for naturally occurring cells to locate effectively into protected niches on the leaf surface [51]. Erkman [52] reported that 10 ppm HOCl (pH 7.0) applied via immersion with agitation for 5 min reduced *E. coli* on lettuce, parsley, and pepper by 1.2, 1.6, and 2.6 log_{10} CFU/mL, respectively. Nevertheless, in produce packing operations, accumulation of organic matter (e.g., field soil, debris, fruit, leaves) in a dump tank or flume water, as well as alkaline pH of wash water, can decrease effectiveness of chlorine [47,53].

In this study, micelle-loaded eugenol produced the highest numerical reductions in naturally contaminating aerobic bacteria, *Enterobacteriaceae*, and fungi, although with the exception of the *Enterobacteriaceae*, these did not differ statistically from reductions achieved by un-encapsulated eugenol and 200 ppm chlorine. It was reported that *Enterobacteriaceae* and pseudomonads are predominant on surfaces of leafy greens [45]. Thus, increased populations of aerobic bacteria and *Enterobacteriaceae* on spinach surfaces in this study could have been due to the ability of these bacteria to metabolize or tolerate SDS [54–56]. Kramer et al. [55] reported that 200 strains of independent isolates of *Enterobacteriaceae* members (e.g., *E. coli*, *Shigella flexneri*, *Shigella sonnei*, *Salmonella* Arizonae, *Klebsiella pneumoniae*, etc.) were highly tolerant to SDS and were able to grow in the presence of ≥5% SDS. In contrast, previous research has indicated that SDS demonstrated antimicrobial activity against foodborne fungal microbes, inhibiting colony development and mycotoxin synthesis [57,58].

Utilization of EOC-encapsulating micelles or emulsions for inactivation of pathogens on fresh produce surfaces has been reported. Park et al. [59] reported clove bud oil (0.02%) + benzothoium chloride (0.002%) emulsion inactivated inoculated *S*. Typhimurium and *Listeria monocytogenes* on fresh-cut pak choi by 1.9 to 2.0 $\log_{10}$ CFU/g, respectively. Kang et al. [22] showed that cinnamon leaf essential oil in cetylpyridinium chloride produced 1.8 and 1.5 $\log_{10}$ CFU/g reductions against *L. monocytogenes* and *E. coli* O157:H7, respectively; quality of kale leaves was not affected during storage. In our previous study, eugenol (1% *w/v*) encapsulated in SDS (1% *w/v*) micelles were used for inhibition of *S*. Saintpaul and *E. coli* O157:H7 as well as native microbiota on tomato skin surfaces during refrigerated and abuse storage [28]. In that study, antimicrobial effects of free and encapsulated eugenol did not differ from those of HOCl and empty SDS micelles during refrigerated storage. However, reductions in pathogen counts to non-detectable levels were only observed with free and encapsulated eugenol [28]. EOC-encapsulated micelles could be used as an alternative to the commonly used sanitizers to reduce pathogens on fresh produce, potentially achieving greater pathogen reductions versus those typically observed by washing in chlorinated water [60].

5. Conclusions

Overall, micelle-encapsulated and eugenol displayed similar efficacies for reducing the enteric bacterial human pathogens *E. coli* O157:H7 and *Salmonella*, as well as for microbial hygiene-indicating microorganisms, on surfaces of spinach leaf samples during a simulated washing and subsequent refrigerated storage. Antimicrobial-loaded micelles may be used as an alternative to conventional antimicrobial technologies for decontaminating surfaces of leafy green produce commodities from microbial pathogens as a means to produce human food safety for consumers of these agricultural commodities.

Author Contributions: Experiment conceptualization, S.R. and T.M.T.; methodology, S.R. and T.M.T.; formal analysis, S.R. and C.R.K.; writing—original draft preparation, T.M.T. and S.R.; writing—review and editing, T.M.T., S.R., and C.R.K.; project administration, T.M.T.; funding acquisition, T.M.T.

Funding: This work is supported by National Integrated Food Safety Initiative Competitive Program [grant no. 2010-51110-21079] from the USDA National Institute of Food and Agriculture. Additional funding for author S.R. assistantship was provided by the Department of Nutrition and Food Science, Texas A&M University, College Station, TX, USA.

Acknowledgments: Authors acknowledge technical suggestions toward research by L. Cisneros-Zevallos, Department of Horticultural Sciences, Texas A&M University, College Station, TX, USA.

Conflicts of Interest: The authors declare no conflict of interest. The funders had no role in the design of the study; in the collection, analyses, or interpretation of data; in the writing of the manuscript; or in the decision to publish the results.

References

1. Centers for Disease Control and Prevention (CDC). Burden of Foodborne Illness: Overview. Available online: http://www.cdc.gov/foodborneburden/estimates-overview.html (accessed on 4 November 2019).
2. Painter, J.A.; Hoekstra, R.M.; Ayers, T.; Tauxe, R.V.; Braden, C.R.; Angulo, F.J.; Griffin, P.M. Attribution of foodborne illnesses, hospitalizations, and deaths to food commodities by using outbreak data, United States, 1998–2008. *Emerg. Infect. Dis.* **2013**, *19*, 407–415. [CrossRef] [PubMed]
3. Scallan, E.; Hoekstra, R.M.; Angulo, F.J.; Tauxe, R.V.; Widdowson, M.A.; Roy, S.L.; Jones, J.L.; Griffin, P.M. Foodborne illness acquired in the United States-major pathogens. *Emerg. Infect. Dis.* **2011**, *17*, 7–15. [CrossRef] [PubMed]
4. Crowe, S.J.; Mahon, B.E.; Vieira, A.R.; Gould, L.H. Vital signs: Multistate foodborne outbreaks—United States, 2010–2014. *Morb. Mortal. Wkly. Rep.* **2015**, *64*, 1221–1225. [CrossRef] [PubMed]
5. CDC. Preliminary Incidence and Trends of Infection with Pathogens Transmitted Commonly through Food—Foodborne Diseases Active Surveillance Network, 10 U.S. Sites, 2006–2017. Available online: https://www.cdc.gov/mmwr/volumes/67/wr/mm6711a3.htm?s_cid=mm6711a3_w (accessed on 5 November 2019).
6. Harris, L.J.; Farber, J.N.; Beuchat, L.R.; Parish, M.E.; Suslow, T.V.; Garrett, E.H.; Busta, F.F. Outbreaks associated with fresh produce: Incidence, growth, and survival of pathogens in fresh and fresh-cut produce. *Compr. Rev. Food Sci. Food Saf.* **2003**, *2*, 78–141. [CrossRef]
7. Centers for Disease Control and Prevention (CDC). Multistate Outbreak of *E. coli* O157:H7 Infections Linked to Fresh Spinach (Final Update). Available online: http://www.cdc.gov/ecoli/2006/september/updates/100606.htm (accessed on 4 November 2019).
8. Centers for Disease Control and Prevention (CDC). Multistate Outbreak of Shiga Toxin-Producing Escherichia coli O157:H7 Infections Linked to Organic Spinach and Spring Mix Blend (Final Update). Available online: https://www.cdc.gov/ecoli/2012/o157h7-11-12/index.html (accessed on 4 November 2019).
9. CDC. Outbreak of *E. coli* Infections Linked to Romaine Lettuce: FINAL UPDATE. Available online: https://www.cdc.gov/ecoli/2018/o157h7-11-18/index.html (accessed on 5 November 2019).
10. Chaves, R.D.; Ruiz Martinez, R.C.; Bortolossi Rezende, A.C.; Rocha, M.D.; Oteiza, J.M.; Sant'Ana, A.S. Salmonella and *Listeria monocytogenes* in ready-to-eat leafy vegetables. In *Food Hygiene and Toxicology in Ready-to-Eat Foods*; Kotzekidou, P., Ed.; Academic Press: New York, NY, USA, 2016; pp. 123–149.
11. Sapers, G.M.; Doyle, M.P. Scope of the produce contamination problem. In *The Produce Contamination Problem*, 2nd ed.; Matthews, K.R., Sapers, G.M., Gerba, C., Eds.; Academic Press: San Diego, CA, USA, 2014; pp. 3–20.
12. Dhifi, W.; Bellili, S.; Jazi, S.; Bahloul, N.; Mnif, W. Essential oils' chemical characterization and investigation of some biological activities: A critical review. *Medicines* **2016**, *3*, 25. [CrossRef] [PubMed]
13. Seow, Y.; Yeo, C.; Chung, H.; Yuk, H.-G. Plant essential oils as active antimicrobial agents. *CRC Crit. Rev. Food Sci. Nutr.* **2014**, *54*, 625–644. [CrossRef]
14. Weiss, J.; Gaysinsky, S.; Davidson, M.; McClements, D.J. Nanostructured encapsulation systems: Food antimicrobials. In *Global Issues in Food Science and Technology*; Barbosa-Cánovas, G., Mortimer, A., Lineback, D., Spiess, W., Buckle, K., Colonna, P., Eds.; Academic Press: San Diego, CA, USA, 2009; pp. 425–479.
15. Bassole, I.H.; Juliani, H.R. Essential oils in combination and their antimicrobial properties. *Molecules* **2012**, *17*, 3989–4006. [CrossRef]
16. National Center for Biotechnology Information. Eugenol, CID=3314. Available online: https://pubchem.ncbi.nlm.nih.gov/compound/Eugenol (accessed on 5 November 2019).
17. Santos, M.I.S.; Martins, S.R.; Veríssimo, C.S.C.; Nunes, M.J.C.; Lima, A.I.G.; Ferreira, R.M.S.B.; Pedroso, L.; Sousa, I.; Ferreira, M.A.S.S. Essential oils as antibacterial agents against food-borne pathogens: Are they really as useful as they are claimed to be? *J. Food Sci. Technol.* **2017**, *54*, 4344–4352. [CrossRef]
18. Park, J.-B.; Kang, J.-H.; Song, K.B. Antibacterial activities of a cinnamon essential oil with cetylpyridinium chloride emulsion against *Escherichia coli* O157:H7 and *Salmonella* Typhimurium in basil leaves. *Food Sci. Biotechnol.* **2018**, *27*, 47–55. [CrossRef]
19. Gaysinsky, S.; Taylor, T.M.; Davidson, P.M.; Bruce, B.D.; Weiss, J. Antimicrobial efficacy of eugenol microemulsions in milk against *Listeria monocytogenes* and *Escherichia coli* O157:H7. *J. Food Prot.* **2007**, *70*, 2631–2637. [CrossRef]

20. Asker, D.; Weiss, J.; McClements, D.J. Formation and stabilization of antimicrobial delivery systems based on electrostatic complexes of cationic-non-ionic mixed micelles and anionic polysaccharides. *J. Agric. Food Chem.* **2011**, *59*, 1041–1049. [CrossRef] [PubMed]
21. Zhang, L.; Critzer, F.; Davidson, P.M.; Zhong, Q. Formulating essential oil microemulsions as washing solutions for organic fresh produce production. *Food Chem.* **2014**, *165*, 113–118. [CrossRef] [PubMed]
22. Kang, J.-H.; Park, S.-J.; Park, J.-B.; Song, K.B. Surfactant type affects the washing effect of cinnamon leaf essential oil emulsion on kale leaves. *Food Chem.* **2019**, *271*, 122–128. [CrossRef] [PubMed]
23. Kang, J.-H.; Song, K.B. Inhibitory effect of plant essential oil nanoemulsions against *Listeria monocytogenes*, *Escherichia coli* O157:H7, and *Salmonella* Typhimurium on red mustard leaves. *Innov. Food Sci. Emerg. Technol.* **2018**, *45*, 447–454. [CrossRef]
24. Gaysinsky, S.; Davidson, P.M.; McClements, D.J.; Weiss, J. Formulation and characterization of phytophenol-carrying antimicrobial microemulsions. *Food Biophys.* **2008**, *3*, 54–65. [CrossRef]
25. Florence, A.T.; Attwood, D. Surfactants. In *Physicochemical Principles of Pharmacy*, 4th ed.; Pharmaceutical Press: Grayslake, IL, USA, 2006; pp. 177–228.
26. McClements, D.J. *Food Emulsions Principles, Practices and Techniques*, 2nd ed.; CRC Press: Boca Raton, FL, USA, 2005; pp. 95–174.
27. Landry, K.S.; Komaiko, J.; Wong, D.E.; Xu, T.; McClements, D.J.; McLandsborough, L. Inactivation of *Salmonella* on sprouting seeds using a spontaneous carvacrol nanoemulsion acidified with organic acids. *J. Food Prot.* **2016**, *79*, 1115–1126. [CrossRef]
28. Ruengvisesh, S.; Oh, J.K.; Kerth, C.R.; Akbulut, M.; Taylor, T.M. Inhibition of bacterial human pathogens on tomato skin surfaces using eugenol-loaded surfactant micelles during refrigerated and abuse storage. *J. Food Saf.* **2019**, *39*, e12598. [CrossRef]
29. Ruengvisesh, S.; Loquercio, A.; Castell-Perez, E.; Taylor, T.M. Inhibition of bacterial pathogens in medium and on spinach leaf surfaces using plant-derived antimicrobials loaded in surfactant micelles. *J. Food Sci.* **2015**, *80*, M2522–M2529. [CrossRef]
30. Calix-Lara, T.F.; Rajendran, M.; Talcott, S.T.; Smith, S.B.; Miller, R.K.; Castillo, A.; Sturino, J.M.; Taylor, T.M. Inhibition of *Escherichia coli* O157:H7 and *Salmonella enterica* on spinach and identification of antimicrobial substances produced by a commercial Lactic Acid Bacteria food safety intervention. *Food Microbiol.* **2014**, *38*, 192–200. [CrossRef]
31. Castillo, A.; Lucia, L.M.; Goodson, K.J.; Savell, J.W.; Acuff, G.R. Use of hot water for beef carcass decontamination. *J. Food Prot.* **1998**, *61*, 19–25. [CrossRef]
32. Devi, K.P.; Nisha, S.A.; Sakthivel, R.; Pandian, S.K. Eugenol (an essential oil of clove) acts as an antibacterial agent against *Salmonella typhi* by disrupting the cellular membrane. *J. Ethnopharmacol.* **2010**, *130*, 107–115. [CrossRef] [PubMed]
33. Burt, S. Essential oils: Their antibacterial properties and potential applications in foods—A review. *Int. J. Food Microbiol.* **2004**, *94*, 223–253. [CrossRef] [PubMed]
34. Yossa, N.; Patel, J.; Millner, P.; Lo, Y.M. Essential oils reduce *Escherichia coli* O157:H7 and *Salmonella* on spinach leaves. *J. Food Prot.* **2012**, *75*, 488–496. [CrossRef] [PubMed]
35. Singer, M.M.; Tjeerdema, R.S. Fate and effects of the surfactant sodium dodecyl sulfate. *Rev. Environ. Contam. Toxicol.* **1993**, *133*, 95–149. [PubMed]
36. Kabara, J.J.; Swieczkowski, D.M.; Conley, A.J.; Truant, J.P. Fatty acids and derivatives as antimicrobial agents. *Antimicrob. Agents Chemother.* **1972**, *2*, 23–28. [CrossRef] [PubMed]
37. Elliot, R.; Singhal, N.; Swift, S. Surfactants and bacterial bioremediation of polycyclic aromatic hydrocarbon contaminated soil—Unlocking the targets. *Crit. Rev. Environ. Sci. Technol.* **2010**, *41*, 78–124. [CrossRef]
38. Arachea, B.T.; Sun, Z.; Potente, N.; Malik, R.; Isailovic, D.; Viola, R.E. Detergent selection for enhanced extraction of membrane proteins. *Protein Expr. Purif.* **2012**, *86*, 12–20. [CrossRef]
39. Terjung, N.; Löffler, M.; Gibis, M.; Hinrichs, J.; Weiss, J. Influence of droplet size on the efficacy of oil-in-water emulsions loaded with phenolic antimicrobials. *Food Funct.* **2012**, *3*, 290–301. [CrossRef]
40. Perez-Conesa, D.; Cao, J.; Chen, L.; McLandsborough, L.; Weiss, J. Inactivation of *Listeria monocytogenes* and *Escherichia coli* O157:H7 biofilms by micelle-encapsulated eugenol and carvacrol. *J. Food Prot.* **2011**, *74*, 55–62. [CrossRef]
41. Ben Arfa, A.; Combes, S.; Preziosi-Belloy, L.; Gontard, N.; Chalier, P. Antimicrobial activity of carvacrol related to its chemical structure. *Lett. Appl. Microbiol.* **2006**, *43*, 149–154. [CrossRef]

42. Zhang, M.; Oh, J.K.; Cisneros-Zevallos, L.; Akbulut, M. Bactericidal effects of nonthermal low-pressure oxygen plasma on *S. typhimurium* LT2 attached to fresh produce surfaces. *J. Food Eng.* **2013**, *119*, 425–432.

43. Zhang, M.; Oh, J.K.; Huang, S.-Y.; Lin, Y.-R.; Liu, Y.; Mannan, M.S.; Cisneros-Zevallos, L.; Akbulut, M. Priming with nano-aerosolized water and sequential dip-washing with hydrogen peroxide: An efficient sanitization method to inactivate *Salmonella* Typhimurium LT2 on spinach. *J. Food Eng.* **2015**, *161*, 8–15. [CrossRef]

44. Wang, H.; Zhou, B.; Feng, H. Surface characteristics of fresh produce and their impact on attachment and removal of human pathogens on produce surfaces. In *Decontamination of Fresh and Minimally Processed Produce*, 1st ed.; Gómez-López, V.M., Ed.; John Wiley & Sons, Inc.: Ames, IA, USA, 2012; pp. 43–57.

45. Lund, B.M. Ecosystems in vegetable foods. *J. Appl. Bacteriol.* **1992**, *73*, 115s–126s. [CrossRef] [PubMed]

46. Heredia, A.; Dominguez, E. The plant cuticle: A complex lipid barrier between the plant and the environment. An overview. In *Counteraction to Chemical and Biological Terrorism in East European Countries*; Dishovsky, C., Pivovarov, A., Eds.; Springer Netherlends: Dordrecht, The Netherlands, 2009; pp. 109–116.

47. Chaidez, C.; Campo, N.C.-d.; Heredia, J.B.; Contreras-Angulo, L.; González–Aguilar, G.; Ayala–Zavala, J.F. Chlorine. In *Decontamination of Fresh and Minimally Processed Produce*; Gomez-Lopez, V.M., Ed.; Wiley-Blackwell: Ames, IA, USA, 2012; pp. 121–133.

48. Zhang, S.; Farber, J.M. The effects of various disinfectants against *Listeria monocytogenes* on fresh-cut vegetables. *Food Microbiol.* **1996**, *13*, 311–321. [CrossRef]

49. Pan, X.; Nakano, H. Effects of chlorine-based antimicrobial treatments on the microbiological qualities of selected leafy vegetables and wash water. *Food Sci. Technol. Res.* **2014**, *20*, 765–774. [CrossRef]

50. Neo, S.Y.; Lim, P.Y.; Phua, L.K.; Khoo, G.H.; Kim, S.-J.; Lee, S.-C.; Yuk, H.-G. Efficacy of chlorine and peroxyacetic acid on reduction of natural microflora, *Escherichia coli* O157:H7, *Listeria monocytogenes* and *Salmonella* spp. on mung bean sprouts. *Food Microbiol.* **2013**, *36*, 475–480. [CrossRef]

51. Vandekinderen, I.; Devlieghere, F.; De Meulenaer, B.; Ragaert, P.; Van Camp, J. Optimization and evaluation of a decontamination step with peroxyacetic acid for fresh-cut produce. *Food Microbiol.* **2009**, *26*, 882–888. [CrossRef]

52. Erkmen, O. Antimicrobial effects of hypochlorite on *Escherichia coli* in water and selected vegetables. *Foodborne Pathog. Dis.* **2010**, *7*, 953–958. [CrossRef]

53. Brecht, J. Chlorine use in produce packing lines. *Am. Veg. Grower* **2002**, *50*, 27.

54. Flahaut, S.; Frere, J.; Boutibonnes, P.; Auffray, Y. Comparison of the bile salts and sodium dodecyl sulfate stress responses in *Enterococcus faecalis*. *Appl. Environ. Microbiol.* **1996**, *62*, 2416–2420.

55. Kramer, V.C.; Nickerson, K.W.; Hamlett, N.V.; O'Hara, C. Prevalence of extreme detergent resistance among the *Enterobacteriaceae*. *Can. J. Microbiol.* **1984**, *30*, 711–713. [CrossRef] [PubMed]

56. Thomas, O.R.; White, G.F. Metabolic pathway for the biodegradation of sodium dodecyl sulfate by *Pseudomonas* sp. C12B. *Biotechnol. Appl. Biochem.* **1989**, *11*, 318–327. [PubMed]

57. Rodriguez, S.B.; Mahoney, N.E. Inhibition of aflatoxin production by surfactants. *Appl. Environ. Microbiol.* **1994**, *60*, 106–110. [PubMed]

58. Wu, D.; Lu, J.; Zhong, S.; Schwarz, P.; Chen, B.; Rao, J. Influence of nonionic and ionic surfactants on the antifungal and mycotoxin inhibitory efficacy of cinnamon oil nanoemulsions. *Foods Funct.* **2019**, *10*, 2817–2827. [CrossRef]

59. Park, J.-B.; Kang, J.-H.; Song, K.B. Clove bud essential oil emulsion containing benzethonium chloride inactivates *Salmonella* Typhimurium and *Listeria monocytogenes* on fresh-cut pak choi during modified atmosphere storage. *Food Cont.* **2019**, *100*, 17–23. [CrossRef]

60. Parish, M.E.; Beuchat, L.R.; Suslow, T.V.; Harris, L.J.; Garrett, E.H.; Farber, J.N.; Busta, F.F. Methods to reduce/eliminate pathogens from fresh and fresh-cut produce. *Compr. Rev. Food Sci. Food Saf.* **2003**, *2*, 161–173. [CrossRef]

 *foods*

Article

Antimicrobial Properties of Encapsulated Antimicrobial Natural Plant Products for Ready-to-Eat Carrots

Yosra Ben-Fadhel, Behnoush Maherani, Melinda Aragones and Monique Lacroix *

Research Laboratories in Sciences Applied to Food, Canadian Irradiation Center, INRS–Armand Frappier, Health and Biotechnology Center, Institute of Nutraceutical and Functionals Foods, 531 Boulevard des Prairies, Laval, QC H7V 1B7, Canada; Yosra.bf@gmail.com (Y.B.-F.); bmaherani@gmail.com (B.M.); melindaragones@gmail.com (M.A.)
* Correspondence: Monique.Lacroix@iaf.inrs.ca; Tel.: +1-450-687-5010 (ext. 4489); Fax: +1-450-686-5501

Received: 24 September 2019; Accepted: 28 October 2019; Published: 1 November 2019

Abstract: The antimicrobial activity of natural antimicrobials (fruit extracts, essential oils and derivates), was assessed against six bacteria species (*E. coli* O157:H7, *L. monocytogenes*, *S.* Typhimurium, *B. subtilis*, *E. faecium* and *S. aureus*), two molds (*A. flavus* and *P. chrysogenum*) and a yeast (*C. albicans*) using disk diffusion method. Then, the antimicrobial compounds having high inhibitory capacity were evaluated for the determination of their minimum inhibitory, bactericidal and fungicidal concentration (MIC, MBC and MFC respectively). Total phenols and flavonoids content, radical scavenging activity and ferric reducing antioxidant power of selected compounds were also evaluated. Based on in vitro assays, five antimicrobial compounds were selected for their lowest effective concentration. Results showed that, most of these antimicrobial compounds had a high concentration of total phenols and flavonoids and a good anti-oxidant and anti-radical activity. In situ study showed that natural antimicrobials mix, applied on the carrot surface, reduced significantly the count of the initial mesophilic total flora (TMF), molds and yeasts and allowed an extension of the shelf-life of carrots by two days as compared to the control. However, the chemical treatment (mix of peroxyacetic acid and hydrogen peroxide) showed antifungal activity and a slight reduction of TMF.

Keywords: natural antimicrobials; encapsulation; shelf-life; microbiological quality

---

1. Introduction

Plants, spices, fruits and vegetable extracts have been exploited since antiquity for their aromas, coloring ability, antioxidant and antimicrobial properties [1]. However, at the beginning of the 19th century, a rapid rise of the use of chemical additives has been observed. Among the chemical additives used in food, nitrites, sulfide dioxide, sulfites, parabens, peroxyacetic acid and hydrogen peroxide are the best known. However, these additives are controversial as many have shown potential health risks, mainly carcinogenic effects, irritation and the appearance of resistant strains [1,2]. There is, therefore, a growing interest in identifying natural antimicrobial extracts which have the advantage of being effective with much less toxic and less allergenic effects. Natural antimicrobial extracts have demonstrated various antiviral, antifungal, antibacterial, anti-parasitic, antioxidant, and even insecticidal activities [3,4]. For example, it was demonstrated that garlic juice and tea extract could inhibit bacteria even those resistant to antibiotics, such as ciprofloxacin, methicillin and vancomycin [5]. In addition to their antimicrobial properties, natural antimicrobials often have functional properties already used as anticancer, radioprotective and hypoglycemic [1]. For example, it was observed that lime juice extract can inhibit the growth of pancreatic cancer cells [6]. Antioxidant properties have also been reported for certain plant extracts like garlic and onion. Antioxidant properties can help in the prevention of meat discoloration, the preservation of vitamin content (B_1 and B_2) and the prevention of lipid oxidation [7]. Some of the active compounds present in plants, herbs, spices, fruits and vegetables are known as secondary metabolites. The main groups of compounds responsible for the antimicrobial activity of plants extracts include phenols (phenolic acids, flavonoids: i.e., flavonols, tannins), quinones, saponins, coumarins, terpenoids and alkaloids [8]. Natural extracts under the form of essential oils are rich in flavonoids, terpenes, terpenoids and aromatic and aliphatic constituents and could be obtained by hydro or steam distillation, solvent extraction, ultrasound, microwave, ohmic heating, supercritical CO_2 extraction or pulsed electric field [3]. Most of their active compounds are found in leaf extract (i.e., rosemary, sage), flowers and flower buds (i.e., cloves), bulbs (i.e., garlic, onion), rhizomes (i.e., asafetida) and fruits (i.e., pepper) [9]. Depending on plant type and bacterial strain, essential oil derivatives could have a high antibacterial activity. Bertoli, et al. [10] reported that 60% of plant essential oils have antifungal activity. Their mode of action on microorganisms has been the object of several studies and demonstrated that essential oils, due to their hydrophobic nature, are able to react with the lipid layer of the bacterial cell membrane, thereby increasing the permeability of membranes inducing leakage of ions and cell contents, lysis and death of bacteria [11]. Their efficiency against several bacteria, molds and yeasts made of the essential oils a good candidate for food industry to insure food safety. Unfortunately, their use in food industry is restricted by a low dose due to their strong sensorial impact and toxicity [12,13]. On the other hand, the hydrophobic nature of essential oils affects their homogeneity and bioavailability on the food surface. Their encapsulation in a more suitable matrix could help to avoid this inconvenient and can prevent volatilization and oxidation of their active compounds. Moreover, encapsulation could mask the strong aroma and prevent the degradation of the active compounds [14].

Carrots have been implicated in several outbreaks in England and Wales during 1992–2005, in the United States during 1973–1997 [15,16] and in 2004 [17]. The most frequent pathogens involved in these outbreaks are *E. coli* O6 (strain that produced the heat-stable and heat labile toxins (O6: NM LT ST), VTEC, *Yersinia pseudotuberculosis* which caused gastrointestinal illness and erythema nodosum among schoolchildren in Finland and *Shigella sonnei* [15,18]. Others studies demonstrated the possibility of growth of *Salmonella spp.* and *Listeria monocytogenes* on carrots [19]. The fungal strains of *Alternaria, Rhizopus, Aspergillus, Stemphylium* and *Botrytis* were also found to contaminate carrots [20,21]. The mechanism of contamination of carrots remains not well known. Monaghan and Hutchison [22] reported inadequate hand hygiene in the field can transfer bacterial contamination to hand-harvested carrots. Direct contact with wildlife feces during storage and cross-contamination of the equipment during washing and peeling could also be contributing factors [16].

The main objective of this study was to assess the antimicrobial activities of 17 antimicrobial agents against nine different microorganisms (Gram negative, Gram positive, molds and yeast) that could affect food products in order to select the most efficient antimicrobial extracts. The total phenols and flavonoids content, the anti-radical and antioxidant activity were assessed for each selected extract. In this study, a strategy was developed in order to reduce the efficient dose of natural antimicrobial extracts by the development of formulation containing a mixture of natural extracts encapsulated in o/w emulsion which could act in synergy. Then, the antimicrobial efficiency of the antimicrobial-loaded emulsion was tested in situ onto pre-cut carrots. Finally, sensorial evaluation was done on the treated carrots.

2. Materials and Methods

2.1. Antimicrobial Extracts

Biosecur F440D (33–39%) was provided by Biosecur Lab, Inc. (Mont St-Hilaire, Québec, QC, Canada). Citral was provided from BSA, Inc. (BSA Ingredients s.e.c/l.p., Montreal, QC, Canada). Cranberry juice (*Vaccinium macrocarpon*) was provided by Atoka Cranberries, Inc. (Manseau, QC, Canada) and was stored at −80 °C until used. Fourteen essential oils from spices, fruits and plants were bought from Biolonreco, Inc. (Dorval, QC, Canada) and their main constituents are presented in Table 1. Biosecur F440D, citral and essential oils were stored at 4 °C.

Table 1. List of organic essential oils (EO) and their composition.

Common Name	Botanic Name	Part	Compositions (%) *
Bergamote EO	*Citrus bergamia*	Zest	Limonene (36.2), Linalyle acetate (29.7), linalool (13.2), γ-terpinene (6.8), β-pinene (5.4)
Pan tropical EO	*Cinnamomum verrum*	Peel	E-cinnamaldehyde (55.1), cinnamyl acetate (9.6), β-caryophyllene (4.0)
Citrus EO	*Cymbopogon winterianus*	Aerial part	Citronellal (35.4), geraniol (20.1), Citronellol (12.2), elemol (4.6), Limonene (3.0), citronellyl acetate (2.9), germacrene D (2.7), geranyl acetate de (2.5), linalool (0.6)
Ginger EO	*Zingiber officinalis*	Rhizome	α-zingiberene (25.4), β-sesquiphellandrene + α-curcumene (13.9), Camphene (10.5), β-phellandrene + 1, 8-cineole (8.3), β-bisabolene + β-selinene (7.7), E,E-α-farnesene (4.2), α-pinene (3.3)
Asian EO	*Cymbopogon flexuosus*	Herb	Geranial (39.1), neral (31.6), geraniol (6.7), geranyl acetate (3.7)
Marjolaine shells EO	*Origanum majorana*	Flower top	Terpinene-4-ol (28.0), γ-terpinene (15.5), α-terpinene (9.5), Cis-thuyanol (7.3), α-terpineol (3.7)
Peppermint EO	*Mentha x piperita*	Aerial part	Menthol (30.6), menthone (29.3), 1,8-cineole + β-phellandrene (5.2), menthyl acetate (4.5), neomenthol (3.1), Isomenthone (4.4), menthofurane (4.2), Limonene (2.4)
Myrte cineole EO	*Myrtus communis*	leaf	α-pinene (51.5), 1,8-cineole (23.9), Limonene (10.4), Linalool (3.0)
Sweet orange EO	*Citrus sinensis*	Zest	Limonene (94.8)
Tea tree EO	*Melaleuca alternifolia*	Leaf	Terpinene-4-ol (37.6), γ-terpinene (21.1), α-terpinene (10.1), Terpinolene (4.8), 1,8-cineole + β-phellandrene (4.2), α-pinene (2.6), α-terpineol (2.5)
Mediterranean EO	*Origanum compactum*	Flower top	Carvacrol (46.1), thymol (17.6), γ-terpinene+ trans-β-ocimene (14.8), p-cymene (8.5)
Thyme leaf savory EO	*Thymus satureioides*	Flower top	Borneol (27.0), α-terpineol (11.9), camphene (10.5), α-pinene + α-thuyene (6.5), β-caryophyllene (5.5), Carvacrol (5.3), p-cymene (3.9), Linalol (3.7), Terpinene-4-ol + methyl carvacrol ether (2.9), 1,8-cineole + β-phellandrene (2.9), Thymol (2.8)
Cloves EO	*Eugenia caryophyllus*	Floral button	Eugenol (81.8), Eugenyl acetate (12.9), β-caryophyllene (3.4)
Thyme thymol EO	*Thymus vulgaris CT6*	Flower top	Thymol (46.6), p-cymene (16.9), γ-terpinene (9.3), Linalool (4.1), Carvacrol (3.5)

* Composition was provided by Biolonreco, Inc. and was determined by CPG-SM Hewlett Packard /CPG- FID; Column: HP Innowax 60-0.5-0.25; Carrier gas Helium: 22 psi.

2.2. Preparation of Bacterial Cultures

Six bacterial strains, four Gram positive: *Listeria monocytogenes* HPB 2812 (Health Canada, Health Products and Food Branch, Ottawa, Canada), *Staphylococcus aureus* ATCC 29213 (American Type Culture Collection, Rockville, MD, USA), *Enterococcus faecium* ATCC 19434 (American Type Culture Collection, Rockville, MD, USA) and *Bacillus subtilis* ATCC 23857 (INRS-Institut Armand-Frappier, Laval, QC, Canada), and two Gram negative: *Escherichia coli* O157:H7 (EDL 933, provided by Pr. Charles Dozois) and *Salmonella* Typhimurium SL 1344 (INRS-Institut Armand-Frappier, Laval, QC, Canada) were used as target bacteria in antimicrobial tests. *Aspergillus flavus* (INRS-Institut Armand-Frappier, Laval, QC, Canada) and *Penicillium chrysogenum* (INRS-Institut Armand-Frappier, Laval, QC, Canada) were used as fungal strains and *Candida albicans* ATCC10231 (INRS-Institut Armand-Frappier, Laval, QC, Canada) as yeast. All the bacteria were stored at −80 °C in Tryptic Soy Broth medium (TSB; BD, Franklin Lakes, NJ, USA) containing glycerol (20% *v/v*). Before each experiment, bacterial stock cultures were propagated through two consecutives 24 h growth cycles in TSB at 37 °C to reach the concentration of approximately 10^9 CFU/mL. The grown cultures were then diluted in sterile peptone water 0.1% (Alpha Biosciences, Inc., Baltimore, MD, USA) to obtain a working culture of approximately 10^6 CFU/mL.

For fungal evaluation, *A. flavus and P. chrysogenum* were propagated through 72 h growth cycle on potato dextrose agar (PDA, Difco, Becton Dickinson) at 28 ± 2 °C. Colonies were isolated from the agar media using sterile platinum loop, suspended in sterile peptone water, and filtrated through sterile cell strainer (Fisher scientific, Ottawa, ON, Canada). *C. albicans* was inoculated in potato dextrose broth (PDB, Difco, Becton Dickinson) for 24 h at 28 °C. The filtrate was adjusted to 10^6 CFU/mL using a microscope before dilution to reach approximately 10^6 CFU/mL for the disk diffusion agar and the minimum inhibitory, bactericidal and fungicidal concentration (MIC, MBC and MFC, respectively) determination [23].

2.3. Preliminary Study

First, 100 µL of the tested microorganisms 10^6 CFU/mL were seeded on sterile Petri dishes containing Muller Hinton Agar (MHA, BD, Franklin Lakes, NJ, USA). Then, 5 µL of each pure antimicrobial compounds were deposited on the surface of a sterile 6-mm filter disk. A negative control was used by deposing 5 µL of sterile water on the surface of the disk. All plates were sealed with parafilm to avoid evaporation and incubated for 24 h at 37 °C for bacteria and for 48 h to 72 h at 28 °C for molds and yeasts followed by the measurement of the diameter zone of the inhibition expressed in mm. On the basis of the disk diffusion results, the most efficient antimicrobial compounds have been selected to determine their MIC, MBC and MFC, their total phenols and flavonoids content and their antioxidant and anti-radical properties and to evaluate the in situ antimicrobial efficiency of the mixture on pre-cut carrot surface.

2.4. Antimicrobial Efficiency

The minimum inhibitory concentration (MIC) and the minimum bactericidal and fungicidal concentration (MBC and MFC) were determined on the emulsion as an encapsulation form composed of essential oils 2.5% (*w/w*), tween 80 2.5% (*w/w*) and 95% (*w/w*) distilled water. The mixture was homogenized by vortex for 1 min and by Ultra-Turrax (IKA T25 digital Ultra-Turrax disperser, IKA Works Inc., Wilmington, NC, USA) for 1 min at 15,000 rpm. Because of its water solubility, Biosecur F440D was prepared at 0.4% (*w/w*) in distilled water. All the prepared solutions were then filtered through 0.2 µm syringe filter.

The MIC value of each antimicrobial compound was determined in sterilized flat-bottomed 96-well microplate according to the serial microdilution method [23]. Briefly, serial dilutions (200:100 µL) of the antimicrobial compounds were made in Mueller Hinton Broth (MHB, Difco, Becton Dickinson) for bacteria and in Potato Dextrose Broth (PDB, Difco, Becton Dickinson) for molds and yeast and dispensed into 96-well microplates to obtain a dilutions range of 2000–15 ppm for Biosecur F440D and 12,500–145 ppm for essential oils. Then, a volume of 15 µL of bacteria, molds and yeast suspension (10^6 CFU/mL) was added. Two control samples were evaluated; the 1st was to control the growth of the evaluated microorganisms where a volume of 100 µL of MHB/PDB was mixed to 15 µL of the selected microorganism. The 2nd control was the blank where a volume of 15 µL of distilled water was added to 100 µL of each antimicrobial dilution. The MIC of tween 80 at 2.5% was also evaluated. The final volume in all the wells was 115 µL. Microplates were sealed with acetate foil to avoid evaporation and then incubated on a shaker (Forma Scientific. Inc., Marietta, OH, USA) at 80 rpm at 37 °C for 24 h and 28 °C for 48 h respectively for bacteria and molds/yeasts to insure a better homogenization. The absorbance was then measured at 595 nm in an absorbance microplate reader (BioTek ELx800®, BioTek Instruments Inc., Winooski, VT, USA). The MIC is considered to be the lowest concentration of the antimicrobial compounds that completely inhibits bacterial and fungal strain growth by showing equal absorbance as blank. Afterwards, to assess the MBC and the MFC, 5 µL of each well were taken from the microplate and were deposit on a Petri dish containing Tryptic Soy Agar (TSA) for bacteria and PDA for molds and yeasts. Finally, Petri dishes were incubated for 24 h at 37 °C for bacteria or 48–72 h at 28 °C for molds and yeasts respectively. The MBC and the MFC were respectively determined as the concentration where no colony was detected.

2.5. Total Phenol Determination

The total phenol content was carried out using a Folin–Ciocalteu colorimetric method according to Dewanto, et al. [24]. Pure essential oils and Biosecur F440D were diluted in anhydrous ethanol and water respectively to obtain suitable dilution within the standard curve ranges of 0–200 μg of gallic acid/mL. Measurements were done at 760 nm versus the blank prepared similarly with water or ethanol. All values were expressed as mean (milligrams of gallic acid equivalents per g of antimicrobial compounds).

2.6. Radical Scavenging Activity (DPPH)

The antioxidant activity of the antimicrobial compounds was determined using 2,2-diphenyl-1-picrylhydrazyl (DPPH) as a free radical [25]. The reaction for scavenging DPPH radicals was performed in polypropylene tubes at room temperature. One milliliter of a 40 μM of methanolic solution of DPPH was added to 25 μL of diluted antimicrobial compounds. The mixture was shaken vigorously and left for 90 min. The absorbance of the resulting solution was measured at 517 nm. Anhydrous methanol was used as a blank solution, and DPPH solution without any sample served as control. The Trolox equivalent antioxidant capacity (TEAC) values were calculated from the equation determined from linear regression after plotting known solutions of Trolox or ascorbic acid with different concentrations (0–1 mM). The DPPH inhibition percentage was calculated using Equation (1) and the antiradical activity was expressed as mM of Trolox or ascorbic acid.

$$\text{Radical scavenging activity (\%)} = (\text{Control OD} - \text{Sample OD}) \times 100/\text{Control OD} \tag{1}$$

2.7. Ferric-Reducing Antioxidant Power (FRAP)

Total antioxidant activity was estimated by FRAP assays [26]. Three aqueous stock solutions containing 0.1 M acetate buffer (pH 3.6), 10 mM TPTZ [2,4,6-tris(2-pyridyl)-1,3,5-triazine] in 40 mM hydrochloric acid solution, and 20 mM ferric chloride were prepared and stored under dark conditions at 4 °C. Stock solutions were combined (10:1:1, *v/v/v*) to form the FRAP reagent just prior to analysis. FRAP reagent was heated in a water bath for 30 min at 37–40 °C. For each assay, 2.8 mL of FRAP reagent and 200 μL of diluted sample were mixed. After 10 min, the absorbance of the reaction mixture was determined at 593 nm. The standard curve was prepared with ascorbic acid (0–2 mM). Results were expressed as equivalent μM of ascorbic acid per gram of antimicrobial.

2.8. Determination of Total Flavonoids Content

Total flavonoids content was determined by using a colorimetric method [24]. Briefly, 0.25 mL of diluted antimicrobial compounds or (+) catechin standard solution was mixed with 1.25 mL of distilled water followed by the addition of 75 μL of a 5% $NaNO_2$ solution. After 6 min, 150 μL of a 10% $AlCl_3$ $6H_2O$ solution was added and allowed to stand for 5 min at room temperature before 0.5 mL of 1 M NaOH was added. The mixture was brought to 2.5 mL with distilled water and mixed well. The absorbance was measured immediately against the blank at 510 nm in comparison with the standards prepared similarly with known (+)-catechin concentrations. The results were expressed as mean (micrograms of catechin equivalents per gram of antimicrobial).

2.9. In Situ Test on Pre-Cut Carrots

2.9.1. Antimicrobial Loaded Emulsion

To encapsulate the natural antimicrobial compounds, an emulsion was prepared by mixing Biosecur F440D® to citrus, Asian, Mediterranean and pan tropically essential oils composed mainly with lemongrass, oregano and cinnamon essential oils respectively [27]. Sunflower lecithin (HLB 7) and sucrose monopalmitate (HLB 18) were used as emulsifiers (180 ppm) to obtain a stable emulsion with a HLB = 12 and an oil phase: emulsifier's ratio of 1:1. The emulsion was magnetically homogenized then mixed with Ultra-Turrax at 10,000 rpm for 1 min.

2.9.2. Samples Preparation

Freeze pre-cut carrots were provided by Bonduelle, Inc. (Sainte-Martine, Canada). Carrot was washed with water then divided into 3 groups: untreated carrots (control), treated carrots with antimicrobial formulation-loaded emulsion (containing a mixture of Biosecur F440D extract and Asian, Mediterranean, citrus and pan tropical essential oils) and treated carrots with commercial chemical antimicrobial (0.03% of Tsunami: a mix of 15.2% of peroxyacetic acid and 11.2% of hydrogen peroxide). For treated samples, carrots were dipped in the antimicrobial solution for 30 s, kept drying under laminar flow hood for 15 min to discard the exceeding solution. Samples were then stored in Whirl-Pak™ Sterile Filter Bags (Nasco, Whilpack®, Fort Atkinson, WI, USA) at 4 °C for 8 days (20 g per bag). Emulsifiers were considered too low to not affect the antimicrobial activity of the emulsion.

2.9.3. Shelf-life Estimation

The total mesophilic bacterial count (TMF) was evaluated during 8 days of storage at 4 °C. The TMF was selected based on previous studies, as TMF contains a complex mix of different autochthonous microorganisms including *Candida* spp. [28], *Entrobacter* spp., *Salmonella* spp. and *S. aureus* [29]. To estimate the initial count of TMF, a bacterial analysis was carried out for the control on day 0. During storage, all treatments and control were evaluated on day 1, 3, 6 and 8. On each day of analysis, 60 g of 0.1% (*w/v*) peptone water (Alpha Biosciences Inc., Baltimore, MD, USA) were added to filter bag containing 20 g of carrots previously prepared. The carrot samples were mixed during 2 min at high speed (260 rpm) in a Lab-blender 400 stomacher (Laboratory Equipment, London, UK), then 100 µL were seeded on TSA for TMF evaluation and on PDA with chloramphenicol for molds and yeasts evaluation. Plates were incubated at 37 °C and 28 °C during 48–72 h for TMF and molds and yeast respectively. Results were expressed as bacterial count and fungal count (log CFU/g) during storage at 4 °C.

Shelf-life limit was considered at the limit of unacceptability, when TMF count and the total molds and yeasts reached the current authorities regulation level of 10^7 CFU/g and 10^4 CFU/g, respectively [30]. Equation (2) was used to describe the growth of bacteria (*Y*) over time during the exponential phase.

$$Y = X \exp{(\mu t)} \tag{2}$$

where *X* is the initial population, μ the growth rate of TMF (Ln CFU/g/day) and *t* the number of storage days.

2.10. Sensory Evaluation

In order to evaluate the effect of the developed antimicrobial formulation on the sensory properties of carrots, the sensory evaluation, was carried out by comparing the control to treated carrots with the developed antimicrobial formulation. The sensorial evaluation of treated and untreated carrots was done using a hedonic test [31]. The level of appreciation was determined using nine points (1 = dislike extremely; 5 = neither like nor dislike; 9 = like extremely). Samples were treated with the antimicrobial formulation-loaded emulsion (containing a mixture of Biosecur F440D and Asian, Mediterranean, citrus and pan tropical essential oils) and kept to dry. The sensorial evaluation was done by a panel of 24 untrained people after 1 day of the treatment application. For each panelist, 3 pieces of carrots were served to evaluate the flavor, the odor and the global appreciation. Treated samples consisted of carrot samples coated with the antimicrobial formulation.

2.11. Statistical Analysis

Each experiment was done in triplicate ($n = 3$). For each replicate 2 samples were analyzed. Analysis of variance (ANOVA), Duncan's multiple range tests for equal variances and Tamhane's test for unequal variances were performed for statistical analysis using SPSS 18.0 software (SPSS Inc., Chicago, IL, USA). Differences between means were considered significant when the confidence interval was lower than 5% ($p \leq 0.05$).

3. Results

3.1. Preliminary Study

Results of the disk diffusion method (Table 2) showed that from 17 evaluated antimicrobial compounds, five antimicrobial agents that showed high inhibitory diameter against all the tested microorganisms were identified. Based on their bioactivity, these antimicrobial compounds could be also grouped into four distinctive groups: Group 1 contains pan tropical, Mediterranean and thyme essential oils which have a large spectral activity against bacteria, yeast and molds with an inhibitory diameter ≥23.7 mm. Their effectiveness was higher against yeast and molds with an inhibitory diameter between 38.3 and 80 mm for *C. albicans*, *P. chrysogenum*, and *A. flavus* as compared to an inhibitory diameter between 23.7 and 44.3 mm for *S.* Typhimurium, *L. monocytogenes*, *B. subtilis*, *E. coli*, *S. aureus* and *E. faecium*. Group 2 contains Asian, cloves, citrus and thyme savory leaves essential oils and citral and was very efficient to inhibit molds and yeasts. Asian essential oil and citral showed an average antibacterial activity against six bacterial strains with an inhibitory diameter ≤22.5 mm and an antifungal activity with an inhibitory diameter between 23.0 mm and 80.0 mm. Citrus and cloves essential oils were efficient to reduce *B. subtilis*, *S. aureus*, *C. albicans*, *A. flavus* and *P. chrysogenum* showing an inhibitory diameter between 22.0 and 68.7 mm. Otherwise, they showed above-average efficiency against the other microorganisms. Group 3 contains Biosecur F440D which possesses a good antimicrobial activity against all the microorganisms. The inhibitory diameter of Biosecur F440D varied from 12.3 mm to 25.4 mm for *E. faecium* and *S. aureus*, respectively, showing a medium antimicrobial activity whether against bacteria molds or yeast. Biosecur F440D was more efficient to inhibit bacteria, molds and yeasts than cranberry juice. Group 4 contains bergamot, marjoram, peppermint, sweet orange, tea tree, myrtle and ginger essential oils and cranberry juice, and showed a very low antimicrobial activity. Pepper mint essential oil was efficient only to inhibit the growth of *C. albicans* showing an inhibitory diameter of 31.3 mm. Results showed that essential oils of bergamot, sweet marjoram, sweet orange, myrtle and ginger with an inhibitory diameter ≤18.3 mm showed a very low antimicrobial activity against bacteria, molds and yeasts.

Based on these results, five antimicrobial extracts were selected to characterize their MIC, MBC, MFC and to determine their total phenols and flavonoids composition and their antiradical and antioxidant properties: citrus and Asian essential oils for their antifungal activity, pan tropical and Mediterranean essential oils for their large spectral activity and Biosecur F440D for its good activity and its hydrophilic properties.

Table 2. Inhibitory diameter of antimicrobials extracts against tested microorganisms ($n = 3$).

| | | Inhibition Diameter: Mean ± std.dev (mm) | | | | | | | | |
| | | Gram Positive | | | | Gram Negative | | Yeast | Molds | |
		L. monocytogenes	*B. subtilis*	*E. faecium*	*S. aureus*	*S.* Typhimurium	*E. coli*	*C. albicans*	*A. flavus*	*P. chrysogenum*
1	**Biosecur F440D**	16.6 ± 1.9	18.9 ± 1.0	12.3 ± 0.7	25.4 ± 2.4	12.5 ± 1.1	13.7 ± 0.9	22.8 ± 1.6	14.1 ± 0.7	13.6 ± 3.0
2	**Cranberry juice**	8.7 ± 0.9	9.3 ± 1.6	6.0 ± 0.0	6.0 ± 0.0	6.0 ± 0.0	7.0 ± 1.2	6.0 ± 0.0	6.0 ± 0.0	6.0 ± 0.0
3	**Bergamote EO**	6.0 ± 0.0	13.7 ± 1.2	6.0 ± 0.0	6.0 ± 0.0	6.0 ± 0.0	6.0 ± 0.0	10.7 ± 0.4	6.0 ± 0.0	8.5 ± 0.3
4	**Citrus EO** [*]	8.4 ± 0.6	68.7 ± 4.9	13.9 ± 0.4	35.6 ± 2.8	13.8 ± 1.0	14.8 ± 2.0	36.2 ± 5.5	22.4 ± 5.2	45.7 ± 4.0
5	**Cloves EO**	14.8 ± 1.2	24.9 ± 3.4	19.0 ± 1.9	22.0 ± 2.5	20.0 ± 0.8	20.1 ± 3.3	27.2 ± 0.7	38.6 ± 1.2	41.7 ± 1.2
6	**Marjoram EO**	13.9 ± 0.9	17.4 ± 3.5	16.1 ± 0.7	17.1 ± 0.8	17.3 ± 1.1	19.1 ± 2.4	13.0 ± 0.2	6.0 ± 0.0	11.2 ± 0.4
7	**Pepper menthe EO**	7.8 ± 0.4	19.1 ± 3.4	15.5 ± 0.9	18.9 ± 3.3	13.3 ± 0.4	14.4 ± 1.9	31.3 ± 1.8	6.0 ± 0.0	9.9 ± 1.1
8	**Sweet orange EO**	6.0 ± 0.0	14.0 ± 1.5	6.0 ± 0.0	6.0 ± 0.0	6.0 ± 0.0	6.0 ± 0.0	11.1 ± 0.9	6.0 ± 0.0	8.5 ± 0.1
9	**Mediterranean EO**	23.8 ± 0.5	44.3 ± 4.1	33.9 ± 4.4	42.7 ± 4.0	28.5 ± 3.6	27.2 ± 2.5	52.0 ± 1.6	59.0 ± 2.6	80.0 ± 0.0
10	**Tea tree EO**	12.2 ± 0.4	17.2 ± 1.6	16.5 ± 0.8	18.3 ± 3.9	16.7 ± 2.2	17.3 ± 1.7	12.3 ± 1.3	6.0 ± 0.0	9.5 ± 0.4
11	**Thyme savory leaves EO**	11.3 ± 0.3	21.3 ± 3.7	16.0 ± 0.9	27.6 ± 2.6	17.7 ± 1.5	19.3±3.6	30.6 ± 1.9	20.5 ± 1.9	33.4 ± 0.5
12	**Myrte EO**	8.6 ± 0.4	11.0 ± 1.7	6.8 ± 0.9	9.3 ± 0.9	10.1 ± 2.4	8.7 ± 0.6	12.8 ± 1.2	6.0 ± 0.0	11.3 ± 1.3
13	**Ginger EO**	6.0 ± 0.0	6.0 ± 0.0	6.0 ± 0.0	7.8 ± 2.9	6.0 ± 0.0	7.1 ± 1.3	12.4 ± 0.6	16.3 ± 1.4	11.4 ± 0.4
14	**Pan tropical EO**	31.1 ± 3.4	30.6 ± 2.1	23.7 ± 0.3	25.4 ± 2.1	32.0 ± 6.4	29.2 ± 1.3	61.0 ± 5.8	70.3 ± 3.4	63.0 ± 0.2
15	**Citral EO**	12.5 ± 1.4	10.2 ± 1.3	11.8 ± 1.6	18.4 ± 0.8	11.5 ± 1.4	10.4 ± 0.6	80.0 ± 0.0	23.0 ± 3.0	80.0 ± 0.0
16	**Asian EO**	8.8 ± 0.3	10.3 ± 2.6	9.2 ± 0.6	22.5 ± 1.2	9.6 ± 1.0	10.2 ± 0.7	42.7 ± 1.6	62.6 ± 6.1	80.0 ± 0.0
17	**Thyme thymol EO**	32.1 ± 2.2	41.4 ± 4.0	26.9 ± 3.4	31.3 ± 4.0	27.2 ± 3.7	30.5 ± 4.0	53.9 ± 2.6	38.3 ± 2.3	44.2 ± 5.3

* EO: Essential oil.

3.2. Determination of MIC, MBC and MFC

The results of MIC, MBC and MFC of the selected antimicrobial compounds are presented in Table 3. Results showed that Biosecur F440D was the most efficient in inhibiting the bacterial growth, showing a MIC and a MBC between 17 and 171 ppm against all evaluated bacterial strains. Pan tropical essential oil was also more efficient in inhibiting the growth of molds and *C. albicans* showing a fungicidal activity against *A. flavus* and *P. chrysogenum* with a MFC between 155 and 621 ppm. Pan tropical and Mediterranean essential oils showed the highest antimicrobial activity against almost all microorganisms tested showing a bactericidal and fungicidal activity. They inhibited the growth of all evaluated microorganisms at a concentration ≤1241 ppm for pan tropical essential oil and ≤2474 ppm for Mediterranean essential oil. These results indicate that these two essential oils have an interesting antimicrobial potential. Asian essential oil showed a high activity in inhibiting the growth of molds and yeast and showed a MFC of 311, 622 and 4979 ppm for *C. albicans*, *A. flavus* and *P. chrysogenum*, respectively. On the other hand, Biosecur F440D had a bactericidal activity against all the evaluated bacterial strains as compared to essential oils which have fungicidal activity.

Table 3. Minimum inhibitory, bactericidal and fungicidal concentrations (MIC, MBC and MFC) of the selected antimicrobial compounds.

		MIC, MBC and MFC Expressed in parts-per-million, PPM (Mean Value ± SD, n =3)					
		Biosecur F440D	Pan Tropical EO	Mediterranean EO	Asian EO	Citrus EO	Tween 80 2.5%
L. monocytogenes	MIC	171 ± 5	621 ± 3	619 ± 2	4974 ± 0	4974 ± 0	> 12500
	MBC	171 ± 0	1241 ± 5	1237 ± 3	4974 ± 0	4974 ± 0	-
B. subtilis	MIC	33 ± 1	1241 ± 6	1237 ± 3	2487 ± 0	4974 ± 0	> 12500
	MBC	33 ± 0	1241 ± 0	2470 ± 0	4974 ± 0	4974 ± 0	-
E. faecium	MIC	142 ± 33	1241 ± 6	2474 ± 7	4979 ± 0	4974 ± 0	> 12500
	MBC	142 ± 28	2488 ± 8	4947 ± 10	4979 ± 7	4974 ± 0	-
S. aureus	MIC	17 ± 0	1050 ± 0	1049 ± 1	1056 ± 0	2474 ± 0	> 12500
	MBC	17 ± 0	2227 ± 0	2224 ± 0	2239 ± 0	2474 ± 0	-
S. Typhimurium	MIC	171 ± 5	621 ± 3	309 ± 1	1245 ± 2	4974 ± 0	> 12500
	MBC	171 ± 4	621 ± 2	619 ± 1	1245 ± 1	4974 ± 0	-
E. coli	MIC	114 ± 3	621 ± 3	619 ± 2	1245 ± 2	2474 ± 0	> 12500
	MBC	114 ± 2	1243 ± 5	619 ± 1	1245 ± 0	2474 ± 0	-
C. albicans	MIC	427 ± 12	155 ± 1	155 ± 0	311 ± 0	1245 ± 0	> 12500
	MFC	628 ± 0	621 ± 3	618 ± 1	311 ± 0	1245 ± 0	-
A. flavus	MIC	836 ± 23	621 ± 3	2474 ± 7	4979 ± 0	4979 ± 0	> 12500
	MFC	1261 ± 26	621 ± 1	4958 ± 5	4979 ± 7	4979 ± 0	-
P. chrysogenum	MIC	552 ± 11	155 ± 1	1237 ± 3	622 ± 1	1245 ± 0	> 12500
	MFC	609 ± 76	155 ± 1	2477 ± 5	622 ± 0	1245 ± 0	-

3.3. Total Phenols and Flavonoids

Results of total phenols and total flavonoids content (Table 4) showed that Mediterranean and pan tropical essential oils were highly concentrated in total phenols (respectively 220.57 and 34.62 mg gallic acid equivalent/ g of antimicrobial) and total flavonoids (respectively 34.62 and 17.63 mg catechin equivalent/g of antimicrobial). Biosecur F440D showed a concentration of 4.38 mg gallic acid equivalent/g of antimicrobial for total phenols content and 1.26 mg catechin equivalent/g of antimicrobial for total flavonoids. Citrus and Asian essential oils showed the least concentration of total phenol and flavonoid content with, respectively, 1.51 and 1.41 mg gallic acid equivalent/g of antimicrobial and 0.06 and 0.56 mg catechin equivalent/g of antimicrobial.

Table 4. Total phenols and total flavonoids content of the antimicrobial extracts.

Natural Antimicrobial Products	Total Phenols (mg gallic acid/g of AM) *	Total Flavonoids (mg catechin/g of AM) *
Biosecur F440D	4.38 ± 0.16 [a]	1.26 ± 0.06 [a]
Pan tropical EO	34.62 ± 3.68 [b]	17.63 ± 1.40 [b]
Mediterranean EO	220.57 ± 17.67 [c]	34.75 ± 2.4 [c]
Asian EO	1.41 ± 0.18 [a]	0.56 ± 0.07 [a]
Citrus EO	1.51 ± 0.03 [a]	0.06 ± 0.03 [a]

* Within each column, means with the same letter are not significantly different ($p > 0.05$); AM: Antimicrobial.

3.4. Radical Scavenging Activity and FRAP

Biosecur F440D and Mediterranean, Asian, pan tropical and citrus essential oils were tested for their ability to scavenge radicals by the DPPH method. Biosecur F440D has the highest radical scavenging activity above all the other compounds with 0.28 mM of Trolox (Table 5). The radical scavenging of Biosecur F440D was two times higher than Mediterranean essential oil (0.18 mM equivalent), three times higher than citrus essential oil (0.07 mM equivalent) and 10 times higher than Asian essential oil (0.02 mM of Trolox equivalent).

The antioxidant activity measured with the ferric reducing power assay revealed similar results to those obtained with the DPPH technique (Table 5). The highest antioxidant activities were obtained with Mediterranean essential oil (0.76 Eq µM of ascorbic acid equivalent/g of extract), followed by pan tropical essential oil and Biosecur F440D (0.43 and 0.30 Eq µM of ascorbic acid equivalent/g of antimicrobial respectively). Asian and citrus essential oils have the lowest values (below 0.04 Eq µM of ascorbic acid equivalent/g of antimicrobial).

Table 5. Ferric reducing antioxidant power (FRAP) and Radical Scavenging Activity of the antimicrobial compounds.

Natural Antimicrobial Products	FRAP *	Radical Scavenging Activity *	
	Eq µM of Ascorbic acid/g of AM	mM Trolox	mM AA
Biosecur F440D	0.30 ± 0.04 [ab]	0.28 ± 0.05 [d]	0.29 ± 0.05 [d]
Pan tropical EO	0.43 ± 0.02 [b]	0.15 ± 0.02 [c]	0.15 ± 0.02 [c]
Mediterranean EO	0.76 ± 0.03 [c]	0.18 ± 0.03 [c]	0.19 ± 0.03 [c]
Asian EO	0.04 ± 0.00 [a]	0.02 ± 0.00 [a]	0.02 ± 0.00 [a]
Citrus EO	0.03 ± 0.00 [a]	0.07 ± 0.01 [b]	0.07 ± 0.01 [b]

* Within each column, means with the same letter are not significantly different ($p > 0.05$).

3.5. In Situ Analysis

Results of the growth of TMF, molds and yeasts (Figure 1) showed that on Day 0, the encapsulation of the antimicrobial formulation in o/w emulsion (containing a mixture of Biosecur F440D and Asian, Mediterranean, citrus and pan tropical essential oils), applied on the surface of carrots, allowed 2 log reductions for TMF and 1 log reduction for molds and yeasts as compared to the control ($p \leq 0.05$). The mix of selected antimicrobial ingredients-loaded emulsion was more effective than the commercial mix (Tsunami 100). A significant reduction of TMF, molds and yeasts counts was also observed during the whole storage period showing a 1 log reduction of TMF on carrots treated with the antimicrobial ingredients-loaded emulsion as compared to the control which signifies a better control of the microbiological growth of TMF on pre-cut carrots. The antimicrobial activity of the commercial mix of peroxyacetic acid and hydrogen peroxide against TMF was also lower than the antimicrobial ingredients-loaded emulsion during the whole storage. The shelf-life of pre-cut carrots was reached on

Day 6 for untreated carrots, treated carrots with the commercial chemical preservatives and on Day 8 for treated carrots with the developed antimicrobial-loaded emulsion (Figure 1a). By considering Days 1, 3 and 6, the growth rate was also lower in treated carrot with the antimicrobial formulation and with Tsunami samples showing a growth rate of 0.1291 and 0.1852 Ln CFU/g/day respectively as compared to 0.2193 Ln CFU/g/day for untreated samples (Table 6).

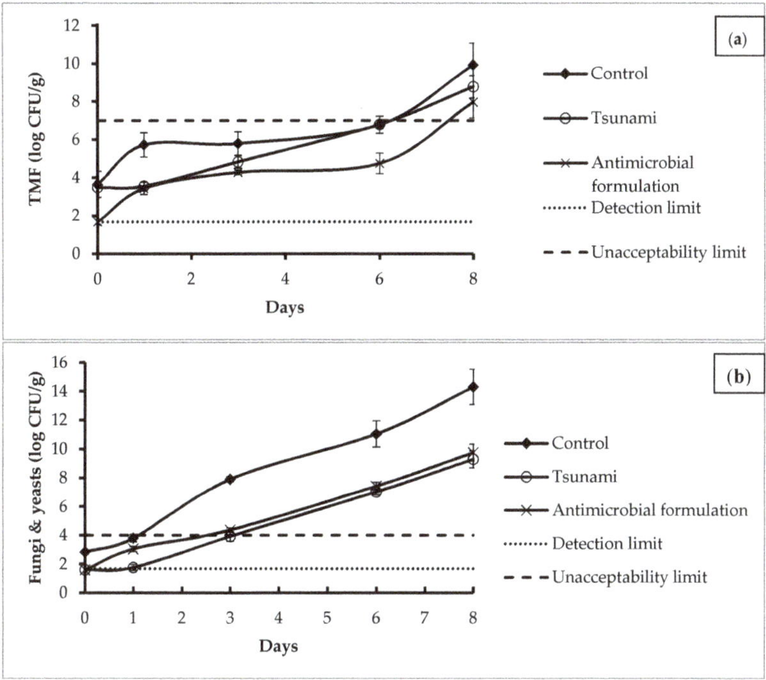

Figure 1. Total mesophilic flora (**a**) and total molds and yeasts (**b**) growth on pre-cut carrots.

Table 6. Growth rate of total mesophilic flora (TMF) in refrigerated pre-cut carrots.

Sample	Growth Rate of TMF (Ln CFU/g/Day)
Control	0.2193
Tsunami	0.1852
Antimicrobial formulation	0.1291

By considering the results of total molds and yeasts (Figure 1b), the shelf-life of pre-cut carrots was reached on Day 1 for untreated carrots and on Day 3 for both treated carrots with the antimicrobial-loaded emulsion and treated carrots with the chemical preservative (Tsunami). The obtained in situ results indicated that the antimicrobial formulation was effective against TMF and molds and yeasts, not only immediately after treatment but also during a mid-term storage.

3.6. Sensory Evaluation

Sensory analysis of pre-cut carrots treated or not with the antimicrobial formulation-loaded emulsion, was done by evaluating its odor, taste and global appreciation, using a nine-point hedonic scale and a panel of 24 untrained people and results are presented in Figure 2. Results showed that the antimicrobial treatment did not have any detrimental effect on the sensorial quality of the coated carrots. The values of the odor, the taste and the global appreciation were 6.8, 6.6 and 6.6 for the carrots treated with the antimicrobial formulation as compared to 6.8, 7.1 and 7.2 for the control samples. The odor was not affected by the applied treatment and a slight reduction on the attributed note was observed on the taste and the global appreciation. Overall, no significant negative effect ($p > 0.05$) was observed.

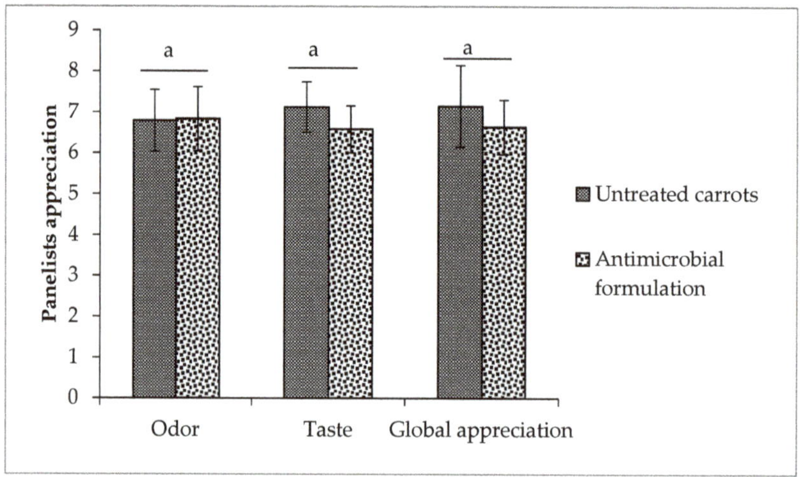

Figure 2. Effect of antimicrobial treatment on sensorial properties of pre-cut carrots.

4. Discussion

Valorization of natural antimicrobials has been extensively investigated during the last decades. In the present study, it was demonstrated that natural antimicrobials have a good antioxidant and antimicrobial activity against a wide range of food pathogens and spoilage microorganisms, and that their combination allows a better control of the microbiological quality of pre-cut carrots without altering their sensory properties.

Using the disk diffusion method, we have identified five antimicrobial compounds that showed a high inhibitory diameter against the tested microorganisms: Biosecur F440D and citrus, Asian, Mediterranean and pan tropical essential oils. Similar results for inhibitory diameter obtained by disk diffusion were also reported by Baser and Buchbauer [13] for cinnamon and citronella against *L. monocytogenes* and *S.* Typhimurium. Despite the medium inhibitory diameter (12.5–25.4 mm) of Biosecur F440D as compared to essential oils, its MIC and MBC was the lowest against all the evaluated bacteria. According to Ghabraie, et al. [32] and Lopez, et al. [33], the antimicrobial activity of essential oils is due to both solid and vapor-phase fractions. The antimicrobial activity of the vapor-phase could be observed only when essential oils are seeded on surface which was the case with the disk diffusion method. With the MIC method, the antimicrobial evaluation was done in liquid medium which reduces significantly the antimicrobial effect of the vapor fraction. However, Biosecur F440D, because of its water solubility, has a bactericidal activity when employed in liquid media and the obtained MIC was similar to the MBC (Table 3). Results obtained with disk diffusion agar confirmed previous observations and showed a higher or similar sensitivity of Gram positive bacteria to essential oils than Gram negative [13,34]. On the other hand, results obtained with MIC and MBC of essential oils showed that overall, essential oils were more efficient to inhibit Gram-negative bacteria than Gram positive as well showing a lowest MIC and MBC. These results suggest that volatile compounds in essential oils (MW < 300) could have a higher efficiency against Gram negative probably due to its various chemical compounds: alcohols, ethers or oxides, aldehydes, ketones, esters, amines, amides, phenols, heterocycles, and mainly the terpenes. It is known that the composition has an impact on the antimicrobial efficiency [35].

The antimicrobial behavior observed in the *in vitro* study of each antimicrobial compound differs mainly due to the difference in their chemical composition and nature. The Mediterranean and the pan tropical essential oils are highly effective antimicrobial compounds, leads to a significant inhibition against almost all evaluated microorganisms.

The Mediterranean essential oil, for example, is rich in total phenols and total flavonoids (Table 4). Similar results were observed by Wogiatzi, et al. [36] where several oregano origins were evaluated. Wogiatzi, Gougoulias, Papachatzis, Vagelas and Chouliaras [36] demonstrated that the total phenol content is also intimately related to the plant area of cultivation (foot/middle mountain). The hydroxyl group (-OH) of the phenolic compounds could interact with the membrane cell of bacteria and reduce the pH gradient through the cytoplasmic membrane which disrupts its structure and causes the loss of intracellular ATP and cell death [37]. The -OH group can also bind to the active site of enzymes (i.e., ATPase, histidine carboxylase), thereby altering the cellular metabolism of microorganisms [37,38]. The presence of phenolic compounds is also responsible for the good antioxidant activity of the Mediterranean essential oil observed, which act as free radical terminators [39]. Mediterranean essential oil is thus able to reduce the redox potential of the culture medium and to reduce the growth of microorganisms.

The antimicrobial activity of pan tropical essential oil is related to its high concentration on cinnamaldehyde. Cinnamaldehyde is capable of modifying the lipid profile of the microbial cell membrane probably due to its high antioxidant activity [40] which allows it to oxidase lipids on the bacterial membrane. Cinnamaldehyde can also inhibit the respiratory tract in certain bacteria by disrupting K^+ and pH homeostasis [38]. In this study, pan tropical essential oil was also characterized by a great antifungal activity probably due to its ability to inhibit b-(1,3)-glucan and chitin synthesis in yeasts and molds which are the major structural compounds of the fungal cell walls [41].

Asian essential oil is highly concentrated on geranial and neral. These two isomers are the main compounds of the monoterpene citral which its antimicrobial activity is well known against several bacteria and molds [42,43]. Despite the antifungal effectiveness of Asian and citrus essential oils with disk diffusion method, the effectiveness in broth media was lower due probably to the ability of some microorganisms to transform citronelal and citral and other of their components to the sole carbon and energy source [13]. The antifungal activity of citral and cinnamaldehyde is the result of perturbation in ergosterol biosynthesis which causes a damage to the intracellular structure, loss of intracellular substance and membrane damage [44].

Citrus essential oil is highly concentrated with citronellol and geraniol, and showed a lower antimicrobial activity when compared to the other antimicrobials mainly due to the presence of only one double bond on its main compounds [37]. Nakahara, et al. [45] showed that citronelal and linalool has antifungal activity at a dose of 112 ppm. The antifungal activity of components found in citrus essential oil (i.e. mono-terpenes) was previously reported to the interference of such compounds with enzymatic reaction of wall, i.e., structure [46,47]. This allows a lack of cytoplasm, damage of integrity and finally the mycelial death [48]. Simic, et al. [49] showed also that the antimicrobial activity of citronella essential oil is intimately related to the association of citronella and citronellol due probably to a synergistic effect of their combination.

Biosecur F440D was efficient to inhibit the growth of Gram positive and Gram-negative bacteria showing a bactericidal activity. According to Álvarez-Ordóñez, et al. [50], citrus extracts at higher concentrations than the MIC, pore formation in the cell membrane is observed inducing leakage of nucleic acids. According to the same authors, to achieve a significant bacterial reduction, the exposure time or the antimicrobial concentration used should be two to four times higher than the MIC. Citrus extract mainly acts on the membrane. It causes conformational damage and/or compositional in some or all components of the cell membrane. It mainly affects the carboxyl groups of membrane fatty acids and thus impairs the macromolecular structure of the bacterial membrane. Several studies have tried to identify the components that are involved in the antimicrobial activity of citrus extract. It possesses strong antioxidant and antimicrobial properties, pleasant aromas and flavors, especially due to the presence of flavonoids. Citrus flavanones include naringenin, hesperidin, hesperitin and prunine and have a broad spectrum of action against many Gram-negative bacteria.

Citrus flavonoids have also a direct role in scavenging reactive oxygen species (ROS) as confirmed by the obtained results of antiradical activity [51]. This suggests that the ROS could be involved in the bactericidal activity observed on citrus extracts. Inoue, et al. [52] supported this suggestion and showed that ROS act in conjunction to induce the strong bactericidal activity. The antiradical activity is also due to the presence of vitamin C at a high concentration in citrus extract which is a natural free radical scavenger.

The obtained results of in vitro study showed a very good antimicrobial and antioxidant properties of the selected natural antimicrobials. As their mode of action against bacteria fungi and yeasts differs, the mix of natural antimicrobial-loaded emulsion applied on carrots as a food model, presented a large spectral activity against targeted microorganisms.

The application of this developed formulation encapsulated in o/w emulsion at a concentration that did not affect the sensory properties of carrots (Figure 2) was efficient to reduce TMF, molds and yeasts growth during storage at 4 °C. The developed formulation was also more effective than the chemical antimicrobial (mix of peroxyacetic acid and hydrogen peroxide) to control TMF and had similar efficiency to control molds and yeasts. Based on previous studies, the developed formulation seems to be also more effective than other chemical methods such as HOCl, 4% H_2O_2 which showed less than 2 log reduction of TMF of carrots [53]. In situ efficiency is mainly due to combined activity of different compounds. The use of such combination could help to better control spoilage of fruits and vegetables. According to Bassolé and Juliani [54], combining cinnamon and oregano yielded in most cases, in a synergistic activity against *E. coli* and *S.* Typhimurium. Monoterpene hydrocarbon (α-pinene) when mixed with limonene or linalool also showed additive and synergistic effects [54]. The obtained results present a new antimicrobial formulation based on natural plant extracts that allowed a better control of initial microflora that could replace the methods presently used in industries such as blanching and ozonized water.

5. Conclusions

This study showed that natural antimicrobial extracts are rich on antioxidant and antiradical compounds. Biosecur F440D has the highest radical scavenging activity and has a bactericidal activity against all evaluated bacteria. Pan tropical essential oil has particularly an antifungal activity. Mediterranean essential oil was highly rich on total phenol and has the highest antioxidant activity. The mixture of natural antibacterial extracts when encapsulated in o/w emulsion and applied on carrot surface showed a better antimicrobial effectiveness than commercial chemical treatment widely used to treat vegetables. The mixture could be used as food treatment to extend the shelf-life of pre-cut carrots by two days without affecting their sensory properties. Finally, this user-friendly antimicrobial formulation-loaded emulsion could be applied in the food industry as a way to fulfill federal regulation requirements.

Author Contributions: Conceptualization, M.L.; methodology, Y.B.-F. and M.A.; software, Y.B.-F. and M.A.; validation, Y.B.-F., B.M. and M.L.; formal analysis, Y.B.-F. and M.A.; investigation, Y.B.-F.; resources, Y.B.-F.; data curation, Y.B.-F. and M.L.; writing—original draft preparation, Y.B.-F.; writing—review and editing, Y.B.-F., B.M. and M.L.; visualization, Y.B.-F.; supervision, B.M.; project administration, M.L.; funding acquisition, M.L.

Funding: This research was funded by the Natural Sciences and Engineering Research Council of Canada (CRSNG) (Project #CRDS-488702-15), by the Consortium de Recherche et Innovations en Bioprocédés Industriels au Québec (CRIBIQ) (Project#2015-023-C16), by the Ministère de l'Agriculture, des Pêcheries et de l'Alimentation du Québec (MAPAQ; Project #IA-115316) and by Biosecur Lab Inc., Foodarom group Inc. and Skjodt-Barrett Foods Inc.

Acknowledgments: The authors appreciate the Biosecur Lab for providing Biosecur products. Bonduelle is also acknowledged for having provided pre-cut carrots and Tsunami treatment. Yosra Ben Fadhel was a fellowship recipient of Fondation INRS-Armand-Frappier.

Conflicts of Interest: The authors declare no conflicts of interest Boca Raton.

References

1. Rai, M. *Natural Antimicrobials in Food Safety and Quality*; CABI: Oxfordshire, UK, 2011.
2. Gagnaire, F.; Marignac, B.; Hecht, G.; Héry, M. Sensory irritation of acetic acid, hydrogen peroxide, peroxyacetic acid and their mixture in mice. *Ann. Occup. Hyg.* **2002**, *46*, 97–102. [CrossRef] [PubMed]
3. Hashemi, S.M.B.; Khaneghah, A.M.; de Souza Sant'Ana, A. *Essential Oils in Food Processing: Chemistry, Safety and Applications*; John Wiley & Sons: West Sussex, UK, 2017.
4. Nollet, L.M.; Rathore, H.S. *Green Pesticides Handbook: Essential Oils for Pest Control*; CRC Press: Boca Raton, FL, USA, 2017.
5. Lee, Y.-L.; Cesario, T.; Wang, Y.; Shanbrom, E.; Thrupp, L. Antibacterial activity of vegetables and juices. *Nutrition* **2003**, *19*, 994–996. [CrossRef] [PubMed]
6. Patil, J.R.; Chidambara Murthy, K.; Jayaprakasha, G.; Chetti, M.B.; Patil, B.S. Bioactive compounds from Mexican lime (Citrus aurantifolia) juice induce apoptosis in human pancreatic cells. *J. Agric. Food Chem.* **2009**, *57*, 10933–10942. [CrossRef] [PubMed]
7. Ben-Fadhel, Y.; Leroy, V.; Dussault, D.; St-Yves, F.; Lauzon, M.; Salmieri, S.; Jamshidian, M.; Vu, D.K.; Lacroix, M. Combined effects of marinating and γ-irradiation in ensuring safety, protection of nutritional value and increase in shelf-life of ready-to-cook meat for immunocompromised patients. *Meat Sci.* **2016**, *118*, 43–51. [CrossRef] [PubMed]
8. Cowan, M.M. Plant Products as Antimicrobial Agents. *Clin. Microbiol. Rev.* **1999**, *12*, 564–582. [CrossRef] [PubMed]
9. Tiwari, B.K.; Valdramidis, V.P.; O'Donnell, C.P.; Muthukumarappan, K.; Bourke, P.; Cullen, P.J. Application of Natural Antimicrobials for Food Preservation. *J. Agric. Food Chem.* **2009**, *57*, 5987–6000. [CrossRef] [PubMed]
10. Bertoli, A.; Cirak, C.; Silva, J. *Hypericum* species as sources of valuable essential oils. *Med. Aromat. Plant Sci. Biotechnol.* **2011**, *5*, 29–47.
11. Ayari, S.; Dussault, D.; Millette, M.; Hamdi, M.; Lacroix, M. Changes in membrane fatty acids and murein composition of Bacillus cereus and Salmonella Typhi induced by gamma irradiation treatment. *Int. J. Food Microbiol.* **2009**, *135*, 1–6. [CrossRef]
12. Tisserand, R.; Young, R. *Essential Oil Safety-E-Book: A Guide for Health Care Professionals*; Elsevier Health Sciences: London, UK, 2013.
13. Baser, K.H.C.; Buchbauer, G. *Handbook of Essential Oils: Science, Technology, and Applications*; CRC Press: Boca Raton, FL, USA, 2015.
14. Santos, M.; Martins, S.; Veríssimo, C.; Nunes, M.; Lima, A.; Ferreira, R.; Pedroso, L.; Sousa, I.; Ferreira, M. Essential oils as antibacterial agents against food-borne pathogens: Are they really as useful as they are claimed to be? *J. Food Sci. Technol.* **2017**, *54*, 4344–4352. [CrossRef]
15. Sivapalasingam, S.; Friedman, C.R.; Cohen, L.; Tauxe, R.V. Fresh Produce: A Growing Cause of Outbreaks of Foodborne Illness in the United States, 1973 through 1997. *J. Food Prot.* **2004**, *67*, 2342–2353. [CrossRef]
16. Jalava, K.; Hakkinen, M.; Valkonen, M.; Nakari, U.-M.; Palo, T.; Hallanvuo, S.; Ollgren, J.; Siitonen, A.; Nuorti, J.P. An Outbreak of Gastrointestinal Illness and Erythema Nodosum from Grated Carrots Contaminated with Yersinia pseudotuberculosis. *J. Infect. Dis.* **2006**, *194*, 1209–1216. [CrossRef]
17. Gaynor, K.; Park, S.; Kanenaka, R.; Colindres, R.; Mintz, E.; Ram, P.; Kitsutani, P.; Nakata, M.; Wedel, S.; Boxrud, D. International foodborne outbreak of Shigella sonnei infection in airline passengers. *Epidemiol. Infect.* **2009**, *137*, 335–341. [CrossRef]
18. Da Silva Felício, M.T.; Hald, T.; Liebana, E.; Allende, A.; Hugas, M.; Nguyen-The, C.; Johannessen, G.S.; Niskanen, T.; Uyttendaele, M.; McLauchlin, J. Risk ranking of pathogens in ready-to-eat unprocessed foods of non-animal origin (FoNAO) in the EU: Initial evaluation using outbreak data (2007–2011). *Int. J. Food Microbiol.* **2015**, *195*, 9–19. [CrossRef] [PubMed]
19. Ruiz-Cruz, S.; Acedo-Félix, E.; Díaz-Cinco, M.; Islas-Osuna, M.A.; González-Aguilar, G.A. Efficacy of sanitizers in reducing Escherichia coli O157:H7, Salmonella spp. and Listeria monocytogenes populations on fresh-cut carrots. *Food Control* **2007**, *18*, 1383–1390. [CrossRef]
20. Banwart, G. *Basic Food Microbiology*; Springer Science & Business Media: New York, NY, USA, 2012.
21. Tournas, V. Spoilage of vegetable crops by bacteria and fungi and related health hazards. *Crit. Rev. Microbiol.* **2005**, *31*, 33–44. [CrossRef] [PubMed]

22. Monaghan, J.; Hutchison, M. Ineffective hand washing and the contamination of carrots after using a field latrine. *Lett. Appl. Microbiol.* **2016**, *62*, 299–303. [CrossRef]
23. Ben-Fadhel, Y.; Saltaji, S.; Khlifi, M.A.; Salmieri, S.; Vu, K.D.; Lacroix, M. Active edible coating and γ-irradiation as cold combined treatments to assure the safety of broccoli florets (Brassica oleracea L.). *Int. J. Food Microbiol.* **2017**, *241*, 30–38. [CrossRef]
24. Dewanto, V.; Wu, X.; Adom, K.K.; Liu, R.H. Thermal processing enhances the nutritional value of tomatoes by increasing total antioxidant activity. *J. Agric. Food Chem.* **2002**, *50*, 3010–3014. [CrossRef]
25. Fattouch, S.; Caboni, P.; Coroneo, V.; Tuberoso, C.I.; Angioni, A.; Dessi, S.; Marzouki, N.; Cabras, P. Antimicrobial activity of Tunisian quince (Cydonia oblonga Miller) pulp and peel polyphenolic extracts. *J. Agric. Food Chem.* **2007**, *55*, 963–969. [CrossRef]
26. Benzie, I.F.; Strain, J. The ferric reducing ability of plasma (FRAP) as a measure of "antioxidant power": The FRAP assay. *Anal. Biochem.* **1996**, *239*, 70–76. [CrossRef]
27. Takala, P.N.; Vu, K.D.; Salmieri, S.; Khan, R.A.; Lacroix, M. Antibacterial effect of biodegradable active packaging on the growth of *Escherichia coli*, *Salmonella typhimurium* and *Listeria monocytogenes* in fresh broccoli stored at 4 °C. *LWT-Food Sci. Technol.* **2013**, *53*, 499–506. [CrossRef]
28. Babic, I.; Hilbert, G.; Nguyen-The, C.; Guiraud, J. The yeast flora of stored ready-to-use carrots and their role in spoilage. *Int. J. Food Sci. Technol.* **1992**, *27*, 473–484. [CrossRef]
29. Itohan, A.M.; Peters, O.; Kolo, I. Bacterial contaminants of salad vegetables in abuja municipal area council, Nigeria. *Malays. J. Microbiol.* **2011**, *7*, 111–114.
30. MAPAQ. Lignes Directrices et Normes Pour L'interprétation des Résultats Analytiques en Microbiologie Alimentaire. Available online: https://www.mapaq.gouv.qc.ca/fr/Publications/recueil.pdf (accessed on 31 October 2019).
31. da Silva, A.C.; de Freitas Santos, P.D.; Palazzi, N.C.; Leimann, F.V.; Fuchs, R.H.B.; Bracht, L.; Gonçalves, O.H. Production and characterization of curcumin microcrystals and evaluation of the antimicrobial and sensory aspects in minimally processed carrots. *Food Funct.* **2017**, *8*, 1851–1858. [CrossRef] [PubMed]
32. Ghabraie, M.; Vu, K.D.; Tata, L.; Salmieri, S.; Lacroix, M. Antimicrobial effect of essential oils in combinations against five bacteria and their effect on sensorial quality of ground meat. *LWT-Food Sci. Technol.* **2016**, *66*, 332–339. [CrossRef]
33. Lopez, P.; Sanchez, C.; Batlle, R.; Nerin, C. Solid-and vapor-phase antimicrobial activities of six essential oils: Susceptibility of selected foodborne bacterial and fungal strains. *J. Agric. Food Chem.* **2005**, *53*, 6939–6946. [CrossRef]
34. Dussault, D.; Vu, K.D.; Lacroix, M. In vitro evaluation of antimicrobial activities of various commercial essential oils, oleoresin and pure compounds against food pathogens and application in ham. *Meat Sci.* **2014**, *96*, 514–520. [CrossRef]
35. Dhifi, W.; Bellili, S.; Jazi, S.; Bahloul, N.; Mnif, W. Essential oils' chemical characterization and investigation of some biological activities: A critical review. *Medicines* **2016**, *3*, 25. [CrossRef]
36. Wogiatzi, E.; Gougoulias, N.; Papachatzis, A.; Vagelas, I.; Chouliaras, N. Greek Oregano Essential Oils Production, Phytotoxicity and Antifungal Activity. *Biotechnol. Biotechnol. Equip.* **2009**, *23*, 1150–1152. [CrossRef]
37. Gyawali, R.; Ibrahim, S.A. Natural products as antimicrobial agents. *Food Control* **2014**, *46*, 412–429. [CrossRef]
38. Nazzaro, F.; Fratianni, F.; De Martino, L.; Coppola, R.; De Feo, V. Effect of essential oils on pathogenic bacteria. *Pharmaceuticals* **2013**, *6*, 1451–1474. [CrossRef] [PubMed]
39. Shahidi, F.; Janitha, P.; Wanasundara, P. Phenolic antioxidants. *Crit. Rev. Food Sci. Nutr.* **1992**, *32*, 67–103. [CrossRef] [PubMed]
40. Schmidt, E.; Jirovetz, L.; Buchbauer, G.; Eller, G.A.; Stoilova, I.; Krastanov, A.; Stoyanova, A.; Geissler, M. Composition and antioxidant activities of the essential oil of cinnamon (Cinnamomum zeylanicum Blume) leaves from Sri Lanka. *J. Essent. Oil Bear. Plants* **2006**, *9*, 170–182. [CrossRef]
41. Bang, K.-H.; Lee, D.-W.; Park, H.-M.; Rhee, Y.-H. Inhibition of fungal cell wall synthesizing enzymes by trans-cinnamaldehyde. *Biosci. Biotechnol. Biochem.* **2000**, *64*, 1061–1063. [CrossRef]
42. Silva, C.D.B.D.; Guterres, S.S.; Weisheimer, V.; Schapoval, E.E. Antifungal activity of the lemongrass oil and citral against Candida spp. *Braz. J. Infect. Dis.* **2008**, *12*, 63–66. [CrossRef]
43. Dorman, H.; Deans, S. Antimicrobial agents from plants: Antibacterial activity of plant volatile oils. *J. Appl. Microbiol.* **2000**, *88*, 308–316. [CrossRef]

44. Wang, Y.; Feng, K.; Yang, H.; Yuan, Y.; Yue, T. Antifungal mechanism of cinnamaldehyde and citral combination against Penicillium expansum based on FT-IR fingerprint, plasma membrane, oxidative stress and volatile profile. *RSC Adv.* **2018**, *8*, 5806–5815. [CrossRef]
45. Nakahara, K.; Alzoreky, N.S.; Yoshihashi, T.; Nguyen, H.T.; Trakoontivakorn, G. Chemical composition and antifungal activity of essential oil from Cymbopogon nardus (citronella grass). *Jpn. Agric. Res. Q. JARQ* **2013**, *37*, 249–252. [CrossRef]
46. Sharma, N.; Tripathi, A. Effects of Citrus sinensis (L.) Osbeck epicarp essential oil on growth and morphogenesis of Aspergillus niger (L.) Van Tieghem. *Microbiol. Res.* **2008**, *163*, 337–344. [CrossRef]
47. Aguiar, R.W.D.S.; Ootani, M.A.; Ascencio, S.D.; Ferreira, T.P.; Santos, M.M.d.; Santos, G.R.d. Fumigant antifungal activity of Corymbia citriodora and Cymbopogon nardus essential oils and citronellal against three fungal species. *Sci. World J.* **2014**, *2014*. [CrossRef]
48. Pereira, F.D.O.; Wanderley, P.A.; Viana, F.A.C.; Lima, R.B.D.; Sousa, F.B.D.; Lima, E.D.O. Growth inhibition and morphological alterations of Trichophyton rubrum induced by essential oil from Cymbopogon winterianus Jowitt ex Bor. *Braz. J. Microbiol.* **2011**, *42*, 233–242. [CrossRef] [PubMed]
49. Simic, A.; Rančic, A.; Sokovic, M.; Ristic, M.; Grujic-Jovanovic, S.; Vukojevic, J.; Marin, P.D. Essential oil composition of Cymbopogon winterianus. and Carum carvi. and their antimicrobial activities. *Pharm. Biol.* **2008**, *46*, 437–441. [CrossRef]
50. Álvarez-Ordóñez, A.; Carvajal, A.; Arguello, H.; Martínez-Lobo, F.; Naharro, G.; Rubio, P. Antibacterial activity and mode of action of a commercial citrus fruit extract. *J. Appl. Microbiol.* **2013**, *115*, 50–60. [CrossRef] [PubMed]
51. Calabro, M.; Galtieri, V.; Cutroneo, P.; Tommasini, S.; Ficarra, P.; Ficarra, R. Study of the extraction procedure by experimental design and validation of a LC method for determination of flavonoids in Citrus bergamia juice. *J. Pharm. Biomed. Anal.* **2004**, *35*, 349–363. [CrossRef]
52. Inoue, Y.; Hoshino, M.; Takahashi, H.; Noguchi, T.; Murata, T.; Kanzaki, Y.; Hamashima, H.; Sasatsu, M. Bactericidal activity of Ag–zeolite mediated by reactive oxygen species under aerated conditions. *J. Inorg. Biochem.* **2002**, *92*, 37–42. [CrossRef]
53. Amanatidou, A.; Slump, R.A.; Gorris, L.G.M.; Smid, E.J. High Oxygen and High Carbon Dioxide Modified Atmospheres for Shelf-life Extension of Minimally Processed Carrots. *J. Food Sci.* **2000**, *65*, 61–66. [CrossRef]
54. Bassolé, I.H.N.; Juliani, H.R. Essential Oils in Combination and Their Antimicrobial Properties. *Molecules* **2012**, *17*, 3989. [CrossRef]

Review

Phytotoxicity of Essential Oils: Opportunities and Constraints for the Development of Biopesticides. A Review

Pierre-Yves Werrie [1,*], Bastien Durenne [2], Pierre Delaplace [3] and Marie-Laure Fauconnier [1]

[1] Laboratory of Chemistry of Natural Molecules, Gembloux Agro-Bio Tech, University of Liège, 5030 Gembloux, Belgium; marie-laure.fauconnier@uliege.be

[2] Soil, Water and Integrated Production Unit, Walloon Agricultural Research Centre, 5030 Gembloux, Belgium; b.durenne@cra.wallonie.be

[3] Plant Sciences, Gembloux Agro-Bio Tech, University of Liège, 5030 Gembloux, Belgium; pierre.delaplace@uliege.be

* Correspondence: pierre-yves.werrie@uliege.be

Received: 27 August 2020; Accepted: 9 September 2020; Published: 14 September 2020

Abstract: The extensive use of chemical pesticides leads to risks for both the environment and human health due to the toxicity and poor biodegradability that they may present. Farmers therefore need alternative agricultural practices including the use of natural molecules to achieve more sustainable production methods to meet consumer and societal expectations. Numerous studies have reported the potential of essential oils as biopesticides for integrated weed or pest management. However, their phytotoxic properties have long been a major drawback for their potential applicability (apart from herbicidal application). Therefore, deciphering the mode of action of essential oils exogenously applied in regards to their potential phytotoxicity will help in the development of biopesticides for sustainable agriculture. Nowadays, plant physiologists are attempting to understand the mechanisms underlying their phytotoxicity at both cellular and molecular levels using transcriptomic and metabolomic tools. This review systematically discusses the functional and cellular impacts of essential oils applied in the agronomic context. Putative molecular targets and resulting physiological disturbances are described. New opportunities regarding the development of biopesticides are discussed including biostimulation and defense elicitation or priming properties of essential oils.

Keywords: essential oils; phytotoxicity; mode of action; biopesticides

---

1. Introduction

Essential oils (EOs) have been used historically in the food and perfume industries and are extracted from various plant organs (flowers, leaves, barks, wood, roots, rhizomes, fruits and seeds) through steam distillation, hydro-distillation and cold expression for citrus. These natural products are mainly composed of volatile organic compounds (VOCs), having a high vapor pressure at room temperature and belonging mainly to the phenylpropanoid and terpenoid families. Briefly, terpenes are classified according to the number of isoprene sub-units: two for monoterpene ($C_{10}H_{16}$) and three for sesquiterpene ($C_{15}H_{24}$). Oxygenated terpenes or terpenoids also contain additional functional groups such as alcohol, carboxylic acid, ester, etc. [1], and phenylpropanoids are produced from L-phenylalanine through deamination by phenylalanine ammonia-lyase [2].

Many research studies have been undertaken on the use of EOs in more sustainable agronomic practices. In this regard, numerous findings have described the strong biopesticidal potential of EOs thanks to their antibacterial [3], antifungal [4], insecticidal [5], acaricidal [6], nematicidal [7] and herbicidal activities [8]. Included under the Generally Recognized as Safe (GRAS) product categories of the United States Food and Drug Administration, the impact of EOs on human health and ecosystems seems to be lower compared to synthetic plant protection products (PPP). Biocidal actions of EOs can be specific, and therefore their use could be compatible with integrated pest management (IPM) [9].

The application of EOs is, however, subject to a major constraint. They may present phytotoxic properties to untargeted plants such as crops. The most effective EOs in pest control are phytotoxic too, and considerable precautions are required regarding product formulation (unless the objective is the formulation of a total herbicide) [10]. Empirical tests for commercial EOs are commonly realized on major crops [11]. However these strategies have led to poor knowledge relating to other biological systems [12]. Many parameters determine this impact, such as the application mode (root watering, aerial spraying or injection in the vascular system), the plant organs targeted, the phenological stage (seed, plantlet or mature plant), the physiological state and product formulation. As illustrated by the opposing claims regarding the presence or absence of phytotoxicity of *Mentha pulegium* (pennyroyal) EOs towards *Cucumis sativus* (cucumber) and *Solanum lycopersicum* (tomato), it is necessary to gain insight into the molecular mechanism involved in order to design suitable biopesticides [13–15].

Phytotoxicity can be defined as a negative impact on plant growth or plant fitness and can be linked to cellular dysfunctions. Physiological impairment can be observed through integrative measurements of stress, for example on the photosynthetic apparatus. However, determination of the primary site of action is much more challenging. Diverse phytochemical products have been demonstrated to influence several physiological processes of growth and development in plant cell division and root elongation [16]. Blends of natural plant compounds often have numerous mechanisms of action, making them very efficient at acting on a plant's primary metabolism. It therefore seems most important to gain an insight into the physiological impact of EOs on plant crops to design proper bioassays and efficient biopesticides. Avoiding residual phytotoxicity, which is currently an underestimated constraint in the field, will allow the broader application of EOs [17]. However even if some processes seem to be inhibited in a dose-dependent manner, a concentration below the phytotoxic threshold could also stimulate the plant, a phenomenon referred to as biostimulation. New opportunities arising from this biostimulation and elicitation of defenses will be discussed in this review.

All the mechanisms involved in the phytotoxicity of EOs cannot be easily interpreted individually [18]. This review aims to discuss the latest putative molecular targets (mode of action) involved in plant metabolism with a physiological approach including water status alteration, membrane interaction/disruption, reactive oxygen/nitrogen species induction, genotoxicity and microtubule disruption, mitochondrial respiration or photosynthesis inhibition and enzymatic or phytohormones regulation. The different mechanisms presented throughout this review have been graphically summarized in Figure 1.

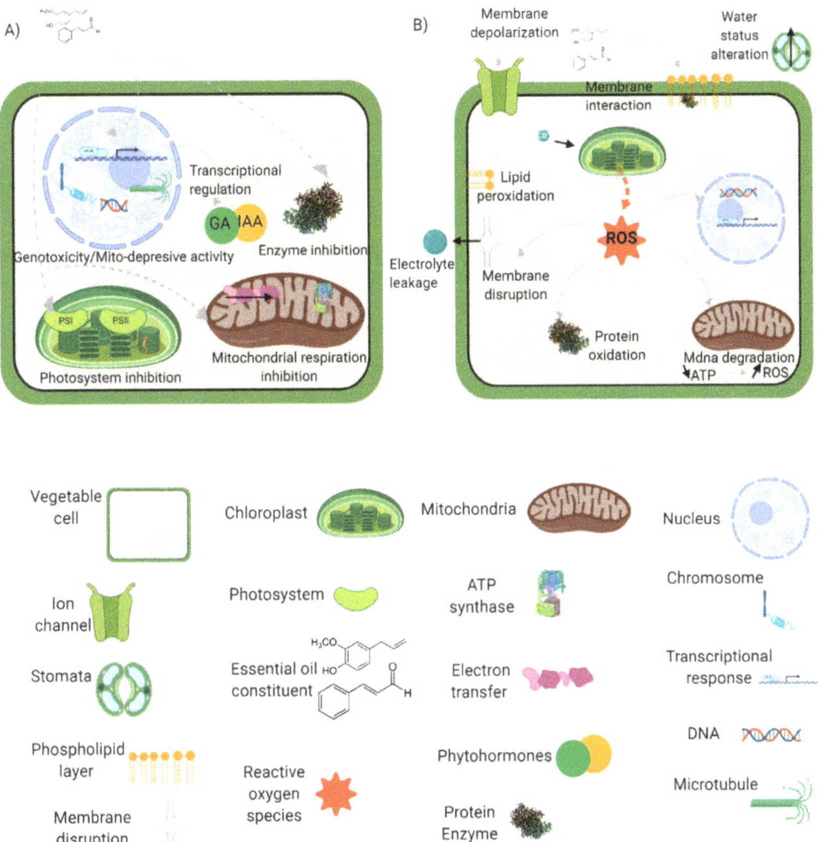

Figure 1. Mode of action of essential oil at the cellular level. (**A**) Photosynthesis and mitochondrial respiration inhibition, microtubule disruption and genotoxicity, enzymatic and phytohormone regulation. (**B**) Water status alteration, membrane properties and interactions, reactive oxygen species induction.

2. Essential Oils' Cellular and Physiological Impacts

2.1. Essential Oils' Translocation

Essential oil constituents (EOC) must access specific targets in order to carry out the physiological impact previously listed within a plant. Numerous publications describe the VOCs released by plants [19–21]. However little is known about their cellular entrance and translocation in plant organisms in the case of a systemic effect.

When sprayed, the first interaction occurs with the cuticular wax components of the leaves. In fact, the cuticle is considered to be the plant's first barrier to molecule penetration. The interaction between monoterpene with epicuticular waxes and stomata will be further described. Briefly, once it has entered through the stomata opening by gas exchange or diffusion through the waxy cuticle, each EOC is partitioned into the gas phase and liquid phase following a defined ratio determined by Henry's law. The liquid phase is materialized by the cell wall in which EOC accumulates. Compounds then diffuse to the cytosol following their oil/water partition coefficients [22]. Finally, active transport should also be considered as has been demonstrated for emissions [23].

Regarding root uptake, a study with radio-labelled thymol demonstrates the translocation of monoterpenes in citrus trees. However, the determination of the mechanism was beyond the scope of the study, although the authors suggest it could be similar to that for ethylenediaminetetraacetic acid (EDTA) [24].

2.2. Water Status Alteration

Depending on the mode of application (aerial or root), two different phenomena have been suggested for disturbing the water status of plants after treatment with EOs.

The deleterious effect of monoterpene (camphor and menthol) on cuticular wax and stomatal closure inhibition has been observed [25]. These two effects act synergistically on plant transpiration leading to guard cell disruption and desiccation. Interestingly, an opposite growth promoting effect is described for *Arabidopsis thaliana* during short vapor exposure to these terpenes. The molecular mechanism responsible for this prevention of stomatal closure is mediated through modification in the cytoskeleton and especially in the actin filament. Furthermore, stress symptoms appear together with a change in gene expression [26]. The amount of leaf epicuticular waxes determines the sensitivity of crop seedlings and weed species [27].

Water status alteration of plants was also observed after root watering application with citral, a mixture of two monoterpene isomers neral and geranial [28]. In a similar study with the sesquiterpene trans-caryophyllene, the authors suggest that this alteration could be responsible for the oxidative burst and a strong proline accumulation due to its osmo-regulative function [29].

2.3. Membrane Properties and Interactions

After entering the intercellular space through the mesh of the cell wall, EOCs directly solubilize within the plasma membrane depending on their physical properties, particularly their vapor pressure and molecular mass. Their specific accumulation was demonstrated to modify the lipid packing density, membrane-bound enzymes and ion flux [30].

This interaction can lead to a reversible depolarization of the membrane potential (Vm) and to membrane disruption [31]. Furthermore, stronger membrane depolarization occurs for more water soluble monoterpenes presenting a low octanol/water partition coefficient (Kow). A change in the polarization state implies ion mobility through the membrane. A drastic entrance of Ca^{2+} in the cytosol is triggered by opening the calcium channel. Ca^{2+} is known to be largely involved in cellular signaling. It performs allosteric regulation of many enzymes and proteins. Moreover, Ca^{2+} is an intracellular second messenger of signal transduction pathways and gene expression. Finally, the increase in Ca^{2+} concentration can lead to an oxidative burst [32].

Studies on artificial monolayer membranes of dipalmitoyl-phosphatildylcholine describe the penetration of monoterpenes such as camphor, cineole, thymol, menthol and geraniol, which affect the vesicles topology [33]. Similar work on model bilayer interactions with related monoterpenes, including limonene, perillyl alcohol and aldehyde, demonstrates the diffusion across the membrane and an ordering effect on the lipid bilayer [34]. More recently, novel molecular techniques of dynamic interaction were applied to study the interaction between citronellal (monoterpene), citronellol (monoterpene) and cinnamaldehyde (phenylpropanoids) with a biomimetic membrane [35]. Briefly, the in silico insertion model predicted different behaviors between the two classes (monoterpenes and phenylpropanoids). These predictions were confirmed using in vitro biophysical assays. Citronellal and citronellol interaction with the model membranes was demonstrated without permeabilizing it, while cinnamaldehyde did not interact with the model membrane. This suggests two different mechanisms of action: (i) the modification of lipid bilayer organization by monoterpenes and (ii) the interaction with membrane receptors for phenylpropanoid pathway metabolites.

Associated with the modification of membrane properties, a change in the membrane's composition also occurs. In fact, an increase in unsaturated fatty acids was demonstrated following application of monoterpenes such as 1,8-cineole, geraniol, thymol, menthol and camphor [36]. Quantitative and qualitative changes in most abundant free and esterified sterols (sitosterol, stigmasterol, and campesterol) and phospholipid fatty acids (16:0, 16:1, 18:0, 18:1, 18:2, 18:3) were also highlighted in a study investigating the effect of the same monoterpenes [37]. This results in an increase in the percentage of unsaturated fatty acid (PLFAs) and stigmasterol. Interestingly, alcoholic monoterpenes seem to have a different mode of action affecting more unsaturated fatty acid and stigmasterol leading to seedling growth interferences.

2.4. Reactive Oxygen and Nitrogen Species Induction

Reactive oxygen species (ROS) are essential in cellular signaling. They can be produced in various locations in plant cells such as in the chloroplast, the peroxisome, the mitochondria and in the endoplasmic reticulum. ROS are very reactive compounds that in excess lead to the degradation of macromolecules such as lipids, carbohydrates, proteins and DNA [38].

Oxidative burst or generation of ROS has long been proposed as one of the main mechanisms of action of phytotoxins [39]. We know that the uncoupling of photosynthesis and respiration leads to the production of superoxide radicals (O^{2-}), which are transformed into hydrogen peroxide (H_2O_2) by the superoxide dismutase. Moreover, the reaction with transition metal triggers a reduction of H_2O_2 to OH·, another very reactive species [40].

Oxidative stress was acknowledged after treatment with α-pinene through hydrogen peroxide, proline and the lipid peroxidation product malondialdehyde (MDA). Moreover, an antioxidant enzyme activity assay (superoxide dismutase, catalase, ascorbate, peroxidase, guaiacol peroxidase and glutathione reductase) was also performed in the roots. The oxidative stress generated by these ROS leads to membrane lipid peroxidation and ultimately to membrane disruption launching the programmed cell death. These membrane disruptions are evidenced via electrolyte leakage (EL) and vital staining [41].

In a similar experiment determining germination and growth inhibition by β-pinene EL, lipid peroxidation and lipoxygenase (LOX) activity were assessed. The result showed a strong increase in EL, dienes and H_2O_2 content and the authors suggest that despite an increase in the activity of ROS scavenging enzymes, root membrane integrity was lost [42]. Later on, they studied the early ROS generation and activity of the antioxidant defense system in the root and shoot of hydroponic wheat. The damaged was more severe in the root and a higher lipoxygenase activity was observed in parallel with accumulation of MDA [43]. The up-regulation of LOX activity has been observed for citronellol as well and the authors suggest that its hydroperoxide derivatives may destroy the membrane [44].

EOs inhibiting the growth of tested plants via ROS overproduction leading to oxidative stress and degradation of membrane integrity was evidenced via increased levels of MDA and EL, and decreased levels of conjugated dienes were demonstrated for other EOs such as *Pogostemon benghalensis* [45], *Monarda didyma* [46] and *Artemisia scoparia* [47].

Secondary effects of ROS generation include depigmentation of cotyledons in *A. thaliana* by *Heterothalamus psiadioides* EOs. The effects are here observed in a dose-dependent manner and in very small amounts. The authors also suggest that alteration on auxin levels occur as a secondary effect. Exogenous addition of antioxidants did not reverse effects on adventitious rooting, indicating that damages were too severe [48].

The generation of ROS, one of the most prevalent plant responses to stress, is described in direct response to the application of EOs. However, it is unlikely to be the main mechanism of toxicity but rather an indirect consequence resulting from LOX activity, chloroplast or mitochondria alteration [38]. The fundamental involvement of ROS in stress signaling as well as their interaction with other signaling components such as transcription factors, plant hormones, calcium, membrane, G-protein and mitogen-activated protein kinases need to be highlighted [49]. These interactions may explain many of the numerous physiological impacts induced by EOs' application in plants. Moreover, after treatment with α-farnesene, they also observed the induction of nitric oxide production, a reactive nitrogen species (RNS) associated with an oxidative burst [38].

2.5. Photosynthesis Inhibition

Photosynthesis inhibition has also been proposed as one of the putative modes of action of EOs. While the impact of certain allelochemicals on photosynthesis is well established, for instance quinone, this is not the case for EOs where numerous mechanisms have been proposed. Direct ROS-mediated disruption through oxidation of photosystem II (PSII) protein has been suggested to inhibit photosynthesis as suggested by the increase in the proline content, whose function is to accept electrons to protect the photosystem [50]. The effect of β-pinene on the chloroplast membrane has long been demonstrated by the inhibition of the electron transport of PSII [51,52].

Numerous studies report a decrease in the photosynthetic pigments namely chlorophylls (a and b) and carotenoids after treatments with EOs in a dose-dependent way [53–55]. This can result from a direct pigment photo-degradation or from a decrease in *de novo* synthesis. Plants have developed a non-photochemical quenching (fluorescence) strategy to avoid the ROS production resulting from this photo-inhibition. The decrease in carotenoid content could explain a higher fluorescence emission and a decrease of the PSII performance due to some damage to the complex antenna via ROS production and lipid peroxidation [56].

Artemisia fragrans EO impacts on the photosynthetic apparatus of perennial weed *Convolvulus arvensis* were studied using the most important chlorophyll fluorescence parameters. Increase in minimal fluorescence level (F0) implies a restriction in the PSII transport chain. The decrease in maximum quantum yield of PSII (Fv/Fm) results from photosystem inactivation (photo-damage) and/or a blockade in electron transport. PSII electron transport chain state (φPSII) reduction in plants treated with EOs restricts the non-cyclic electron transport chain. The last two parameters represent energy used in photochemical quenching (qP) and non-photochemical quenching (NPQ). qP decreases following concentration of EOs whereas NQP increases. Taken altogether, these results imply that the excited energy was not used in photosynthesis due to photosystem degradation by EO treatment [57].

Two specific fluorescence parameters QYmax (a maximum quantum yield of PSII photochemistry) and Rfd (a fluorescence decrease ratio) have even been proposed as early predictors of broccoli plant response treatment to clove oil [58].

Moreover, in a study of photo respiratory pathway alteration by *Origanum vulgare* EOs in *A. thaliana*, Araniti et al. [59] suggested that alteration of glutamate and aspartate metabolism leads to leaf chlorosis and necrosis. Glutamine synthetase is crucial to incorporate ammonia in organic compounds and may be a molecular target of *O. vulgare* EO. Finally, ammonia accretion has direct inhibiting properties on PSI and PSII due to its bonding with the oxygen-evolving complex. In addition, the decrease in pH gradients across membranes is able to uncouple photophosphorylation.

2.6. Mitochondrial Respiration Inhibition

Mitochondrial respiration inhibition is another putative target in the cellular mode of action of EOs. Monoterpene treatment has long been reported to decrease respiratory oxygen consumption in whole plants, dissected organs and isolated mitochondria for 1,8-cineole [60] and juglone [61].

The effect of monoterpenes has been well documented on isolated mitochondria, on germination and on primary root growth of maize [62]. Briefly, the authors demonstrated that α-pinene triggers two different mechanisms which are the uncoupling of oxidative phosphorylation and the inhibition of electron transfer. This action drastically decreases adenosine triphosphate (ATP) production and the authors suggest it occurs following unspecific disruption in the inner mitochondrial membrane [63,64]. The mode of action of other monoterpenes such as camphor and limonene have been investigated. They respectively cause mitochondrial uncoupling and act on ATP synthase or on adenine nucleotide translocase complexes [63,65].

Accessibility to mitochondria in vivo can strongly affect phytotoxicity. A study performed using soy hypocotyl showed that the effect on mitochondria alone did not fully explain the resulting phytotoxic effect. Absence of correlation between respiratory inhibition in mitochondria and seed germination or root growth treated with α-pinene and limonene suggest that their inhibition properties are probably dependent on their ability to permeate intracellular compartments [65].

Furthermore, the description of the cytochrome-oxidase pathway inhibition highlights the fact that this inhibition is likely to increase mitochondrial reactive oxygen species and membrane lipoperoxidation as demonstrated by increased concentrations of lipoperoxide products, activation of lipoxygenase and antioxidant enzymes [66].

Microscopic evaluation highlights the drastic reduction in the number of intact organelles among which mitochondria and membranes disrupt nuclei, mitochondria and dictyosomes [67]. This mitochondrial membrane deleterious effect leads to a decrease in energy production and ROS generation affecting numerous biochemical processes and cellular activities as observed for tobacco BY-2 cells treated with 1,8-cineole [68,69].

2.7. Microtubule Disruption and Genotoxicity

Vapor exposure of citral at µmolar concentrations completely depolymerizes microtubules without any damage to the plasma membrane [70]. Results suggest an in vitro dose/time relationship for microtubule disruption whereas the actin filament remained intact. Finally mitotic microtubules were more damaged than the cortical ones, leading to impairment in the mitosis process [71].

To determine whether the microtubule impact results from direct depolymerization or from indirect phytohormones balance modification, Graña et al. [72] studied the short- and long-term effects of citral application in the plant model *A. thaliana*. Auxins (indole 3-acetic acid) polar transport is rapidly inhibited and ethylene content increases. These two hormones have numerous points of interaction and are essential for microtubule organization, which leads to a long-term disorganization of cell ultra-structure. Citral-treated samples present a large number of Golgi complexes together with a thickening of the cell wall. Those phenomena affect cell division and intracellular communication in the long term.

More recently, Chaimovitsh et al. [73] studied microtubule and membrane damages for a large number of terpenes and further demonstrated the difference in their mechanisms of action. In fact, they observed strong microtubule depolarization for limonene and (+)-citronellal and moderate microtubule depolarization for citral, geraniol, (−)-menthone, (+)-carvone and (−)-citronellal. Moreover, many compounds lacked antitubular activity such as pulegone, (−)-carvone, carvacrol, nerol, geranic acid, (+)/(−)-citronellol and citronellic acid. Furthermore, they demonstrated enantioselectivity of microtubule disruption for citronellal and carvone, the (+) enantiomers being more effective. They compared this antitubular activity with the membrane disrupting properties and found that citral did not cause membrane disruption. Carvacrol induced membrane leakage, and limonene both depolymerized microtubules and induced membrane leakage. Finally, through in vivo quantification of applied monoterpene they discover the biotransformation of citral (i) and limonene (ii) to (i) nerol and geraniol and (ii) carvacrol, respectively. This conversion explains the dual mode of action of limonene in both the membrane and microtubule. Dual mode of action was recently highlighted for menthone in tobacco BY-2 plant cells and seedlings of *A. thaliana* [74].

Concerning direct genotoxicity, numerous chromosome abnormalities have been observed, such as sticky chromosome, chromosome bridges, spindle disturbance, c-mitosis and bi-nucleated cells in root tip cells after treatment with EOs of *Schinus terebinthifolius*, *Citrus aurantiifolia*, *Lectranthus amboinicus*, *Mentha longifolia* and *Nepeta nuda*. The damaging reaction of EOs on the chromatin organization could lead to chromosome bridges or sickness and ultimately to apoptosis. Interestingly, different results for EOs with the same principal terpene suggest that there is a synergic interaction between major and minor compounds [75–79].

Another mito-depressive activity of EOs could be mediated by the inhibition of DNA synthesis. It was effectively demonstrated by Nishida et al. [80] that monoterpenes are able to hinder organelle and nuclear DNA synthesis. Direct damage to DNA has been highlighted through the effect of EOs on head and tail DNA. Although the mechanisms behind this are still vague, authors suggest that ROS following EO treatments may be responsible for the genotoxic effect [81].

2.8. Enzymatic Inhibition and Regulation

Beside glutamine synthetase as a particular enzymatic target of EOs, studies suggest direct or indirect inhibition of specific enzymes as a putative mode of action. For example, a first case is related to the long known potato tuber bud dormancy inhibition using peppermint oil. A decrease in the activity of 3-hydroxy-3-methylglutaryl Coenzyme A reductase (HMGR; E.C. 1.1.1.34), a key-enzyme in the mevalonate pathway, was observed but without explanation at the transcriptional level [82,83].

Rentzsch et al. [84] demonstrated a specific monoterpene interaction with gibberellin (GAs) signaling at the dose-, tissue- and gene-level during dormancy release and sprout growth. They also described a typical case of biostimulation. At low concentrations, peppermint essential oil and carvone promote bud sprouting and dormancy release, whereas at high concentrations they completely inhibit it. They demonstrated that dormancy release is associated with tissue-specific α- and β-amylase modulation and that EOs could affect this modulation. Indeed, at low concentration, amylase expressions were modulated by carvone through specific enhancement of a-AMY2 gene transcription by interacting with its transcription factor. This was not the case for peppermint EOs, for which they proposed interaction with specific components of the GAs signaling pathway that enhanced the GAs-mediated responses [84].

These enzyme modulating activities have been reported for other compounds such as β-pinene reduction of hydrolyzing enzyme (protease, α- and β-amylase) in rice seedlings. At the same time, peroxidases and polyphenol oxidase activity increases, suggesting their role in resistance against β-pinene-induced oxidative stress [53].

Strict inhibition phenomena have been proposed for cinmethylin, which is a synthetic analogue of 1,4 and 1,8-cineole through asparagine synthetase inhibition. Authors have suggested that benzyl ether moiety cleaved to generate toxophore that inhibits the enzyme. However due to an inability to reproduce these results in vivo afterwards, the authors decided to retract the paper. This illustrates well the difficulties in rigorously establishing a single molecular target [85].

Later another target was proposed for the herbicide cinmethylin, the tyrosine aminotransferase (TAT; EC 2.6.1.5). Indeed, TAT provides quinones for the prenylquinones pathway in the inner chloroplast membrane. Furthermore, plastoquinone is a cofactor in the carotenoid pathway. Therefore, the decrease in carotenoid resulting from this inhibition may trigger photo-oxidative degradation of chlorophyll and photosynthetic membranes, disturbing chloroplast function [86].

More recently, Abdelgaleil, Gouda and Saad [87] postulated that phytotoxicity of EOs could be mediated through carbonic anhydrase inhibition. Indeed, this enzyme plays a key role in the (de)carboxylation reaction involved in both respiration and photosynthesis and contributes to the movement of inorganic carbon to photosynthetic cells. Thus, CO_2 content in these cells would decrease, leading to the formation of ROS by diverting a photosynthetic electron from CO_2 [87].

2.9. Phytohormones and Priming of Plant Defence

A first evidence of the interaction with phytohormones has already been developed previously concerning the gibberellin (GAs). Two other interconnected hormones have been suggested as main targets, auxins and ethylene. Indeed, citral impacts the polar auxins transport, resulting in an alteration of its content, cell division and ultrastructure of *A. thaliana* root meristem seedlings cell [72]. Concentration balance between auxin and ethylene is responsible for root growth, radicle elongation and root hair formation. Citral was suggested as a promising herbicide with strong short term and long lasting toxicity. Similar results on polar auxin transportation were obtained with farnesene [88], which affects specific PIN-FORMED (PIN) protein. Furthermore, modification in PIN gene expression leads to a decrease in meristem size and a left-handed phenotype. Interestingly, a previous study reported an increase in the auxin content [56]. This loss of gravitropism was suggested to result from an alteration in the hormonal balance and stimulation of oxidative stress via ROS and RNS production interfering with cell division and cytokinesis through microtubule disruption altering root morphology.

Phytohormone balance is also involved in priming and plant defense induction mechanisms. Monoterpenoids are able to activate defense genes by signaling processes and Ca^{2+} influx causes by membrane depolarization, protein phosphorylation/dephosphorylation and the action of ROS [89]. This gene expression can either lead to priming (an accelerated gene-response to biotic stress) or direct defense elicitations.

Priming of plant defenses has already been acknowledged in agricultural practices, as for example exposure to mint volatiles, which enhanced transcripts levels of defense genes in soy through histone acetylation within the promoter regions [90]. This priming was stronger at mid-distance, implying a nonlinear relationship to concentration. Recently, priming against bacteria was observed in apple using thyme oil. Indeed, the authors noted a much stronger expression of pathogenesis-related (PR) genes PR-8 following *Botrytis cinerea* application [91].

Regarding elicitation of plant defense, resistance can either be constitutive with the systemic acquired resistance (SAR) or induced with the induced systemic resistance (ISR). There is large cross-talk between the two systems which rely on salicylic acid (SA) and jasmonate (JA) hormones.

Transcriptomic study following exposure to volatile monoterpenes myrcene and ocimene demonstrated that plants develop a similar response to that induced by methyl jasmonate (MeJA) [92]. Microarray profiling revealed the induction of several hundreds of transcripts annotated as stress or defense genes or transcription factor. Multiple stages of the octadecanoid pathway were present, and metabolite analysis demonstrates an increased level of MeJA in *A. thaliana* tissues.

The induction of SAR has also been acknowledged when using *Gaultheria procumbens* essential oil, which is composed almost only of methyl salicylate. To demonstrate the effectiveness of the EO, they inoculated GFP-labelled fungal pathogens and showed a strong reduction in its development, similar to commercial solution [93]. Thyme EO also triggers constitutive defense in tomato against grey mold and fusarium as demonstrated by phenolic compounds and peroxidase activity measurements. Furthermore, root application is more effective than foliar. The authors also suggest that an increase in peroxidase activity resulting from oxidative burst (ROS) is a precursor of phenolic compound accumulation. It seems that activation of a plant defense gene and secondary metabolite production can be attributed to Peroxidase-Mediated Reactive Oxygen Species production [94]. Moreover, induction of defense enzymes associated with SAR such as β-1,3-glucanase, chitinase and peroxidase activity, have been observed for different essential oil/constituents namely *Cinnamomum zeylanicum* oil/trans-cinnamaldehyde [95], Indian clove EO/eugenol [96] and citronella EO/citronellal [97].

3. Mechanism of Detoxification

Plants have evolved pathways to decrease the toxicity of allelochemicals released from neighbors and xenobiotics. These mechanisms can be summarized as the metabolization of phytotoxins or conjugation/sequestration followed by compartmentalization or emissions.

Reduction and esterification of aldehydes to their alcohols have been demonstrated for green leaf volatiles such (GLV) as (Z)-3-hexenal [98], but also as previously mentioned for monoterpenes such as citral to nerol and geraniol and limonene to carvacrol [73]. Similar reaction pathways were mentioned for citronellal by *Solanum aviculare* suspension cultures to menthane-3,8-diol, citronellol and isopulegol [99]. Wheat seeds exposed to EOs were also able to oxidize and reduce different terpenes, namely neral, geranial, citronellal, pulegone and carvacrol, to the corresponding alcohol and acids using non-specific enzyme systems. The authors have suggested that the reduction activity was catalyzed by non-specific dehydrogenase and oxidation by P-450-type enzymes [100]. Interestingly, part of the applied compound is degraded, as demonstrated by the impossibility to account for all the compounds supplied to the germinated seeds. Moreover, derivates are less toxic compared to parent compounds [100]. *Anethum graveolens* hairy root cultures biotransform two oxygen-containing monoterpene substrates, menthol or geraniol in 48 h to menthyl acetate, linalool, α-terpineol, citronellol, neral, geranial, citronellyl, neryl, geranyl acetates and nerol oxides [101].

Other detoxifying mechanisms rely on conjugation with carbohydrates, or glycosylation, to sequestrate VOC. Compared to the free aglycones, they present a higher solubility in water and a smaller reactivity, which facilitates their storage in the vacuoles and protects from aglycones toxicity [102]. Numerous studies demonstrate this glycosylation by *Eucalyptus perriniana* culture cell which converts thymol, carvacrol and eugenol into the corresponding β-glucosides and β-gentiobiosides [103]. Biotransformation products were isolated following administration of 1,8-cineole as well. Following the administration of camphor, seven new mono-glucoside products were isolated. Interestingly, the oxygen function was introduced before the glycosylation and ketone group reduction was observed [104]. (−)-fenchone administration delivered six new biotransformation products with specific regio- and stereoselectivity for the hydroxylation reaction [105]. Similar results were obtained for sesamol [106] and vanillin [107] as well.

Cell suspension of *Achillea millefolium* administrated with geraniol, borneol, menthol, thymol and farnesol converts these into several products and glycosylate, both the substrates and the biotransformation products. The decrease in glycosylated compounds afterwards implies that this glycolization mechanism is both used for detoxification and to convert VOC in readily usable forms to incorporate them in the metabolism [108].

This mechanism was also acknowledged *in planta* as demonstrated for (Z)-3-hexenol produced by plants under insect attack [109]. This glycolized form acts as a defense molecule against herbivores, and is accumulated for the sake of prevention of the next attack. A large number of plant families use glycolization as a common pathway of exogenous VOC plant perception. Similar results are observed for other types of alcohols including aromatic, aliphatic and terpene compounds [110].

Another sequestrating reaction consisted in the glutathionylation of GLV, which has been demonstrated for methacrolein whose gluthation conjugates have been isolated from vapor-exposed tomato [111]. α, β-unsaturated aldehydes also react with gluthation [112]. Overall, various processes have been developed by plants to detoxify and they are summarized in Figure 2.

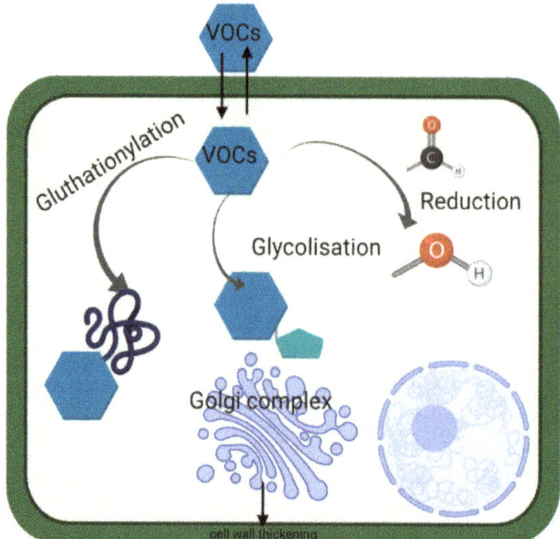

Figure 2. Sequestration and biotransformation of exogenous volatile organic compounds (VOCs) in plant.

4. Discussion and Conclusions

EOs physiological impacts have been and can be studied at the metabolomic [113], proteomic [114] and transcriptomic [115] levels and large amounts of untargeted data will emerge by grouping these techniques of research together. As phytotoxicity is either a goal (herbicide) or a constraint (other biopesticidal application or biostimulation), both parts will be discussed separately.

Regarding herbicidal application, cellular metabolism reactions are clearly involved in the phytotoxic properties of essential oils. The scientific community is making progress in identifying the cellular functions affected, such as photosynthesis, respiration, etc., and research is advancing in molecular target identification. Nevertheless, due to the many interconnecting pathways that are involved simultaneously, no clear distinction has appeared between the diverse chemical classes of EOs compounds. Most of them are grouped within one EO, which makes the unravelling of the specific mode of action a complex process. However, their effects can be distinguished between a general stress type response (ROS or osmotic related) compared to a more specific target (microtubule for example) leading to cellular impairment at a much lower concentration.

To demonstrate persistence and efficiency in the targeted biological system, medium- and long-term effects are most important. To answer these questions, it seems most interesting to deepen the study on the dynamics of the compounds and their fate in plant metabolism in regards to the capacity of the plant to metabolize, detoxify, sequestrate and compartmentalize. Phytotoxicity towards weeds without affecting the crop is essential to develop selective bio-herbicides. In this regard, the identification of other molecular mechanisms such as sugar and amino acid accumulation to prevent EOs stress seems promising as demonstrated in maize [113].

The last point relates to the composition of the EOs. High complexity of EOC needs to be characterized properly as hundreds of compounds sometimes occur [116]. Moreover, variability within the same genus or plant has been frequently observed depending on many parameters such as chemotype, climate, soil, exposure from one year to the next [117,118], sometimes leading to fundamentally different compositions [119]. However, even if fundamental interaction cannot be studied properly for hundreds of compounds, their diverse mechanisms of action can constitute a strong opportunity for synergistic effects and prevent adaptation by weed species. Interaction between different EOC can allow a reduction in the application, while still effectively preventing germination and weed growth [120].

On the other hand, the phytotoxicity of essential oil has long been considered as its main constraint regarding the development of other biopesticides (insecticides, fungicides, etc.) Phytotoxic consideration is currently often limited to the trade-offs of efficiency against the targeted pest versus visual innocuousness to the protected crop. As illustrated in Table 1, large variation occurs regarding the phytotoxic properties of EOs or their constituents depending on the application systems and mode of action considered.

Bioassays should ideally provide a range of toxic concentrations according to the mechanism involved in the toxicity process. Standardized methodologies/protocols to define the toxicity level of individual compounds as well as their blends are needed at the macroscopic or remote level and on a specific scale to allow prediction. It is always a question of targeting an applied plant model and then defining the toxicity levels in those specific application conditions. In this regard, in vivo redox and osmotic status sensor should be used as a specific marker of toxicity levels.

Table 1. Phytotoxic properties of essential oils or constituent solutions in diverse application mode/rate.

Mode of Action	Essential Oils or Constituents (Concentration)	Application Mode (Time)	Plant Target	Observation	Ref
-	Camphor (10 mg/L) menthol (5 mg/L)	Vapor exposure (for 24 to 96 h)	*A. thaliana*	Scanning electron microscopy, transpiration, PCR, western blot	[25]
	Camphor (10 mg/L)	Vapor exposure (for 24 to 96 h)	*A. thaliana*	Real time PCR, in vivo cytoskeleton visualization	[26]
	Clove oil (2.5%) eugenol (1.5%)	Sprayed at 50 mL/m^2	Broccoli, lambsquarte, pigweed	Membrane integrity (EL), spray solution retention	[27]
Water status alteration	Citral (1200–2400 μM)	Watered every 2 day (25 mL per pot)	*A. thaliana*	Water/osmotic potentials (Ψw/Ψs), pigment, protein, anthocyanin, stomata density	[28]
	Trans-caryophyllene (450–1800 μM)	Watering (25 mL/pot) or spraying (15 mL/pot)	*A. thaliana*	Chlorophyll a fluorescence, osmotic potential, MDA, pigment, proline, protein and element content	[29]
	Mentha piperita (5–900 ppm)	Perfusion	*Cucumis sativus*	Root segment membrane potential determination	[31]
	C. zeylanicym *C. winterianus* (3%)	Sprayed (10 L/m^2)	*A. thaliana*	Herbicide tests + in silico approach	[35]
Membrane properties and interaction	1,8-cineole, thymol, menthol, geraniol, camphor (21.7, 2.0, 1.9, 2.5, 7.4 mg/L)	Vapor exposure	*Zea mays*	Lipid, peroxide and lipid peroxidation	[36]
				Sterols and phospholipid fatty acid (PLFA) composition	[37]

Table 1. *Cont.*

Mode of Action	Essential Oils or Constituents (Concentration)	Application Mode (Time)	Plant Target	Observation	Ref
Reactive oxygen and nitrogen species induction	α-pinene (1.36–136 mg/mL)	Vapor exposure in petri dish for 3, 5 and 7 days	*C. occidentalis, A. viridis, T. aestivum, Pisum sativum, Cicer arietinum*	EL, MDA, H_2O_2, proline, ROS scavenging enzymes (SOD, APX, GPX, CAT, GR)	[50]
	β-Pinene (0.02–0.80 mg/mL)				[42]
	β-pinene (1.36–13.6 µg/mL)	Vapor exposure for 4 to 24 h	Wheat seed	H_2O_2, O^{2-}, MDA, ROS scavenging enzymes, LOX	[43]
	Citronellol (50–250 µM)	Watered for 24, 48 and 72 h	Wheat seed	MDA, EL, CDs, LOX, In situ histochemical analyses	[44]
	P. benghalensis (0.25–2.5 mg/mL)	Vapor exposure	*Avena fatua Phalaris minor*	H_2O_2, O^{2-}, MDA, CDs, EL, ROS scavenging enzymes	[45]
	Monarda didyma (0.06–1.25 µg/mL)	Vapor exposure for 5 days	Weed seed	H_2O_2, MDA	[46]
	Artemisia scoparia (0.14–0.70 mg/mL)	Vapor exposure for 5 days	Wheat seed	O^{2-}, H_2O_2, proline, root oxidizability, cell death	[47]
	Heterothalamus psiadioides (1–5 µL)	Vapor exposure in petri dish for 7 days	*A. thaliana*	Histochemical detection of H_2O_2	[48]
Photosynthesis inhibition	β-pinene (135 µM)	Applied to organelles suspension	Chloroplast (*Spinacia oleracea*)	O_2, protein, chlorophyll, electron microscopy	[51]
	β-pinene (945 µM)	Applied to organelles suspension	Chloroplast (*Cucurbita pepo*)	O_2, protein, chlorophyll, Gel electrophoresis and immunoblotting	[52]
	β-pinene (0.02–0.80 mg/mL)	Vapor exposure for 3, 5 and 7 days	*Oryza sativa*	Chlorophyll, protein, carbohydrate, proteases, α- and β-amylases, POD, PER	[53]
	Cymbopogon citratus (1.25–10% (*v/v*))	Foliar sprayed at 1000 L ha^{-1}	Barnyardgrass	Chlorophyll a, b and carotenoid, EL, MDA	[54]

Table 1. *Cont.*

Mode of Action	Essential Oils or Constituents (Concentration)	Application Mode (Time)	Plant Target	Observation	Ref
Photosynthesis inhibition	*Hyptis suaveolens* (1–5% (*v/v*))	Foliar sprayed (10 mL/plant)	*Oryza sativaE. crus-galli*	Total chlorophyll content, cell viability, Cytogenetic analysis	[55]
	Farnesene (0–1200 µM)	Grown in medium for 14 days	*A. thaliana*	Root gravitropism, structural studies, electron microscopy, O^{2-}, H_2O_2, microtubule, ethylene, auxin	[56]
	Artemisia fragrans (0.5, 1, 2 and 4%)	Spraying (100 mL/ pot) for 5 days	*Convolvulus arvensis*	Chlorophyll a fluorescence, chlorophyll, ROS scavenging enzymes, H_2O_2, MDA	[57]
	Clove oil (2.5%), eugenol (1.95%)	Covered by solutions	Broccoli	Chlorophyll a fluorescence imaging at 20, 40 and 60 min	[58]
	Origanum vulgare (0–500 µL/L)	Grown in medium for 10 days	*A. thaliana*	Chlorophyll a fluorescence, chlorophyll, protein, MDA, Ionomic, metabolomic	[59]
Mitochondrial respiration inhibition	1,8-cineole (6 mM)	Apply to organelle	*A. fatua*	O_2 consumption	[60]
	Juglone (10 mM)	Bathed in dark for 30 min	Soybean cotyledons	O_2 consumption and isotope fractionation	[61]
Mitochondrial respiration inhibition	α-pinene, camphor, eucalyptol and limonene (0.1–10 mM)	Vapor exposure/apply to organelle	Maize	Protein, seed germination, growth test and oxygen uptake	[62]
	α-pinene (50–500 µM)	Grown in medium for 10 days	Coleoptiles and primary roots of maize	O_2 consumption, mitochondrial ATP production	[63]

Table 1. *Cont.*

Mode of Action	Essential Oils or Constituents (Concentration)	Application Mode (Time)	Plant Target	Observation	Ref
	Pulegone, menthol, menthone (0–1500 ppm)	Foliar sprayed	Cucumber seeds (roots segments, mitochondria)	O_2 uptake, mitochondrial respiration	[64]
	Camphor, 1,8-Cineole, Limonene, α–pinene (0–500 μM)	Apply to organelle suspension	Corn and soybean	Mitochondrial respiration	[66]
	1,8-cineole (0–2000 μM)	Vapor exposure	*N. tabacum* (seeds)	Growth, protoplasts proliferation, starch accumulation of BY-2	[68]
	Citral (0–1.0 μL)	Vapor exposure	*A. thaliana*	Microscopy, in vitro polymerization of microtubules	[70]
	Citral (0–1.200 μM)	Grown inmedium for 14 days	*A. thaliana*	Ultra-structural, pectin and callose staining, mitotic indices, ethylene, auxin	[71]
	Limonene, citral, carvacrol, pulegone (4.6–9.2 μmol/20 mL)	Vapor exposure for 0, 15, 30 and 60 min	*A. thaliana*	Membrane, microtubules, F-actin, (confocal microscopy), *in Planta* monoterpene concentrations	[73]
	Menthone	Vapor exposure	*Tobacco BY-2A. thaliana*	GFP-tagged markers for microtubules and actin filaments	[74]
Microtubule disruption and genotoxicity	*Schinus molle Schinus terebinthifolius*	Vapor exposure 0.1 mL for 72 h	*Allium cepa, Lactuca sativa*	Cytogenetic assay	[75]
	Citrus aurantiifolia (0.10–1.50 mg/mL)	Vapor exposure (10 mL) for 3–24 h	*Avena fatua, E. crus-galli, Phalaris minor*	Phytotoxicity: dose-response assay, cytotoxicity (*Allium cepa*)	[76]
	Plectrantus amboinicus (0–0.120% *w/v*)	Vapor exposure for 48 h	*Lactuca sativa Sorghum bicolor*	Germination speed index, percentage of germination	[77]
	Mentha longifolia (10–250 μg/mL) (0.5–5%)	Vapor exposure Foliar sprayed (5 mL/pot)	*Cyperus rotundus, E.crus-galli, Oryza sativa*	Germination, root length, coleoptile length, chlorophyll, cytotoxicity assay (*Allium cepa*)	[78]

Table 1. *Cont.*

Mode of Action	Essential Oils or Constituents (Concentration)	Application Mode (Time)	Plant Target	Observation	Ref
Microtubule disruption and genotoxicity	*Nepeta nuda* (0.1–0.8 µL/mL)	Vapor exposure (10 mL) for 7 days	*Zea mays*	Randomly amplified polymorphic DNA, quantitative analysis of proteins	[79]
	Salvia leucophylla (0–1300 µM)	Vapor exposure for 4 days	*Brassica campestris*	DAPI-fluorescence microscopy, immunofluorescence microscopy, DNA Synthesis Activities	[80]
	Vitex negundo (0.1–2.5 mg/mL)	Vapor exposure (12 mL)	*Avena Fatua, E. crus-galli*, Onion bulbs	Phytotoxicity, cytoxicity	[81]
	S-carvone (125 µL)	Vapor exposure (several days)	*Solanum tuberosum*	Potato sprout growth, HMGR activity, membrane protein composition, transcription activity	[82]
Phytohormones	*R/S*-carvone (25–125 µL)	Vapor exposure (several days)	*Solanum tuberosum*	Growth inhibition, carvone and conversion products in potato sprouts	[83]
	Peppermint oil (0.1% (*v/v*))	Vapor exposure	*Solanum tuberosum*	Potato sprout growth, protein extraction, enzyme activity, semi quantitative RT-PCR for potato α–amylase	[84]
	Ten monoterpenes (0.5–2 mM)	Vapor exposure (6 mL) for 9 days	*Silybum marianum*	carbonic anhydrase activity	[87]
	Farnesene (250 µM)	Grown in medium for 14 days	*A. thaliana*	Root anatomy/meristem size, mitotic indices, quantitative PCR, auxin gradient and polar transport	[88]

Other opportunities seem to arise at low concentrations far below the toxicity threshold, such as biostimulation [121] and priming or elicitation of defense mechanisms [91]. This elicitation of the systemic defense mechanism can also result in broader abiotic pest protection and be a pertinent agronomical strategy. However, limitations arise in regard to the allocation of resources (growth-defense trade-off) and reduced efficiency compared to a synthetic product. The same essential oils/constituents are sometimes mentioned to be phytotoxic at high concentrations and beneficial at low ones following a dose response concept. It has been proposed that these low doses simulate mild stress [122]. However, such threshold models as hormesis are still debated in biology and very little is known about the underlying mechanisms [123].

An additional consideration concerns the kinetic release of EOs. Indeed, their persistence and application methods are limited due to their low molecular weight, hydrophobicity and high volatility. To overcome these limitations, much work has been done regarding formulation techniques to allow a control release profile. A recent promising domain is the formulation of nano-emulsion using bio-based surfactants [124] as well as other encapsulation techniques [125].

A final constraint is the market approval by the different regulatory agencies throughout the world as well as economic considerations. Even if procedures are sometimes available for plant-based products such as GRAS, list 25b of the EPA [12] or the European Pesticide Regulation (EC) No. 1107/2009 [126], few active substances have been registered so far. Easier registration also leads to misevaluation regarding efficacy and safety for consumers. Indeed, in high concentrations, their use may be economically disadvantageous and exhibit undesirable phytotoxicity [127]. In fact, the mammalian toxicity (LD50) is >1000 mg kg^{-1} except for some EOs that are moderately toxic to very toxic such as boldo, cedar and pennyroyal with LD50 values of 130, 830 and 400 mg kg^{-1} [128]. Reports of allergenic potential have been made regarding the use of cinnamon and citronella oil [129,130]. Regarding economic considerations, areas of production are increasing every year and decreasing the prohibitive cost of EOs. With controversial products being removed from the market, such as the sprout-preventing chemical chlorpropham (CIPC), alternative products such as EOs are expected to rise. Techno-economic assessments are still lacking regarding a large number of applications. These evaluations combining efficacy, plant safety and social and environmental impacts should clarify many opportunities for the application of EOs [131].

To conclude, the use of EOs for sustainable agricultural practices seems promising, and extensive research will probably clarify or deny their relevance in diverse applications. Due to their inherent characteristics, the pest control properties are usually very transitory and less effective than synthetic products. However, EOs can be an efficient alternative to conventional plant protection products when properly formulated and integrated with other pest management strategies.

Funding: This research was funded by the Department of Research and Technological Development of the Walloon regions of Belgium (DG06) through the TREE-INJECTION project R. RWAL-3157 and by the Education, Audio-visual and Culture Executive Agency (EACEA) through the EOHUB project 600873-EPP-1-2018-1ES-EPPKA2-KA.

Acknowledgments: The authors thank Thierry Hance, Guillaume Le Goff and Patrick du Jardin for their contribution through numerous discussions. All the figures were created with BioRender.com.

Abbreviations

PPP	plant protection product
EO(s)	essential oil(s)
VOCs	volatile organic compounds
EOC	essential oil constituents
IPM	integrated pest management
ATP	adenosine triphosphate
ROS	reactive oxygen species
RNS	reactive nitrogen species
H_2O_2	hydrogen peroxide
MDA	Malondialdehyde
LOX	lipoxygenase
EL	electrolyte leakage
PS	photosystem
GAs	gibberellins
TAT	tyrosine aminotransferase
PR	pathogenesis related
SAR	systemic acquired resistance
ISR	induced systemic resistance
SA	salicylic acid
JA	jasmonic acid
GLV	green leaf volatiles

References

1. Bakkali, F.; Averbeck, S.; Averbeck, D.; Idaomar, M. Biological effects of essential oils—A review. *Food Chem. Toxicol.* **2008**, *46*, 446–475. [CrossRef] [PubMed]
2. Dixon, R.A.; Achnine, L.; Kota, P.; Liu, C.J.; Reddy, M.S.S.; Wang, L. The phenylpropanoid pathway and plant defence-a genomics perspective. *Mol. Plant. Pathol.* **2002**, *3*, 371–390. [CrossRef] [PubMed]
3. O'Bryan, C.A.; Pendleton, S.J.; Crandall, P.G.; Ricke, S.C. Potential of plant essential oils and their components in animal agriculture—In vitro studies on antibacterial mode of action. *Front. Vet. Sci.* **2015**, *2*, 35. [CrossRef] [PubMed]
4. Cavanagh, H. Antifungal activity of the volatile phase of essential oils: A brief review. *Nat. Prod. Commun.* **2007**, *2*, 1297–1302. [CrossRef]
5. Jankowska, M.; Rogalska, J.; Wyszkowska, J.; Stankiewicz, M. Molecular targets for components of essential oils in the insect nervous system—A review. *Molecules* **2018**, *23*, 34. [CrossRef]
6. Peixoto, M.G.; Costa-Junior, L.M.; Blank, A.F.; Lima, A.D.S.; Menezes, T.S.A.; Santos, D.D.A.; Alves, P.B.; Cavalcanti, S.C.D.H.; Bacci, L.; Arrigoni-Blank, M.D.F. Acaricidal activity of essential oils from Lippia alba genotypes and its major components carvone, limonene, and citral against *Rhipicephalus microplus*. *Vet. Parasitol.* **2017**, *210*, 118–122. [CrossRef]
7. Andrés, M.F.; González-Coloma, A.; Sanz, J.; Burillo, J.; Sainz, P. Nematicidal activity of essential oils: A review. *Phytochem. Rev.* **2012**, *11*, 371–390. [CrossRef]
8. Tworkoski, T. Herbicide effects of essential oils. *Weed Sci.* **2002**, *50*, 425–431. [CrossRef]
9. Koul, O.; Walia, S.; Dhaliwal, G. Essential oils as green pesticides: Potential and constraints. *Biopestic. Int.* **2008**, *4*, 63–84.
10. Isman, M.B. Plant essential oils for pest and disease management. *Crop. Prot.* **2000**, *19*, 603–608. [CrossRef]
11. Ibáñez, M.D.; Blázquez, M.A. Phytotoxic effects of commercial essential oils on selected vegetable crops: Cucumber and tomato. *Sustain. Chem. Pharm.* **2020**, *15*, 100209. [CrossRef]
12. Isman, M.B.; Zheljazkov, V.; Cantrell, C.L. Pesticides based on plant essential oils: Phytochemical and practical considerations. In *ACS Symposium Series*; American Chemical Society: Washington, DC, USA, 2016; Volume 1228, pp. 13–26.
13. Domingues, P.M.; Santos, L. Essential oil of pennyroyal (*Mentha pulegium*): Composition and applications as alternatives to pesticides—New tendencies. *Ind. Crop. Prod.* **2019**, *139*, 111534. [CrossRef]

14. Rolli, E.; Marieschi, M.; Maietti, S.; Sacchetti, G.; Bruni, R. Comparative phytotoxicity of 25 essential oils on pre- and post-emergence development of *Solanum lycopersicum* L.: A multivariate approach. *Ind. Crop. Prod.* **2014**, *60*, 280–290. [CrossRef]

15. Topuz, E.; Madanlar, N.; Erler, F. Chemical composition, toxic and development- and reproduction-inhibiting effects of some essential oils against *Tetranychus urticae* Koch (Acarina: *Tetranychidae*) as fumigants. *J. Plant. Dis. Prot.* **2018**, *125*, 377–387. [CrossRef]

16. Yan, Z.Q.; Wang, D.-D.; Ding, L.; Cui, H.Y.; Jin, H.; Yang, X.Y.; Yang, J.S.; Qin, B. Mechanism of artemisinin phytotoxicity action: Induction of reactive oxygen species and cell death in lettuce seedlings. *Plant. Physiol. Biochem.* **2015**, *88*, 53–59. [CrossRef]

17. Singh, P.; Pandey, A.K. Prospective of essential oils of the genus *Mentha* as biopesticides: A review. *Front. Plant. Sci.* **2018**, *9*, 1295. [CrossRef]

18. Grana, E.; Sánchez-Moreiras, A.M.; Reigosa, M.J. Mode of action of monoterpenes in plant-plant interactions. *Curr. Bioact. Compd.* **2012**, *8*, 80–89. [CrossRef]

19. Widhalm, J.R.; Jaini, R.; Morgan, J.A.; Dudareva, N. Rethinking how volatiles are released from plant cells. *Trends Plant. Sci.* **2015**, *20*, 545–550. [CrossRef]

20. Maffei, M.E. Sites of synthesis, biochemistry and functional role of plant volatiles. *S. Afr. J. Bot.* **2010**, *76*, 612–631. [CrossRef]

21. Dong, F.; Fu, X.; Watanabe, N.; Su, X.; Yang, Z. Recent advances in the emission and functions of plant vegetative volatiles. *Molecules* **2016**, *21*, 124. [CrossRef]

22. Sugimoto, K.; Matsui, K.; Takabayashi, J.; Blande, J.D.; Glinwood, R. *Uptake and Conversion of Volatile Compounds in Plant–Plant Communication*; Springer: Cham, Switzerland, 2016; pp. 305–316.

23. Adebesin, F.; Widhalm, J.R.; Boachon, B.; Lefèvre, F.; Pierman, B.; Lynch, J.H.; Alam, I.; Junqueira, B.; Benke, R.; Ray, S.; et al. Emission of volatile organic compounds from petunia flowers is facilitated by an ABC transporter. *Science* **2017**, *356*, 1386–1388. [CrossRef] [PubMed]

24. Wong, C.R.; Coats, J.R. Movement of thymol in citrus plants. In *Roles of Natural Products for Biorational Pesticides in Agriculture*; American Chemical Society: Washington, DC, USA, 2018; Volume 1294, pp. 23–32.

25. Schulz, M.; Kussmann, P.; Knop, M.; Kriegs, B.; Gresens, F.; Eichert, T.; Ulbrich, A.; Marx, F.; Fabricius, H.; Goldbach, H.; et al. Allelopathic monoterpenes interfere with *Arabidopsis thaliana* cuticular waxes and enhance transpiration. *Plant. Signal. Behav.* **2007**, *2*, 231–239. [CrossRef] [PubMed]

26. Kriegs, B.; Jansen, M.; Hahn, K.; Peisker, H.; Šamajová, O.; Beck, M.; Braun, S.; Ulbrich, A.; Baluška, F.; Schulz, M. Plant Signaling & Behavior Cyclic monoterpene mediated modulations of *Arabidopsis thaliana* phenotype Effects on the cytoskeleton and on the expression of selected genes. *Plant Signal. Behav.* **2010**, *5*, 832–838.

27. Bainard, L.D.; Isman, M.B.; Upadhyaya, M.K. Phytotoxicity of clove oil and its primary constituent eugenol and the role of leaf epicuticular wax in the susceptibility to these essential oils. *Weed Sci.* **2006**, *54*, 833–837. [CrossRef]

28. Grana, E.; Díaz-Tielas, C.; López-González, D.; Martínez-Peñalver, A.; Reigosa, M.J.; Sánchez-Moreiras, A.M. The plant secondary metabolite citral alters water status and prevents seed formation in *Arabidopsis thaliana*. *Plant. Biol.* **2016**, *18*, 423–432. [CrossRef]

29. Araniti, F.; Sánchez-Moreiras, A.M.; Graña, E.; Reigosa, M.J.; Abenavoli, M.R. Terpenoid trans-caryophyllene inhibits weed germination and induces plant water status alteration and oxidative damage in adult *Arabidopsis*. *Plant Biol.* **2017**, *19*, 79–89. [CrossRef]

30. Griffin, S.; Markham, J.L.; Dennis, G.; Grant Wyllie, S. Using atomic force microscopy to view the effects of terpenoids on the stability and packing of phosphatidylcholine supported lipid bilayers. In Proceedings of the 31st International Symposium on Essential Oils, Hamburg, Germany, 10–13 September 2000.

31. Maffei, M.E.; Camusso, W.; Sacco, S. Effect of Mentha x piperita essential oil and monoterpenes on cucumber root membrane potential. *Phytochemistry* **2001**, *58*, 703–707. [CrossRef]

32. Marcec, M.J.; Gilroy, S.; Poovaiah, B.; Tanaka, K. Mutual interplay of Ca^{2+} and ROS signaling in plant immune response. *Plant. Sci.* **2019**, *283*, 343–354. [CrossRef]

33. Turina, A.V.; Nolan, M.; Zygadlo, J.; Perillo, M. Natural terpenes: Self-assembly and membrane partitioning. *Biophys. Chem.* **2006**, *122*, 101–113. [CrossRef]

34. Witzke, S.; Duelund, L.; Kongsted, J.; Petersen, M.; Mouritsen, O.G.; Khandelia, H. Inclusion of terpenoid plant extracts in lipid bilayers investigated by molecular dynamics simulations. *J. Phys. Chem. B* **2010**, *114*, 15825–15831. [CrossRef]

35. Lins, L.; Maso, S.D.; Foncoux, B.; Kamili, A.; Laurin, Y.; Genva, M.; Jijakli, M.H.; Fauconnier, M.L.; Fauconnier, M.L.; Deleu, M. Insights into the relationships between herbicide activities, molecular structure and membrane interaction of cinnamon and citronella essential oils components. *Int. J. Mol. Sci.* **2019**, *20*, 4007. [CrossRef] [PubMed]

36. Zygadlo, J.A. Effect of monoterpenes on lipid oxidation in maize. *Planta* **2004**, *219*, 303–309. [CrossRef] [PubMed]

37. Zunino, M.P.; Zygadlo, J.A. Changes in the composition of phospholipid fatty acids and sterols of maize root in response to monoterpenes. *J. Chem. Ecol.* **2005**, *31*, 1269–1283. [CrossRef] [PubMed]

38. Gniazdowska, A.; Krasuska, U.; Andrzejczak, O.; Sołtys-Kalina, D. Allelopathic compounds as oxidative stress agents: Yes or no. In *Signaling and Communication in Plants*; Springer: Cham, Switzerland, 2015; pp. 155–176.

39. Dayan, F.E.; Romagni, J.G.; Duke, S.O. Investigating the mode of action of natural phytotoxins. *J. Chem. Ecol.* **2000**, *26*, 2079–2094. [CrossRef]

40. Mittler, R. Oxidative stress, antioxidants and stress tolerance. *Trends Plant. Sci.* **2002**, *7*, 405–410. [CrossRef]

41. Sunohara, Y.; Baba, Y.; Matsuyama, S.; Fujimura, K.; Matsumoto, H. Screening and identification of phytotoxic volatile compounds in medicinal plants and characterizations of a selected compound, eucarvone. *Protoplasma* **2015**, *252*, 1047–1059. [CrossRef]

42. Chowhan, N.; Singh, H.P.; Batish, D.R.; Kaur, S.; Ahuja, N.; Kohli, R.K. β-Pinene inhibited germination and early growth involves membrane peroxidation. *Protoplasma* **2013**, *250*, 691–700. [CrossRef]

43. Chowhan, N.; Bali, A.S.; Singh, H.P.; Batish, D.R.; Kohli, R.K. Reactive oxygen species generation and antioxidant defense system in hydroponically grown wheat (*Triticum aestivum*) upon β-pinene exposure: An early time course assessment. *Acta Physiol. Plant.* **2014**, *36*, 3137–3146. [CrossRef]

44. Kaur, S.; Rana, S.; Singh, H.P.; Batish, D.R.; Kohli, R.K. Citronellol disrupts membrane integrity by inducing free radical generation. *Z. Naturforsch. Sect. C J. Biosci.* **2011**, *66*, 260–266. [CrossRef]

45. Dahiya, S.; Batish, D.R.; Singh, H.P. *Pogostemon benghalensis* essential oil inhibited the weed growth via causing oxidative damage. *Braz. J. Bot.* **2020**, *43*, 1–11. [CrossRef]

46. Ricci, D.; Epifano, F.; Fraternale, D. The essential oil of *Monarda didyma* L. (*Lamiaceae*) exerts phytotoxic activity in vitro against various weed seed. *Molecules* **2017**, *22*, 222. [CrossRef] [PubMed]

47. Kaur, S.; Singh, H.P.; Batish, D.R.; Kohli, R.K. Artemisia scoparia essential oil inhibited root growth involves reactive oxygen species (ROS)-mediated disruption of oxidative metabolism: In vivo ROS detection and alterations in antioxidant enzymes. *Biochem. Syst. Ecol.* **2012**, *44*, 390–399. [CrossRef]

48. Lazarotto, D.C.; Pawlowski, Â.; Da Silva, E.R.; Schwambach, J.; Soares, G.L.G. Phytotoxic effects of *Heterothalamus psiadioides* (*Asteraceae*) essential oil on adventitious rooting. *Acta Physiol. Plant.* **2014**, *36*, 3163–3171. [CrossRef]

49. Sewelam, N.; Kazan, K.; Schenk, P.M. Global plant stress signaling: Reactive oxygen species at the cross-road. *Front. Plant. Sci.* **2016**, *7*, 187. [CrossRef] [PubMed]

50. Singh, H.P.; Batish, D.R.; Kaur, S.; Arora, K.; Kohli, R.K. α-Pinene inhibits growth and induces oxidative stress in roots. *Ann. Bot.* **2006**, *98*, 1261–1269. [CrossRef]

51. Pauly, G.; Douce, R.; Carde, J.P. Effects of β-pinene on spinach chloroplast photosynthesis. *Z. Pflanzenphysiol.* **1981**, *104*, 199–206. [CrossRef]

52. Klingler, H.; Frosch, S.; Wagner, E. In vitro effects of monoterpenes on chloroplast membranes. *Photosynth. Res.* **1991**, *28*, 109–118. [CrossRef]

53. Chowhan, N.; Singh, H.P.; Batish, D.R.; Kohli, R.K. Phytotoxic effects of β-pinene on early growth and associated biochemical changes in rice. *Acta Physiol. Plant.* **2011**, *33*, 2369–2376. [CrossRef]

54. Poonpaiboonpipat, T.; Pangnakorn, U.; Suvunnamek, U.; Teerarak, M.; Charoenying, P.; Laosinwattana, C. Phytotoxic effects of essential oil from *Cymbopogon citratus* and its physiological mechanisms on barnyardgrass (*Echinochloa crus-galli*). *Ind. Crop. Prod.* **2013**, *41*, 403–407. [CrossRef]

55. Sharma, A.; Singh, H.P.; Batish, D.R.; Kohli, R.K. Chemical profiling, cytotoxicity and phytotoxicity of foliar volatiles of *Hyptis suaveolens*. *Ecotoxicol. Environ. Saf.* **2019**, *171*, 863–870. [CrossRef]

56. Araniti, F.; Grana, E.; Krasuska, U.; Bogatek, R.; Reigosa, M.J.; Abenavoli, M.R.; Sánchez-Moreiras, A.M. Loss of gravitropism in farnesene-treated *Arabidopsis* is due to microtubule malformations related to hormonal and ROS unbalance. *PLoS ONE* **2016**, *11*, e0160202. [CrossRef] [PubMed]

57. Pouresmaeil, M.; Nojadeh, M.S.; Movafeghi, A.; Maggi, F. Exploring the bio-control efficacy of *Artemisia fragrans* essential oil on the perennial weed *Convolvulus arvensis*: Inhibitory effects on the photosynthetic machinery and induction of oxidative stress. *Ind. Crop. Prod.* **2020**, *155*, 112785. [CrossRef]

58. Synowiec, A.; Możdżeń, K.; Skoczowski, A. Early physiological response of broccoli leaf to foliar application of clove oil and its main constituents. *Ind. Crop. Prod.* **2015**, *74*, 523–529. [CrossRef]

59. Araniti, F.; Landi, M.; Lupini, A.; Sunseri, F.; Guidi, L.; Abenavoli, M.R. *Origanum vulgare* essential oils inhibit glutamate and aspartate metabolism altering the photorespiratory pathway in *Arabidopsis thaliana* seedlings. *J. Plant. Physiol.* **2018**, *231*, 297–309. [CrossRef]

60. Muller, W.H.; Lorber, P.; Haley, B.; Johnson, K. Volatile growth inhibitors produced by *Salvia leucophylla*: Effect on oxygen uptake by mitochondrial suspensions. *Bull. Torrey Bot. Club* **1969**, *96*, 89. [CrossRef]

61. Peñuelas, J.; Ribas-Carbo, M.; Giles, L. Effects of allelochemicals on plant respiration and oxygen isotope fractionation by the alternative oxidase. *J. Chem. Ecol.* **1996**, *22*, 801–805. [CrossRef] [PubMed]

62. Abrahim, D.; Braguini, W.L.; Kelmer-Bracht, A.M.; Ishii-Iwamoto, E.L. Effects of four monoterpenes on germination, primary root growth, and mitochondrial respiration of maize. *J. Chem. Ecol.* **2000**, *26*, 611–624. [CrossRef]

63. Abrahim, D.; Francischini, A.C.; Pergo, E.M.; Kelmer-Bracht, A.M.; Ishii-Iwamoto, E.L. Effects of α-pinene on the mitochondrial respiration of maize seedlings. *Plant. Physiol. Biochem.* **2003**, *41*, 985–991. [CrossRef]

64. Mucciarelli, M.; Camusso, W.; Bertea, C.M.; Maffei, M. Effect of (+)-pulegone and other oil components of *Mentha x Piperita* on cucumber respiration. *Phytochemistry* **2001**, *57*, 91–98. [CrossRef]

65. Weir, T.L.; Park, S.-W.; Vivanco, J.M. Biochemical and physiological mechanisms mediated by allelochemicals. *Curr. Opin. Plant. Biol.* **2004**, *7*, 472–479. [CrossRef]

66. Luiza Ishii-Iwamoto, E.; Marusa Pergo Coelho, E.; Reis, B.; Sebastiao Moscheta, I.; Moacir Bonato, C. Effects of monoterpenes on physiological processes during seed germination and seedling growth. *Curr. Bioact. Compd.* **2012**, *8*, 50–64. [CrossRef]

67. Lorber, P.; Muller, W.H. Volatile growth inhibitors produced by *Salvia leucophylla*: Effects on seedling root tip ultrastructure. *Am. J. Bot.* **1976**, *63*, 196–200. [CrossRef]

68. Yoshimura, H.; Sawai, Y.; Tamotsu, S.; Sakai, A. 1,8-cineole inhibits both proliferation and elongation of BY-2 cultured tobacco cells. *J. Chem. Ecol.* **2011**, *37*, 320–328. [CrossRef] [PubMed]

69. Atsushi, S.; Hiroko, Y. Monoterpenes of salvia leucophylla. *Curr. Bioact. Compd.* **2012**, *8*, 90–100. [CrossRef]

70. Chaimovitsh, D.; Abu-Abied, M.; Belausov, E.; Rubin, B.; Dudai, N.; Sadot, E. Microtubules are an intracellular target of the plant terpene citral. *Plant. J.* **2010**, *61*, 399–408. [CrossRef]

71. Chaimovitsh, D.; Stelmakh, O.R.; Altshuler, O.; Belausov, E.; Abu-Abied, M.; Rubin, B.; Sadot, E.; Dudai, N. The relative effect of citral on mitotic microtubules in wheat roots and BY2 cells. *Plant. Biol.* **2012**, *14*, 354–364. [CrossRef]

72. Grana, E.; Sotelo, T.; Díaz-Tielas, C.; Araniti, F.; Krasuska, U.; Bogatek, R.; Reigosa, M.J.; Sánchez-Moreiras, A.M. Citral induces auxin and ethylene-mediated malformations and arrests cell division in *Arabidopsis thaliana* roots. *J. Chem. Ecol.* **2013**, *39*, 271–282. [CrossRef]

73. Chaimovitsh, D.; Shachter, A.; Abu-Abied, M.; Rubin, B.; Sadot, E.; Dudai, N.; Shechter, A.; Abu-Abeid, M. Herbicidal activity of monoterpenes is associated with disruption of microtubule functionality and membrane integrity. *Weed Sci.* **2016**, *65*, 19–30. [CrossRef]

74. Sarheed, M.M.; Rajabi, F.; Kunert, M.; Boland, W.; Wetters, S.; Miadowitz, K.; Kaźmierczak, A.; Sahi, V.P.; Nick, P. Cellular base of mint allelopathy: Menthone affects plant microtubules. *Front. Plant Sci.* **2020**, *11*, 1320.

75. Pawlowski, A.; Kaltchuk-Santos, E.; Zini, C.A.; Caramão, E.; Soares, G. Essential oils of *Schinus terebinthifolius* and *S. molle* (*Anacardiaceae*): Mitodepressive and aneugenic inducers in onion and lettuce root meristems. *S. Afr. J. Bot.* **2012**, *80*, 96–103. [CrossRef]

76. Fagodia, S.K.; Singh, H.P.; Batish, D.R.; Kohli, R.K. Phytotoxicity and cytotoxicity of *Citrus aurantiifolia* essential oil and its major constituents: Limonene and citral. *Ind. Crop. Prod.* **2017**, *108*, 708–715. [CrossRef]

77. Pinheiro, P.F.; Costa, A.V.; Alves, T.D.A.; Galter, I.N.; Pinheiro, C.A.; Pereira, A.F.; Oliveira, C.M.R.; Fontes, M.M.P. Phytotoxicity and cytotoxicity of essential oil from leaves of *Plectranthus amboinicus*, carvacrol, and thymol in plant bioassays. *J. Agric. Food Chem.* **2015**, *63*, 8981–8990. [CrossRef] [PubMed]

78. Singh, N.; Singh, H.P.; Batish, D.R.; Kohli, R.K.; Yadav, S.S. Chemical characterization, phytotoxic, and cytotoxic activities of essential oil of *Mentha longifolia*. *Environ. Sci. Pollut. Res.* **2020**, *27*, 13512–13523. [CrossRef] [PubMed]

79. Bozari, S.; Agar, G.; Aksakal, O.; Erturk, F.A.; Yanmis, D. Determination of chemical composition and genotoxic effects of essential oil obtained from *Nepeta nuda* on *Zea mays* seedlings. *Toxicol. Ind. Health* **2013**, *29*, 339–348. [CrossRef] [PubMed]

80. Nishida, N.; Tamotsu, S.; Nagata, N.; Saito, C.; Sakai, A. Allelopathic effects of volatile monoterpenoids produced by *Salvia leucophylla*: Inhibition of cell proliferation and DNA synthesis in the root apical meristem of *Brassica campestris* seedlings. *J. Chem. Ecol.* **2005**, *31*, 1187–1203. [CrossRef]

81. Issa, M.; Chandel, S.; Singh, H.P.; Batish, D.R.; Kohli, R.K.; Yadav, S.S.; Kumari, A. Appraisal of phytotoxic, cytotoxic and genotoxic potential of essential oil of a medicinal plant *Vitex negundo*. *Ind. Crop. Prod.* **2020**, *145*, 112083. [CrossRef]

82. Oosterhaven, K.; Hartmans, K.J.; Huizing, H.J. Inhibition of potato (*Solanum tuberosum*) sprout growth by the monoterpene *S*-carvone: Reduction of 3-hydroxy-3-methylglutaryl coenzyme a reductase activity without effect on its mRNA level. *J. Plant. Physiol.* **1993**, *141*, 463–469. [CrossRef]

83. Oosterhaven, K.; Hartmans, K.J.; Scheffer, J.J.C. Inhibition of potato sprout growth by carvone enantiomers and their bioconversion in sprouts. *Potato Res.* **1995**, *38*, 219–230. [CrossRef]

84. Rentzsch, S.; Podzimska, D.; Voegele, A.; Imbeck, M.; Müller, K.; Linkies, A.; Leubner-Metzger, G. Dose- and tissue-specific interaction of monoterpenes with the gibberellin-mediated release of potato tuber bud dormancy, sprout growth and induction of α-amylases and β-amylases. *Planta* **2012**, *235*, 137–151. [CrossRef]

85. Romagni, J.G. Inhibition of plant asparagine synthetase by monoterpene cineoles. *Plant. Physiol.* **2000**, *123*, 725–732. [CrossRef]

86. Grossmann, K.; Hutzler, J.; Tresch, S.; Christiansen, N.; Looser, R.; Ehrhardt, T. On the mode of action of the herbicides cinmethylin and 5-benzyloxymethyl-1, 2-isoxazolines: Putative inhibitors of plant tyrosine aminotransferase. *Pest. Manag. Sci.* **2011**, *68*, 482–492. [CrossRef] [PubMed]

87. Abdelgaleil, S.A.M.; Gouda, N.A.; Saad, M.M.G. Phytotoxic properties of monoterpenes on *Silybum marianum* (L.) gaertn. and their inhibitory effect on carbonic anhydrase activity. *Aust. J. Basic Appl. Sci.* **2014**, *8*, 356–363.

88. Araniti, F.; Bruno, L.; Sunseri, F.; Pacenza, M.; Forgione, I.; Bitonti, M.B.; Abenavoli, M.R. The allelochemical farnesene affects *Arabidopsis thaliana* root meristem altering auxin distribution. *Plant. Physiol. Biochem.* **2017**, *121*, 14–20. [CrossRef]

89. Massimo, E. Maffei monoterpenoid plant-plant interactions upon herbivory. *Curr. Bioact. Compd.* **2012**, *8*, 65–70. [CrossRef]

90. Sukegawa, S.; Shiojiri, K.; Higami, T.; Suzuki, S.; Arimura, G. ichiro Pest management using mint volatiles to elicit resistance in soy: Mechanism and application potential. *Plant J.* **2018**, *96*, 910–920. [CrossRef]

91. Banani, H.; Olivieri, L.; Santoro, K.; Garibaldi, A.; Gullino, M.L.; Spadaro, D. Thyme and savory essential oil efficacy and induction of resistance against botrytis cinerea through priming of defense responses in apple. *Foods* **2018**, *7*, 11. [CrossRef]

92. Godard, K.A.; White, R.; Bohlmann, J. Monoterpene-induced molecular responses in *Arabidopsis thaliana*. *Phytochemistry* **2008**, *69*, 1838–1849. [CrossRef]

93. Vergnes, S.; Ladouce, N.; Fournier, S.; Ferhout, H.; Attia, F.; Dumas, B. Foliar treatments with *Gaultheria procumbens* essential oil induce defense responses and resistance against a fungal pathogen in *Arabidopsis*. *Front. Plant. Sci.* **2014**, *5*, 477. [CrossRef] [PubMed]

94. Ben-Jabeur, M.; Ghabri, E.; Myriam, M.; Hamada, W. Thyme essential oil as a defense inducer of tomato against gray mold and *Fusarium* wilt. *Plant. Physiol. Biochem.* **2015**, *94*, 35–40. [CrossRef]

95. Perina, F.J.; De Andrade, C.C.L.; Moreira, S.I.; Nery, E.M.; Ogoshi, C.; Alves, E. *Cinnamomun zeylanicum* oil and trans-cinnamaldehyde against Alternaria brown spot in tangerine: Direct effects and induced resistance. *Phytoparasitica* **2019**, *47*, 575–589. [CrossRef]

96. Lucas, G.C.; Alves, E.; Pereira, R.B.; Zacaroni, A.B.; Perina, F.J.; de Souza, R.M. Indian clove essential oil in the control of tomato bacterial spot. *J. Plant Pathol.* **2012**, *94*, 45–51.

97. Borges Pereira, R.; Lucas, G.C.; Perina, J.; Martins, P.; Júnior, R.; Alves, E. Citronella essential oil in the control and activation of coffee plants defense response agains rust and brown eye spot. *Cienc. Agrotec.* **2012**, *36*, 383–390. [CrossRef]

98. Matsui, K.; Sugimoto, K.; Mano, J.; Ozawa, R.; Takabayashi, J. Differential metabolisms of green leaf volatiles in injured and intact parts of a wounded leaf meet distinct ecophysiological requirements. *PLoS ONE* **2012**, *7*, e36433. [CrossRef] [PubMed]

99. Vaněk, T.; Novotný, M.; Podlipná, R.; Šaman, D.; Valterová, I. Biotransformation of citronellal by *Solanum aviculare* suspension cultures: Preparation of p-menthane-3,8-diols and determination of their absolute configurations. *J. Nat. Prod.* **2003**, *66*, 1239–1241. [CrossRef] [PubMed]

100. Dudai, N.; Larkov, O.; Putievsky, E.; Lerner, H.R.; Ravid, U.; Lewinsohn, E.; Mayer, A.M. Biotransformation of constituents of essential oils by germinating wheat seed. *Phytochemistry* **2000**, *55*, 375–382. [CrossRef]

101. Faria, J.M.S.; Nunes, I.S.; Figueiredo, A.C.; Pedro, L.G.; Trindade, H.; Barroso, J.G. Biotransformation of menthol and geraniol by hairy root cultures of *Anethum graveolens*: Effect on growth and volatile components. *Biotechnol. Lett.* **2009**, *31*, 897–903. [CrossRef]

102. Rivas, F.; Parra, A.; Martinez, A.; García-Granados, A. Enzymatic glycosylation of terpenoids. *Phytochem. Rev.* **2013**, *12*, 327–339. [CrossRef]

103. Shimoda, K.; Kondo, Y.; Nishida, T.; Hamada, H.; Nakajima, N.; Hamada, H. Biotransformation of thymol, carvacrol, and eugenol by cultured cells of *Eucalyptus perriniana*. *Phytochemistry* **2006**, *67*, 2256–2261. [CrossRef]

104. Orihara, Y.; Furuya, T. Biotransformation of 1,8-cineole by cultured cells of *Eucalyptus perriniana*. *Phytochemistry* **1994**, *35*, 641–644. [CrossRef]

105. Orihara, Y.; Furuya, T. Biotransformation of (+)- and (−)-fenchone by cultured cells of *Eucalyptus perriniana*. *Phytochemistry* **1994**, *36*, 55–59. [CrossRef]

106. Shimoda, K.; Ishimoto, H.; Kamiue, T.; Kobayashi, T.; Hamada, H.; Hamada, H. Glycosylation of sesamol by cultured plant cells. *Phytochemistry* **2009**, *70*, 207–210. [CrossRef] [PubMed]

107. Sato, D.; Eshita, Y.; Katsuragi, H.; Hamada, H.; Shimoda, K.; Kubota, N. Glycosylation of vanillin and 8-nordihydrocapsaicin by cultured *Eucalyptus perriniana* cells. *Molecules* **2012**, *17*, 5013–5020. [CrossRef] [PubMed]

108. Figueiredo, A.C.; Scheffer, J.J.C. Biotransformation of monoterpenes and sesquiterpenes by cell suspension cultures of *Achillea millefolium* L. ssp. millefolium. *Biotechnol. Lett.* **1996**, *18*, 863–868. [CrossRef]

109. Sugimoto, K.; Matsui, K.; Takabayashi, J. Conversion of volatile alcohols into their glucosides in Arabidopsis. *Commun. Integr. Biol.* **2015**, *8*, 992731. [CrossRef] [PubMed]

110. Yamauchi, Y.; Kunishima, M.; Mizutani, M.; Sugimoto, Y. Reactive short-chain leaf volatiles act as powerful inducers of abiotic stress-related gene expression. *Sci. Rep.* **2015**, *5*, 8030. [CrossRef]

111. Muramoto, S.; Matsubara, Y.; Mwenda, C.M.; Koeduka, T.; Sakami, T.; Tani, A.; Matsui, K. Glutathionylation and reduction of methacrolein in tomato plants account for its absorption from the vapor phase. *Plant. Physiol.* **2015**, *169*, 1744–1754. [CrossRef]

112. Galindo, J.C.G.; Hernández, A.; Dayan, F.E.; Tellez, M.R.; Macías, F.A.; Paul, R.N.; Duke, S.O. Dehydrozaluzanin C, a natural sesquiterpenolide, causes rapid plasma membrane leakage. *Phytochemistry* **1999**, *52*, 805–813. [CrossRef]

113. Synowiec, A.; Możdżeń, K.; Krajewska, A.; Landi, M.; Araniti, F. *Carum carvi* L. essential oil: A promising candidate for botanical herbicide against *Echinochloa crus-galli* (L.) P. Beauv. in maize cultivation. *Ind. Crop. Prod.* **2019**, *140*, 111652. [CrossRef]

114. Araniti, F.; Miras-Moreno, B.; Lucini, L.; Landi, M.; Abenavoli, M.R. Metabolomic, proteomic and physiological insights into the potential mode of action of thymol, a phytotoxic natural monoterpenoid phenol. *Plant. Physiol. Biochem.* **2020**, *153*, 141–153. [CrossRef]

115. Rienth, M.; Crovadore, J.; Ghaffari, S.; Lefort, F. Oregano essential oil vapour prevents *Plasmopara viticola* infection in grapevine (*Vitis Vinifera*) and primes plant immunity mechanisms. *PLoS ONE* **2019**, *14*, e0222854. [CrossRef] [PubMed]

116. Brokl, M.; Fauconnier, M.L.; Benini, C.; Lognay, G.C.; Du Jardin, P.; Focant, J.F. Improvement of ylang-ylang essential oil characterization by GCxGC-TOFMS. *Molecules* **2013**, *18*, 1783–1797. [CrossRef] [PubMed]

117. Benini, C.; Ringuet, M.; Wathelet, J.P.; Lognay, G.C.; Jardin, P.; Fauconnier, M.L. Variations in the essential oils from ylang-ylang (*Cananga odorata* [Lam.] Hook f. & Thomson forma genuina) in the Western Indian Ocean islands. *Flavour Fragr. J.* **2012**, *27*, 356–366. [CrossRef]

118. Tanoh, E.A.; Boué, G.B.; Nea, F.; Genva, M.; Wognin, E.L.; LeDoux, A.; Martin, H.; Tonzibo, F.Z.; Frédérich, M.; Fauconnier, M.L. Seasonal effect on the chemical composition, insecticidal properties and other biological activities of *Zanthoxylum leprieurii* guill. & perr. essential oils. *Foods* **2020**, *9*, 550. [CrossRef]

119. Nea, F.; Tanoh, E.A.; Wognin, E.L.; Kemene, T.K.; Genva, M.; Saive, M.; Tonzibo, Z.F.; Fauconnier, M.L. A new chemotype of *Lantana rhodesiensis* Moldenke essential oil from Côte d'Ivoire: Chemical composition and biological activities. *Ind. Crop. Prod.* **2019**, *141*, 111766. [CrossRef]

120. Mirmostafaee, S.; Azizi, M.; Fujii, Y. Study of allelopathic interaction of essential oils from medicinal and aromatic plants on seed germination and seedling growth of lettuce. *Agronomy* **2020**, *10*, 163. [CrossRef]

121. Souri, M.K.; Bakhtiarizade, M. Biostimulation effects of rosemary essential oil on growth and nutrient uptake of tomato seedlings. *Sci. Hortic.* **2019**, *243*, 472–476. [CrossRef]

122. Hara, M. Potential use of essential oils to enhance heat tolerance in plants. *Z. Naturforsch. Sect. C J. Biosci.* **2020**, *75*, 225–231. [CrossRef]

123. Calabrese, E.J. The emergence of the dose–Response concept in biology and medicine. *Int. J. Mol. Sci.* **2016**, *17*, 2034. [CrossRef]

124. Uthappa, U.T.; Kurkuri, M.D.; Kigga, M. Nanotechnology Advances for the Development of Various Drug Carriers. In *Nanobiotechnology in Bioformulations*; Springer: Cham, Switzerland, 2019; pp. 187–224.

125. Maes, C.; Bouquillon, S.; Fauconnier, M.L. Maes Encapsulation of essential oils for the development of biosourced pesticides with controlled release: A review. *Molecules* **2019**, *24*, 2539. [CrossRef]

126. Villaverde, J.J.; Sevilla-Morán, B.; Sandín-España, P.; López-Goti, C.; Alonso-Prados, J.L. Biopesticides in the framework of the European Pesticide Regulation (EC) No. 1107/2009. *Pest Manag. Sci.* **2014**, *70*, 2–5. [CrossRef]

127. Cloyd, R.A.; Galle, C.L.; Keith, S.R.; Kalscheur, N.A.; Kemp, K.E. Effect of commercially available plant-derived essential oil products on arthropod pests. *J. Econ. Entomol.* **2009**, *102*, 1567–1579. [CrossRef] [PubMed]

128. Regnault-Roger, C.; Vincent, C.; Arnason, J.T. Essential oils in insect control: Low-risk products in a high-stakes world. *Annu. Rev. Entomol.* **2012**, *57*, 405–424. [CrossRef] [PubMed]

129. Barceloux, N.G. *Medical Toxicology of Natural Substances: Foods, Fungi, Medicinal Herbs, Plants, and Venomous Animals*; Wiley: Hoboken, NJ, USA, 2008; pp. 1–1157.

130. Environmental Protection Agency. US EPA—Pesticides—Fact Sheet for Oil of Citronella. Available online: https://www3.epa.gov/pesticides/chem_search/reg_actions/reregistration/fs_PC-021901_1-Feb-97.pdf?rel=outbound (accessed on 5 September 2020).

131. Mishra, J.; Tewari, S.; Singh, S.; Arora, N.K. Biopesticides: Where we stand? In *Plant Microbes Symbiosis: Applied Facets*; Springer: New Delhi, India, 2015; pp. 37–75.

MDPI
St. Alban-Anlage 66
4052 Basel
Switzerland
Tel. +41 61 683 77 34
Fax +41 61 302 89 18
www.mdpi.com

Foods Editorial Office
E-mail: foods@mdpi.com
www.mdpi.com/journal/foods